Wastewater Treatment

*Edited by Muharrem Ince
and Olcay Kaplan Ince*

Published in London, United Kingdom

Wastewater Treatment
http://dx.doi.org/10.5772/intechopen.97964
Edited by Muharrem Ince and Olcay Kaplan Ince

Contributors

Andres Abin-Bazaine, Alfredo Campos Trujillo, Mario Olmos-Marquez, Bahareh Pirzadeh, Ifeanyi Michael Smarte Anekwe, Jeremiah Adedeji, Stephen Okiemute Akpasi, Sammy Lewis Kiambi, Canisius Mpala, Lubelihle Gwebu, Bilal Essaid, Najla Lassoued, Gonzalo Montes-Atenas, Sabeur Khemakhem, John G. Murnane, Thomas F. O'Dwyer, J. Tony Pembroke, Bashir Ghanim, Ronan Courtney, Ashlene Hudson, Opololaoluwa Oladimarun Ogunlowo, Chourouk Ibrahim, Salah Hammami, Eya Ganmi, Abdennaceur Hassen, Tahira Mahmood, Saima Momin, Rahmat Ali, Abdul Naeem, Afsar Khan, Tatheer Fatima, Tanzeela Fazal, Nusrat Shaheen, Omar M. Waheeb, Mohanad Mahmood Salman, Rand Qusay Kadhim

Notice
Statements and opinions expressed in the chapters are these of the individual contributors and not necessarily those of the editors or publisher. No responsibility is accepted for the accuracy of information contained in the published chapters. The publisher assumes no responsibility for any damage or injury to persons or property arising out of the use of any materials, instructions, methods or ideas contained in the book.

First published in London, United Kingdom, 2022 by IntechOpen
IntechOpen is the global imprint of INTECHOPEN LIMITED, registered in England and Wales, registration number: 11086078, 5 Princes Gate Court, London, SW7 2QJ, United Kingdom

British Library Cataloguing-in-Publication Data
A catalogue record for this book is available from the British Library

Additional hard and PDF copies can be obtained from orders@intechopen.com

Wastewater Treatment
Edited by Muharrem Ince and Olcay Kaplan Ince
p. cm.
Print ISBN 978-1-80355-846-2
Online ISBN 978-1-80355-847-9
eBook (PDF) ISBN 978-1-80355-848-6

Meet the editors

Professor Muharrem Ince received his Ph.D. in Analytical Chemistry from Firat University, Turkey in 2008. From 2009 to 2012, he worked as a research analytical chemist at Mus Alparslan University, Turkey. He has been working at Munzur University since 2012 and served as head of the university's Department of Chemical Engineering from 2013 to 2016. He is currently a professor at the same university. He is an editorial board member of several international journals as well as an author and co-author of more than forty journal papers. His expertise is in analytical method development, spectroscopic and chromatographic techniques, environmental sciences, water pollution identification and prevention, food analysis and toxicology, green and sustainable chemistry, nanoscience and nanotechnology, and smart bio-nano-carrier synthesis for drug delivery.

Prof. Dr. Olcay Kaplan Ince received her BS from Hacettepe University, Turkey, and Ph.D. in Analytical Chemistry from Firat University, Turkey. She is a research analytical chemist and former head of the Food Engineering Department, Munzur University, Turkey. She is also editor-in-chief of the International Journal of Pure and Applied Sciences. Dr. Kaplan is the author of more than forty journal papers. Her research interests include trace and toxic element analysis, analytical chemistry, instrumental analysis, problem-solving in analytical chemistry, food science and chromatography, nanoscience and cytotoxicology, deep eutectic solvents, and smart bio-nanocarrier synthesis for drug delivery.

Contents

Preface

This book provides an overview of the theory and practice of wastewater treatment, wastewater treatment technologies, biological and chemical processes in wastewater treatment, membrane technologies for wastewater treatment and resource recovery, challenges for wastewater treatment, and recent advances in wastewater treatment applications. Over six sections, it provides a broad overview of wastewater treatment processes, starting at a more elementary level and gradually incorporating more advanced concepts. It also explains and clarifies important studies in the field and highlights and compares new and groundbreaking techniques of wastewater treatment. It is a useful resource for a variety of readers, including undergraduate and graduate students, chemists, and professionals interested in wastewater treatment technologies and membrane separation processes.

Professor Muharrem Ince and Professor Olcay Kaplan Ince
Munzur University,
Tunceli, Turkey

Theory and Practice of Wastewater Treatment

Chapter 1

Physical Wastewater Treatment

Bahareh Pirzadeh

Abstract

Water is a valuable material. Water used to dispose of nature or enter the consumption cycle requires disinfection and purification to conserve water resources as well as to provide drinking water. Different processes are carried out on the water to increase water quality as much as possible. In general, the filtration process can be divided into two general categories. In the first process, harmful substances are removed from the water. In the second group, the processes are specifically designed to improve the quality and control parameters such as the pH value. The stages of water purification can be divided into different steps more in detail, which physical purification is one of these steps and has been discussed in this chapter.

Keywords: wastewater treatment, physical treatment, water quality, sewage dumping, active pharmaceutical ingredients, corrugated plate interceptor

1. Introduction

All societies, both solid and liquid, produce waste. Wastewater can be considered as a combination of waste produced by water from residential, administrative, commercial, and industrial facilities and drained into groundwater or surface water. Untreated wastewater contains pathogenic microorganisms and organic matter. Degradation of untreated wastewater organic matter produces stinking gases. Therefore, wastewater treatment is one of the essential measures that must be taken before discharge into the environment. Wastewater treatment is a practical solution to speed up the process of providing safe and transparent reusable water.

The pollutants can be removed from wastewater using a variety of ways that are divided into three main categories: physical, chemical, and biological processes. A purification process generally consists of five successive steps as described in **Figures 1** and **2**: (1) preliminary treatment or pre-treatment (physical and mechanical); (2) primary treatment (physicochemical and chemical); (3) secondary treatment or purification (chemical and biological); (4) tertiary treatment (physical and chemical); and (5) treatment of the sludge formed (supervised tipping, recycling or incineration) [1, 3].

Physical wastewater treatment is the first step in the treatment of industrial and sanitary effluents, which in addition to increasing the efficiency of other steps, prevents damage to the equipment used in chemical and biological treatment. The equipment and processes used in physical wastewater treatment vary according to the type of effluents and the quality desired for wastewater. In other words, since the wastewaters are mostly very colored and contain high biological and chemical oxygen, they have high electrical conductivity and are considered chemically alkaline.

IntechOpen

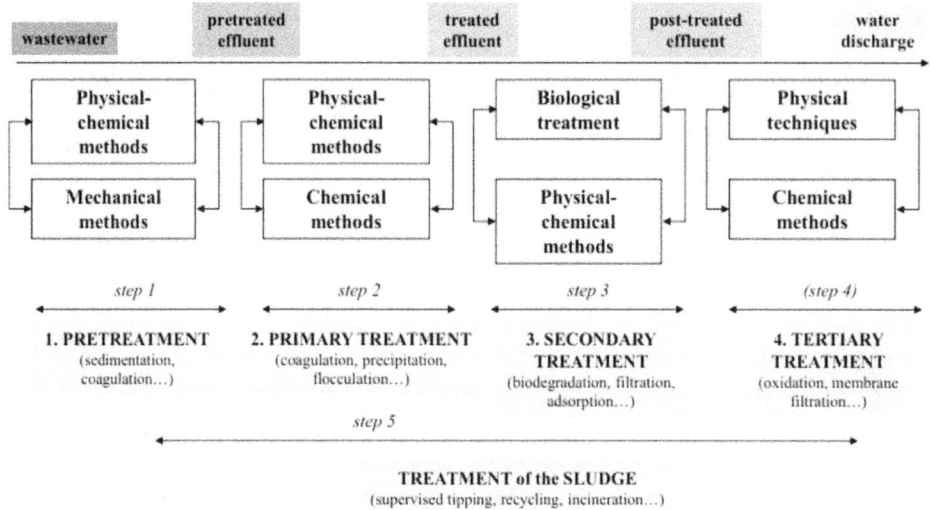

Figure 1.
Main processes for the decontamination of industrial wastewaters [1].

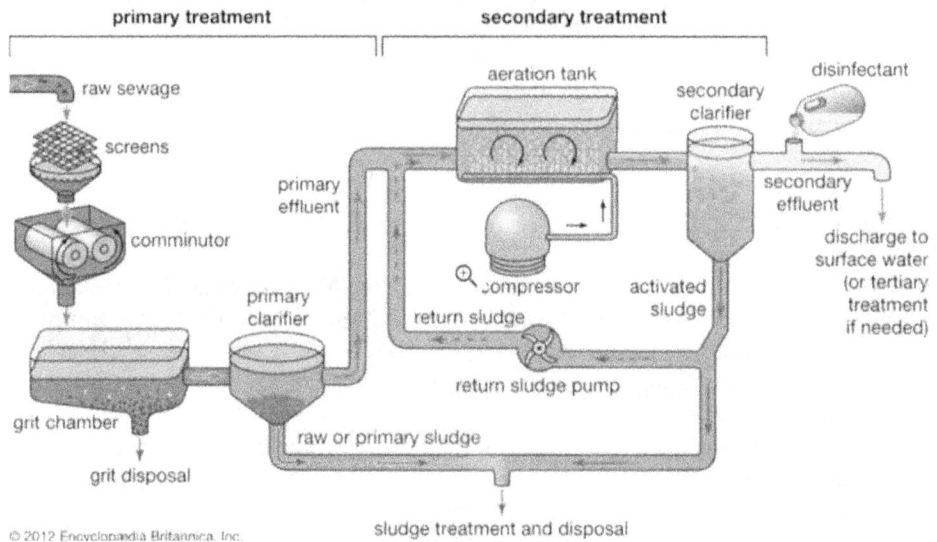

Figure 2.
Primary and secondary treatment of sewage [2].

In other words, since the wastewaters are mostly very colored and contain high biological and chemical oxygen, they have high electrical conductivity and are considered chemically alkaline. Accordingly, different parameters affect the cost of selecting a wastewater treatment method. Factors such as the type of pollutants to be treated, the chemical composition of wastewater, the cost of chemicals required, the operating cost, the cost of collecting waste generated by the treatment process, affect selecting a wastewater treatment method [4].

Based on the various processes and steps mentioned for the physical treatment of industrial and sanitary effluents mentioned above, equipment such as the following is required:

- Types of garbage (manual, mechanical and even grating)

- Settling pools and sludge bridges for collecting deposited sludge

- Classifier and a variety of granular to remove fine grains such as sand

In this chapter, we will explain each of the above.

2. What is physical wastewater treatment

The separation process of particulate matter and solids in industrial and sanitary effluents is called physical purification. Depending on the type of sewage, there may be pieces of fabric, the foliage of trees, sand and plastic parts, etc. in the fluid entering the treatment plants. Entering these particles into wastewater treatment equipment such as pumps, pipes, and fittings may cause damage. In addition, the failure to remove these items causes a lot of pressure on the equipment in chemical and biological wastewater treatment, and their output quality decreases. For this purpose, with practical and simple equipment, physical treatment processes of industrial and sanitary wastewaters are implemented.

3. The most important steps of physical wastewater treatment

3.1 Sewage dumping

Considering the capacity of the treatment plant and the speed of entering fluid, and the size and amount of dissolved and suspended solids in the wastewater, it is necessary to install appropriate littering in the inlet of the sewage canal. The litters used in this step of physical wastewater treatment are varied, and each one has specific features, the most important of which are the following options:

3.1.1 Mechanical garbage collector

Other types of littering used for large refineries include mechanical garbage collectors. The most important feature of this equipment is that the engine is on top of it. So, after accumulating solid particles in the sewage on the blocker screen, the collection process is done. The mechanical garbage collector is divided into the bar, lattice or lace, strapping, round-trip rod, chain, and cylindrical movement groups in terms of appearance structure. In this type of garbage collector, human resources are not used. It can remove suspended solids up to 2–3 mg. It should be noted that the necessary force for motor movement in mechanical garbage is supplied from electricity and therefore has higher energy consumption than the manual type (**Figure 3**) [5].

3.1.2 Handheld sewage collector

A handheld screen consisting of several bars is located at certain distances and prevents particles with a size greater than two centimeters from entering the treatment plant. As the name of this equipment is known, after accumulating particles and solid patches on the plate, human operators perform the collection process. Garbage

Figure 3.
Bar rack and traveling screen [5].

Figure 4.
Preliminary Treatment of Sewage [6].

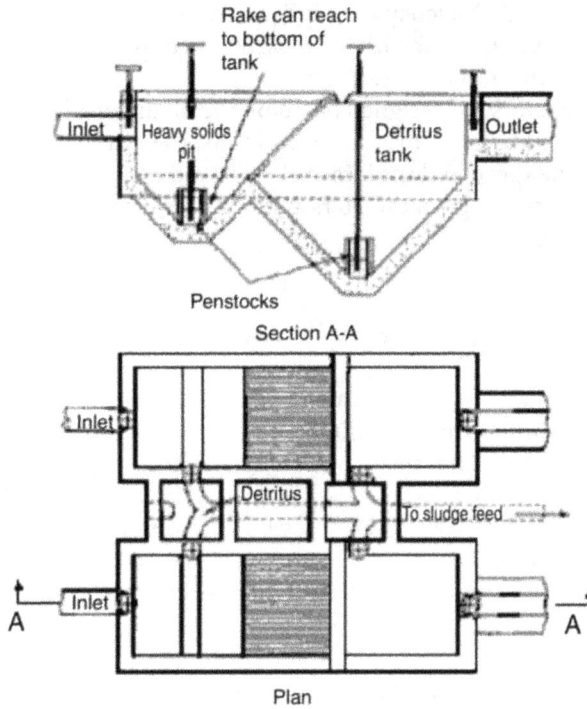

Figure 5.
Bar screen in a detritus tank [5].

Parameter	Mechanically cleaned	Manually cleaned
Bar Size		
Width (mm)	5–20	5–20
Thickness (mm)	20–80	20–80
Bars cleaner spacing (mm)	20–50	15–80
Slope from vertical (degrees)	30–45	0–30
Approach velocity (m s^{-1})	0.3–0.6	0.6–1.0

Table 1.
Design parameters and criteria for Bar screens [5].

collectors have different types of fine and coarse grain. Due to the lack of electricity, it is cost-effective for the physical treatment of wastewater. It should also be noted that the material used to make the body and screen is resistant to acidic and corrosive materials and has a long lifespan (**Figures 4** and **5**) [5].

3.1.3 Mechanical sewage collector

The grating pair is composed of two lattice plates and nets that block the passage of sewage. This type of littering is located in the group of mechanical garbage collectors and without the use of a human operator, the collection of accumulated materials on the plate is done. Due to the use of two lace plates in the overall structure of this screen, its efficiency and efficiency for physical treatment of wastewater and removal of suspended solids are high and with the destruction of one plate, it is possible to continue working with another plate [5].

The garbage collectors are designed based on the diameter of the seeds to be removed, the width and depth of the canal and rods, the distance between the rods, the vertical slope, the speed and the loss of the allowed head.

Table 1 shows some design parameters and criteria for mechanically and hand-cleaned screens.

4. Deposition of particulate matter in wastewater

Industrial and sanitary wastewaters are not completely cleaned of suspended materials despite passing through multiple filters (littering). Many of these small particles will settle on the floor of the treatment plant, depending on their weight if they have the opportunity. The second important process in the physical treatment of industrial and sanitary wastewaters is to allocate time for settling suspended materials in special pools. By doing so, largely fine and particulate matter is transferred to the bottom of the pool and gradually converted into sludge, which will be periodically drained (**Figure 6**) [7].

5. Removal of sand and fine particles in wastewater

In the last steps of the physical treatment of wastewater, it is necessary to provide a solution for the removal of very fine sand. One of the best solutions in these conditions is floating using equipment such as a classifier. Grit Classifier is placed in a simple but functional wastewater treatment equipment group that has different parts such as body, cochlear conveyor, and Electra Gearbox. Water enters the grit classifier body from the inlet valve. The cochlear conveyor flows the strip and creates centrifugal force. Fine particles and sand grains are separated from the water due to the creation of this force. They move to the device output and exit from it. Finally, wastewater without sand particles and similar items will be obtained.

Figure 6.
Sewage settling pond.

6. Active pharmaceutical ingredient (API)

Active pharmaceutical ingredient (API) is one of the most widely used and oldest systems for removing fat and oil from water, wastewater, and wastewater. These systems tend to remove free oil down to less than 15 mg L^{-1} [7]. The construction of water and oil separators by API method is similar to a rectangular clearinghouse, although they have different sizes and design details.

These grease traps are used to remove free oil as well as solid particles from wastewaters of refineries, petrochemicals and chemical plants and other industries. These grease traps are designed according to the standards published by the American Petroleum Institute. The basis of this type of separator is the difference in specific gravity. Less specifically-weighted liquid (usual oil) is collected from the surface, while the fluid with more specific gravity remains in the lower part. Wastewater may contain insoluble oil, sludge, and some soluble components. In common API degreasers, wastewater is initially collected in a pretreatment section for sludge collection. Baffles allow the sewage flow to move slowly towards the outlet, and the oil is separated from the water and effluent.

Usually, in the process of API grease traps, the oil layer, which may contain amounts of water and suspended solids, must be continuously emptied. This layer of isolated oil may be reprocessed or destroyed to recover valuable products. The sedimentary layer containing solid particles formed at the bottom of the degreaser is removed by a remover (or similar device) and a sludge pump.

In the design of API grease traps and other similar gravity tanks, if the input current meets the following conditions, the performance of the grease traps will be difficult:

- The average size of oil droplets in the feed is less than 150 microns,

- Oil density is greater than 925 (Kg m^{-3}),

- Water temperature is less than 5°C,

- The amount of hydrocarbon in the inlet flow is high.

7. Corrugated plate interceptor (CPI)

In addition to the grit classifier at this stage, different types of grease traps can also be used. For example, CPI degreaser is one of the most applicable equipment in this field. It is a kind of gravity separator and causes colloidal particles suspended in wastewater with a size of at least 50 microns to be separated from wastewater. This type of degreaser, sometimes called Tilted Plate Interceptor, has widely replaced API separators and primary settling tanks [8].

In CPI fat, the inlet wastewater is transferred to the machine to diagonal plates at an angle of 45 to 60 degrees. These plates have resin-reinforced layers, eventually causing the fat particles to settle above the surface of the sewage and the rest that weighs heavier. The simple structure of this method reduces the cost of construction and maintenance of facilities. One of the most important uses of CPI separators is oil and gas refineries, oil terminals and ports, repair shops, food industry, detergent, and chemical production industries, automobile manufacturing, and petrochemical industries. The advantages of this method are:

- High efficiency of oil and fat removal

- Low energy consumption

- Simple operation

- Using corrosion-resistant plates, acid, alkalinity

- The dense structure raises concerns about deformation.

8. Dissolved air flotation (DAF)

The removal of pollutants and colloidal particles by injecting compressed air into the effluent is called "Liquid Air Flotation" or "Degradation" of DAF wastewater. As a result of this action, finely released bubbles cling to colloidal particles suspended in the sewer, causing the particles to float on the surface where they are collected by the skimmer and discharged into a sludge storage tank (**Figure 7**). Due to the high efficiency of the DAF degradation system, this equipment is widely used in various industries such as oil and gas refineries, food factories, and chemicals [7]. DAF is an alternative to sedimentation [9].

In the design process of the DAF wastewater degradation machine, several steps are considered for the removal of oil and colloidal particles suspended in water, including [10]:

- Compressed air enters the reservoir and storage of sewage and water, causing bubbles in the fluid.

- Bubbles created in the liquid cling to colloidal particles and transfer to the highest level of the liquid.

- After oily and colloidal particles are placed on the surface of the water, they are collected using skimmers.

To increase the efficiency of the DAF wastewater degradation system, it is better to use coagulants according to the fluid type and particulate matter in it. To do this,

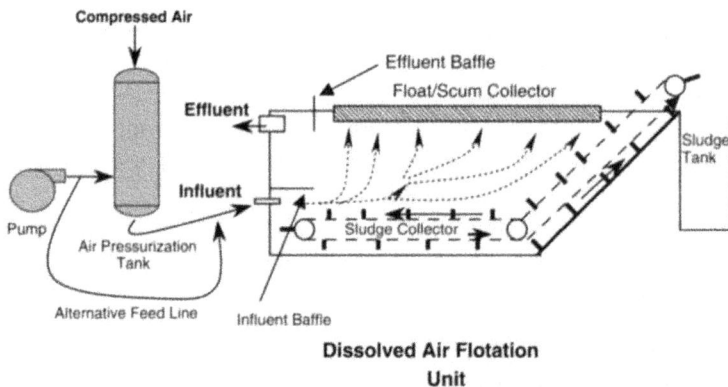

Dissolved Air Flotation Unit

Figure 7.
Dissolved air flotation system configuration [7].

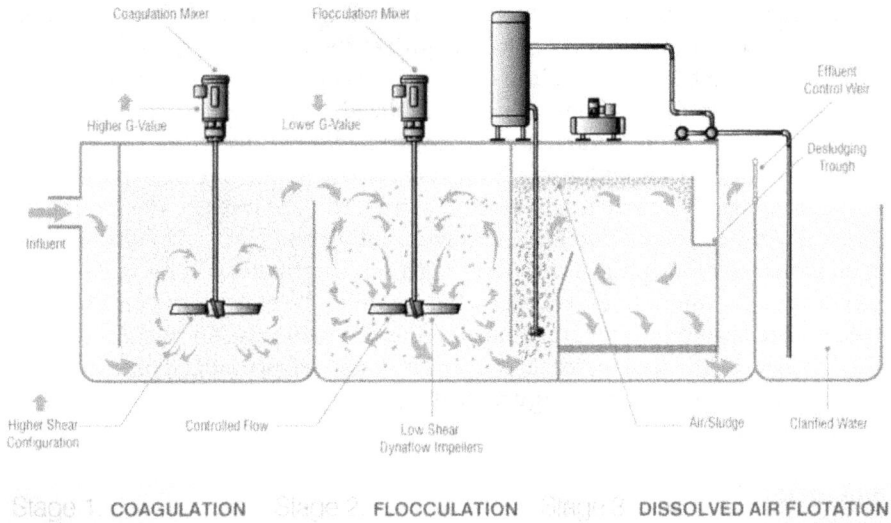

Figure 8.
Coagulation and Flocculation Process and Mixing Essentials [11].

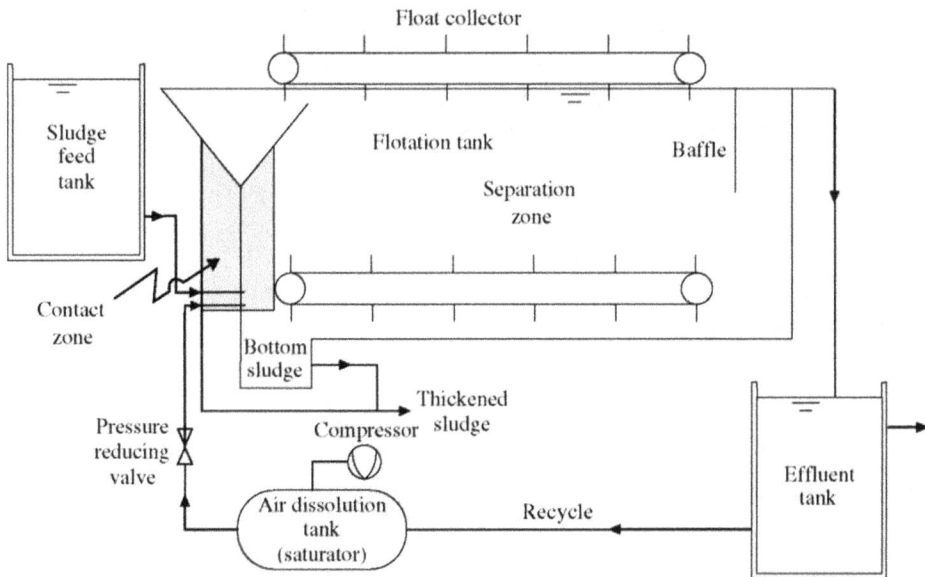

Figure 9.
Dissolved air flotation thickener [9].

add to the inlet effluent of coagulants and polymers to make the load of the particles. For this purpose, spiral tubes are used, which on the one hand, coagulants are injected in different parts, and on the other hand, an air mixture with pressurized liquid is injected into these tubes for better mixing.

In the DAF grease trap system, some of the cleaned and cleared water is inserted into the flocculants pipes and part of it is transferred directly to the floating unit.

Excess treated water is also removed from the DAF system. Clearness of the outlet water from the DAF unit is one of the criteria for the proper operation of this system.

In some cases, coagulation and flocculation tanks are used instead of flocculants tubes. In this case, the coagulants are mixed with effluent in the coagulation tank for about 5 minutes and then injected into the DAF. This process is superior to tubular flocculants when the input effluent is not neutralized and there is enough time to neutralize and on the other hand it is easily controlled by the user.

The objective of the Coagulation Process is to create micro-floc from particulate matter, agglomerate the particles, so that they can settle and move to the sedimentation process. At the Flocculation Process stage, another chemical can be added, a polymer, which helps keep the macro flocs together. Essentially it helps make the bridge between the micro flocs stronger when forming the macro-flocs. A Flocculator is three or more basins with separate mixers. At each stage there are various sizes of flocs, as the size of the flocs increase, the power on the mixers is decreased, to prevent shear, and not damage the flocs (**Figures 8** and **9**).

9. Conclusion

Wastewater treatment is a process in which wastewater (water that is no longer needed or suitable for use) is converted into water that can be restored to the environment and drained in the wild. Wastewater treatment procedures are carried out in 3 ways: chemical treatment, physical treatment and biological treatment.

Physical treatment of wastewater is when physical and mechanical properties and separation and removal of external particulate matter from wastewater are used. Littering, grading, floating, sequestration and other processes are examples of these.

Physical wastewater treatment is the first step in industrial and sanitary wastewater treatment and not only improves the effectiveness of subsequent processes, but also protects chemical and biological treatment equipment. The equipment and techniques used in physical wastewater treatment vary depending on the type of wastewater and the quality of the wastewater.

Physical methods of water and wastewater treatment include different stages of settling, filtration, and aeration that were discussed in this chapter.

Author details

Bahareh Pirzadeh
University of Sistan and Baluchestan, Zahedan, Iran

*Address all correspondence to: b_pirzadeh@eng.usb.ac.ir

IntechOpen

References

[1] Sanctis M, Moro G, Chimienti S, Ritelli P, Levantesi C, Di Iaconi C. Mint: Removal of pollutants and pathogens by a simplified treatment scheme for municipal wastewater reuse in agriculture. Science of The Total Environment. 2017;**580**:17-25. DOI:10.1016/j.scitotenv.2016.12.002

[2] Encyclopedia Baritannica, Inc. 2012. Available from: www.baritannica. com/technology/secondary-treatment [Accessed: January 13, 2022]

[3] Crini G, Lichtfouse E. Mint: Advantages and disadvantages of techniques used for wastewater treatment. Environmental Chemistry Letters. 2019;**17**:145-155. DOI: 10.1007/s10311-018-0785-9

[4] Kalbar P, Karmakar S, Asolekar S. Mint: Selection of an appropriate wastewater treatment technology: A scenario-based. Journal of Environmental Management. 2012;**113**:158-169. DOI: 10.1016/j. jenvman.2012.08.025

[5] Sincero Sr AP, Sincero GA. Physical-chemical treatment of water and wastewater. IWA. 2003

[6] DAS Environmental Expert [Internet]. 2022. Available from: Chemical-Physical Wastewater Treatment Technologies | DAS (das-ee.com). [Accessed: January 13, 2022]

[7] David L, Russell PE. Practical Wastewater Treatment. Hoboken, New Jersey: John Wiley & Sons, Inc.; 2006

[8] World Technologies. Water & Waste Water Expert. CPI/TPI/DAF/NSF/IGF. Available from: https://worldtechnologies. sg/ [Accessed: January 17, 2022]

[9] Droste R, Gehr R. Theory and Practice of Water and Wastewater Treatment. USA: John Wiley & Sons, Inc.; 2019

[10] Nicholas PC. Handbook of Water and Wastewater Treatment Technologies. Boston Oxford Auckland Johnnesburg Melbourne New Delhi: Butterworth-Heinemann; 2002

[11] Dynamix Inc. 2020. Available from: dynamixinc.com/coagulation-and-flocculation-process-and-mixing-essentials [Accessed: January 13, 2022]

Chapter 2

Adsorption Isotherms: Enlightenment of the Phenomenon of Adsorption

Andres Abin-Bazaine, Alfredo Campos Trujillo and Mario Olmos-Marquez

Abstract

Adsorption is a process in which a substance that is in a liquid phase accumulates on a solid surface and is then removed from the liquid phase. An adsorption isotherm describes the equilibrium of adsorption of a substance on a surface at a constant temperature. It represents the amount of material bound to the surface as a function of the material present in the solution. In the adsorption process, the compound to be removed is called the adsorbate and the solid on which the compound is adsorbed is called the adsorbent. The affinity of the adsorbate for the adsorbent is quantified using adsorption isotherms. Adsorption isotherms are mathematical equations that describe the relationship between the amount of adsorbate adsorbed on an adsorbent and the concentration of adsorbate in solution when equilibrium has been reached at constant temperature. Adsorption isotherms are performed by giving a volume-determined solution containing a known amount of adsorbate along with various dosages of the adsorbent. The mixture is held at constant temperature with stirring until it reaches equilibrium. When this is the case, the concentration of the adsorbate in the aqueous phase is measured and the adsorption capacity at equilibrium for each experiment is calculated from the mass balance.

Keywords: isotherms, Henry's, Langmuir, Freundlich, Temkin, Dubinin–Radushkevich, Redlich–Peterson, sips, Halsey, Harkin–Jura, Elovich, Flory–Huggins, fowler–Guggenheim, Jovanovic and Kiselev

1. Introduction

Wastewater treatment has become one of the most important issues for the development of nations. During treatment, contamination can be found that cannot be treated by biological means, but there are also other recalcitrant compounds such as heavy metals that, in certain concentrations, can make biological treatment of wastewater difficult. Treatment of these compounds requires a different type of treatment: chemical precipitation, reverse osmosis, filtration, ion exchange, electrochemical degradation, photocatalytic degradation, nanofiltration, irradiation [1], and

15

adsorption, to name a few. There are processes that are very efficient in removing impurities at trace concentrations, such as reverse osmosis. However, they have the disadvantage that the initial cost is very high and the consumables, such as membranes and maintenance, are very expensive. One of the most promising methods for removing recalcitrant compounds such as heavy metals is adsorption. This is because it is cheap, easy to implement, and environmentally friendly [2]. To understand adsorption mechanisms, it is important to obtain equilibrium data, commonly known as adsorption isotherms. The adsorption isotherms are important for describing how the adsorbate molecules or ions interact with the surface adsorption sites. Therefore, correlation of the equilibrium data using a theoretical or empirical equation is essential for interpretation and prediction of adsorption [3]. An adsorption isotherm is an equation expressing the relationship between the amount of solute adsorbed by the adsorbent and the solute concentration in the liquid phase [4–14]. They are important because they describe how an adsorbate interacts with an adsorbent, and they are critical to the design of an adsorption process. Several equilibrium models have been developed to describe the relationship of isotherms at equilibrium. The models used are those of Henry, Langmuir, Freundlich, Temkin, Dubinin–Radushkevich (DR), Redlich–Peterson (RP), Sips, Halsey, Harkin–Jura, Elovich, Flory–Huggins, Fowler–Guggenheim, Jovanovic, and Kiselev.

2. Adsorption isotherms

2.1 Calculation of isotherms

Batch adsorption experiments are performed. A constant concentration of adsorbent is added and mixed with a constant volume of solution, progressively varying the concentration of the solute to be removed. Stir at a constant rate until equilibrium is reached. The percentage of adsorption is calculated using the following expression:

$$\%\text{Adsorption} = \frac{C_i - C_f}{C_i} \times 100 \tag{1}$$

$$q_e = \frac{C_i - C_f}{m} \times V \tag{2}$$

where C_i and C_f are the initial and final concentrations, respectively, of the ion to be removed (mg L^{-1}), q_e is the adsorption capacity of the ion (adsorbate) by the adsorbent (mg g^{-1}), m is the mass of the adsorbent (g), and V is the volume of the ion solution to be removed (L) [6–12, 15–17].

Once the values of mass removal capacity (q_e) and equilibrium concentration (C_f) have been calculated using a Cartesian coordinate system to represent the two-dimensional distribution, the dispersion diagram provides a set of known points whose analysis allows a qualitative study of the relationship between the two variables. The next step is to determine the functional dependence between the two graphically represented variables that best fits the two-dimensional distribution. It is called linear regression if the function is linear.

The correlation coefficient is a technique used to study the two-dimensional distribution, indicating the intensity or degree of dependence between the variables X and Y. The correlation coefficient r is a number obtained using the following formula:

$$r = \frac{\sum[(x_i - \bar{x})(y_i - \bar{y})]}{\sqrt{\sum(x_i - \bar{x})^2 + \sum(y_i - \bar{y})^2}} \tag{3}$$

where r is correlation coefficient, x_i value of variable of x, $\bar{x}$ is the mean of the variable x, y_i value of variable y, $\bar{y}$ is the mean value of variable y, $\sum(x_i - \bar{x})^2$ sum of the square of the deviation x, $\sum(y_i - \bar{y})^2$ sum of the square of the deviation y. The values of the correlation coefficient vary in the range from -1 to $+1$. A value of -1 indicates a perfect inverse linear correlation while a value of $+1$ indicates a positive linear correlation.

This last procedure can easily be done with the help of software that uses spreadsheets.

2.2 Henry's isotherm

This isothermal model adequately describes the adsorption process at low concentrations so that all adsorbate molecules are without interaction with the neighboring molecules [18]. The concentrations in the phases are associated with a linear expression. It is expressed as:

$$q_e = K_{HE}C_e \tag{4}$$

where q_e is the adsorption capacity in equilibrium (mg g^{-1}); K_{HE} is the equilibrium constant of Henry's; and C_e is the equilibrium concentration of the metal ions in the solution (mg L^{-1}). From the straight line adjusted to graph C_e versus q_e, the K_{HE} coefficient was calculated, which is represented by the slope.

2.3 Langmuir isotherm model

The Langmuir equation assumes that the maximum adsorption corresponds to a monosaturated layer of adsorbate molecules on the surface of the adsorbent, the adsorption energy is constant, and there is no transmigration of the adsorbate on the surface of the adsorbent [8]. All adsorption sites are energetically identical, and the intermolecular forces decrease with increasing distance from the adsorption surface [19]. The Langmuir model is expressed by the following equation:

$$q_e = \frac{q_m K_L C_e}{[1 + K_L C_e]} \tag{5}$$

where q_m is the maximum adsorption capacity (mg g^{-1}) and K_L is the Langmuir constant (L mg^{-1}) [6–8, 10, 12, 16, 20, 21]. The linear form of the Langmuir model is:

$$\frac{C_e}{q_e} = \frac{1}{q_m}C_e + \frac{1}{q_m K_L} \tag{6}$$

The values of the constants q_m and K_L are determined by the slope and the intersection of the fitted line on the abscissa C_e and the ordinate C_e/q_e, respectively. The characteristic of the Langmuir isotherm is that they can express a dimensionless constant called Balance Parameter or Separation Factor, which is expressed with the following equation:

$$R_L = \frac{1}{(1 + (K_L \times C_i))} \tag{7}$$

where R_L is the equilibrium parameter (dimensionless), and C_i is the initial concentration (mg L^{-1}). The R_L values indicate what type of adsorption can be expected. $R_L > 1$ is unfavorable, $R_L = 1$ the adsorption is linear, $R_L = 0$ is irreversible, and if $0 < R_L < 1$, adsorption is favorable [19, 22, 23].

2.4 Freundlich isotherm model

The Freundlich model takes into account the heterogeneity of the surface of the adsorbent and an exponential distribution of the active sites and their energies. The Freundlich model is expressed by the following equation:

$$q_e = K_F + C_e^n \tag{8}$$

where K_F is Freundlich's constant in (mg g^{-1}), and n is the Freundlich exponent related to the intensity of the adsorption; it is dimensionless. The equation of this model can be linearized as follows:

$$\log q_e = \frac{1}{n} \log C_e + \log K_F \tag{9}$$

The values of the constants K_F and n are determined by the intercept and the slope of the graphical line on the ordinate ln q_e and the abscissa ln C_e, respectively. The range of values of $1/n$ is between 0 and 1 showing the degree of nonlinearity between the concentration of the solution and the adsorption. If the value of $1/n$ is equal to 1, the adsorption is linear [6–8, 10, 12, 16, 21, 24]. High values of n indicate a relatively uniform surface, where low values mean high adsorption at low concentrations of solution. In addition, low values of n indicate the existence of a high proportion of high-energy active sites [25].

2.5 Temkin isotherm model

The Temkin isotherm model contains a factor that explicitly considers the adsorption interactions between the species and the adsorbate [14, 26]. The Temkin model is described with the following equation:

$$q_e = BLn(K_T C_e) \tag{10}$$

where $B = RT/b$, b is Temkin's constant, K_T is the Bound Equilibrium Constant (L g^{-1}), and B is related to the heat of adsorption (J mol^{-1}) [27]. The Temkin isotherm model is presented in linear form with the following equation:

$$q_e = BLnC_e + BLnK_T \tag{11}$$

To obtain the constants B and K_T defined by the slope and intersection of the lines, respectively, plot q_e on the ordinate and Ln C_e on the abscissa.

2.6 Dubinin–Radushkevich isotherm model (D–R)

The model of the Dubinin–Radushkevich isotherm (D–R) not assumes a homogeneous surface or a constant adsorption potential. The model is described with the following equation:

$$q_e = q_s e^{-K_{DR}\varepsilon^2} \tag{12}$$

where q_s is the adsorption capacity (mol g^{-1}), K_{DR} is the Dubinin–Radushkevich constant (mol^2/KJ^2), and ε is correlated with the following expression:

$$\varepsilon = RTLn\left(1 + \frac{1}{C_e}\right) \tag{13}$$

where ε is the potential of Polanyi, C_e is the concentration in equilibrium (mol L^{-1}), q_e is the equilibrium concentration in the adsorbent (mol g^{-1}), R is the universal gas constant (8.314×10^{-3} KJ mol^{-1} K^{-1}), and T is the absolute temperature °K. The linear form of this isotherm is expressed with the following equation:

$$q_e = -K_{DR}\varepsilon^2 + Lnq_s \tag{14}$$

where q_s is the saturation capacity of the isotherm (mg g^{-1}). The isotherm constants q_s and K_{DR} are obtained from the intercept and the slope, respectively, from plotting on the ordinate Ln q_e and on the abscissa ε^2. The constant K_{DR} gives the mean of the free energy, E, is the adsorption per molecule of the sorbate when it is transferred to the surface of the solid from infinity in the solution and can be calculated using the following relationship:

$$E = \frac{1}{\sqrt{-2K_{DR}}} \tag{15}$$

The magnitude of E is used to estimate the type of adsorption process. The adsorption process is a chemical absorption if the magnitude of E is greater than 16 KJ mol^{-1}. In turn, when the magnitude of E is less than 8 KJ mol^{-1}, the type of absorption can be defined as a physical process [14, 27, 28]. There is talk of an ion exchange if 8 < E < 16 KJ mol^{-1} [29].

2.7 Redlich–Peterson isotherm model (R–P)

This model is a combination of the Langmuir and Freundlich isotherms. It is used to describe the adsorption on both homogeneous and heterogeneous surfaces. It is considered as a comparison between these two models [30]. This isotherm is expressed with the following expression:

$$q_e = \frac{K_R C_e}{(1 + aRC_e^{\beta})} \tag{16}$$

where K_R is the constant of Redlich–Peterson (L g^{-1}), Redlich–Peterson constant αR (L mg^{-1}), β constant of Redlich-Peterson (dimensionless). The values of β fluctuate between 0 and 1. At low concentrations, the Redlich–Peterson isotherm approximates Henry's law [31]. When the constant β is very close to 1, it is the same as the Langmuir equation and in high concentrations its behavior approaches that of the Freundlich isotherm, since the β exponent tends to zero [32, 33]. From the transformation of the original equation, two linear forms are obtained. One of the linear forms of this isotherm is expressed with the following equation:

$$log\left[\left(K_R\frac{C_e}{q_e}\right) - 1\right] = \beta log C_e + log\,(\alpha R) \tag{17}$$

The values of αR and β for the above equation can be determined from the intercept and the slope, respectively, of the straight line of graphing $log\left[\left(K_R\frac{C_e}{q_e}\right) - 1\right]$ versus $log\,C_e$ [34]. Several values of the constants must be tested before obtaining the optimal line, in order to obtain the values of these. The range of values of these constants is very wide, ranging from 0.01 to several hundred, so it is not easy to obtain the correct values [31]. Another linear form of this equation is:

$$\frac{C_e}{q_e} = \frac{1}{K_R} + \frac{\alpha R}{K_R}C_e^\beta \tag{18}$$

The constants of the Redlich–Paterson isotherm can be determined from the graph between C_e^β and C_e/q_e. However, its application is very complex since it includes three unknown parameters αR, K_R, and β. Therefore, a minimization procedure is adopted to obtain the maximum value of the coefficient of determination R^2, between the theoretical data for q_e obtained from the linearized form of the isotherm equation of Redlich–Peterson and the experimental data [30]. By trial and error, values of β are adopted to obtain an optimal line. In the specific range, the values of b are limited, and it is easy to obtain the correct value [31].

2.8 Sips isotherm model

The Sips isotherm is a combined form of Langmuir and Freundlich isotherms applied for the prediction of heterogeneous adsorption systems. The Sips model avoids the inconveniences and limitations of the Langmuir or Freundlich models. At low concentrations, the adsorbate becomes Freundlich's isotherm and, therefore, does not obey Henry's law [35]. While in high concentrations, the isothermal formula of Langmuir is reduced [36]. The equation of the Sips isotherm is characterized by containing a dimensionless heterogeneity factor, βs. If $\beta s = 1$, the Sips equation is reduced to the Langmuir equation, which indicates that the adsorption process is homogeneous. The Sips isotherm constant (β_s) confirms that the surface of the adsorbent is heterogeneous or not [33]. The Sips isotherm is expressed with the following equation:

$$q_e = \frac{K_S C_e^{\beta_s}}{1 + as C_e^{\beta_s}} \tag{19}$$

where K_S is the equilibrium constant of the Sips isotherm (L mg^{-1}), as is the maximum adsorption capacity (mg g^{-1}), and β_s is the model exponent (dimensionless) [35, 37, 38]. The linear form of the Sips isotherm is [39]:

$$Ln(q_e) = \beta_s Ln(C_e) + Ln(K_S - a_s \times q_e) \tag{20}$$

The coefficients of the Sips isotherm are calculated from graphing $\ln(C_e)$ versus $\ln(q_e)$, where β_s is slope.

2.9 Halsey isotherm model

This model is used to evaluate multilayer adsorption in a system where metal ions are located relatively far from the surface of the adsorbent. The model is expressed with the following Equation [40]:

$$q_e = exp\left(\frac{\ln K_H - \ln C_e}{n_H}\right) \tag{21}$$

The linear form from Halsey isotherm is [41, 43]:

$$Ln(q_e) = -\frac{1}{n_H}Ln\frac{1}{C_e} + \left[\left(\frac{1}{n_H}\right)Ln(K_H)\right] \tag{22}$$

where n_H is the constant of the equation, and K_H is Halsey's equilibrium constant. The constants of the isotherm may be calculated graphing $\ln(1/C_e)$ versus $\ln(q_e)$ and from the straight line obtained, the slope is n_H, and the intercept represents K_H [41].

2.10 Harkins–Jura isotherm model

The Harkins–Jura model describes a multilayer adsorption and the existence of a heterogeneous distribution of the pores of the adsorbent. The model is defined with the following expression:

$$\frac{1}{q_e^2} = -\frac{1}{A_{HJ}} \log(Ce) + \frac{B_{HJ}}{A_{HJ}} \tag{23}$$

where B_{HJ} is a model constant, and A_{HJ} is another model constant. Graphing $1/q_e^2$ versus $\log(C_e)$, the model constants are calculated with slope A_{HJ} and intercept B_{HJ} [41].

2.11 Elovich isotherm model

It assumes that the adsorption sites increase exponentially with adsorption, which implying a multilayer adsorption. This is expressed with [18, 41–43]:

$$\frac{q_e}{q_m} = K_E C_e \, exp\left(-\frac{q_e}{q_m}\right) \tag{24}$$

The linear form is expressed:

$$Ln\frac{q_e}{C_e} = -\frac{1}{q_m}q_e + LnK_E q_m \tag{25}$$

where q_m is the maximum adsorption capacity of Elovich (mg g^{-1}), and K_E is the equilibrium constant of Elovich (L mg^{-1}). K_E and q_m are calculated from the intercept and slope, respectively, of the straight line of $\ln(q_e/C_e)$ versus q_e [18, 41–43].

2.12 Flory-Huggins isotherm model

This model assumes the degree of coverage of the adsorbate on the adsorbent and expresses the degree of feasibility and spontaneity of the adsorption process. It included a parameter indicating the degree of coverage of the surface of the adsorbent, expressed as θ. The general form is given by the following Equation [43]:

$$\frac{\theta}{C_0} = K_{FH}(1 - \theta)^n \tag{26}$$

The linear form of it is expressed:

$$log \frac{\theta}{C_0} = nlog(1 - \theta) + log K_{FH} \tag{27}$$

$$\theta = \left(1 - \frac{C_e}{C_0}\right) \tag{28}$$

where K_{FH} is the equilibrium constant of Flory–Huggins (L mg^{-1}), n is the exponent of the model, and θ is the coverage parameter of the adsorbent surface. For the calculation of the isotherm parameters, $log\ \theta/C_0$ versus $log(1 - \theta)$ should be plotted, where the slope and the intercept represent n_{FH} and K_{FH}, respectively. The constant K_{FH} can be used to calculate the spontaneity of Gibbs free energy [32, 43]. For its calculation is used the following Eq. (32):

$$\Delta G^{\circ} = RTlnK_{FH} \tag{29}$$

where ΔG° Gibbs free energy change (KJ mol^{-1}). The negative values of ΔG° indicate that the adsorption is thermodynamically spontaneous and feasible [44].

2.13 Fowler–Guggenheim isotherm model

It is one of the simplest equations that takes into account the lateral interaction of adsorbate molecules. Its general form is expressed below:

$$K_{FG}C_e = \frac{\theta}{1 - \theta} exp \left(\frac{2\theta W}{RT}\right) \tag{30}$$

The linear form of it is expressed:

$$ln \left[\frac{C_e(1 - \theta)}{\theta}\right] = \frac{2W}{RT}\theta - ln K_{FG} \tag{31}$$

where K_{FG} is the equilibrium constant of Fowler–Guggenheim (L mg^{-1}), and W is the energy of interaction between the molecules of the adsorbate (KJ mol^{-1}) [37, 42]. The parameters of the equation are calculated by graphing $\ln[C_e(1 - \theta)/\theta]$ versus θ.

From this straight line the intercept and the slope represent K_{FG} and W, respectively. The charge and heat of adsorption vary linearly. When the values of W are greater than zero, it indicates that the interaction between the adsorbate molecules is attractive, but if the values of W are negative, the interaction is of the repulsion type, and if $W = 0$, there is no interaction [42].

2.14 Jovanovic isotherm model

This model assumes a superficial adsorption, it is an approximation of adsorption in a monolayer as expressed in the Langmuir model, but it assumes that there is no lateral interaction between the molecules. This model tolerates surface vibration of an adsorbed species [42] and allows some mechanical contact between the adsorbate and the adsorbent [18]. This model is expressed with the following expression [42]:

$$q_e = q_m \left(1 - e^{K_J C_e}\right) \tag{32}$$

Its linear form is [18]:

$$q_t = K_J t^{\frac{1}{2}} \tag{33}$$

where K_J is the equilibrium constant of Jovanovic (L mg^{-1}). When plotting ln q_e versus C_e, the slope and the intercept are K_J and q_m, respectively.

2.15 Kiselev isotherm

This model is known as the model of the localized monomolecular layer and is only valid when $\theta \geq 0.68$. Its linear expression is [18]:

$$\frac{1}{C_e \times (1 - \theta)} = K_i \frac{1}{\theta} + K_i K_n \tag{34}$$

where K_i is Kiselev's constant (L mg^{-1}), and K_n is the equilibrium constant of complex formation between the molecules of the adsorbate. Constants are calculated by plotting $1/[C_e \times (1 - \theta)]$ versus $1/\theta$ where the slope and the intercept represent K_i and $K_i \times K_n$, respectively [18].

2.16 Evaluating the suitability of the isothermal equations using experimental data

Statistical tools are used to evaluate the suitability of the values calculated with the selected isotherm equation in comparison with the experimentally determined values. To measure the differences between the experimentally observed values and the values calculated with the isothermal model, the so-called goodness of fit is used. Among the best known are the following:

2.16.1 Average relative error (%ARE)

To evaluate the goodness of fit of the isothermal equations against the experimentally obtained data, the average relative error (% ARE) can be used. It is explained by the following equation:

$$\%ARE = \frac{100}{n} \sum_{i=1}^{n} \left\| \frac{q_{i,cal} - q_{i,exp}}{q_{i,exp}} \right\| \tag{35}$$

where n is the data number, $q_{i,cal}$ are the equilibrium values calculated with the mathematical isotherm expression (mg g^{-1}), and $q_{i,exp}$ are values obtained experimentally (mg g^{-1}) [14, 28, 43]. An average relative error lower than or equal to 5% is considered adequate.

2.16.2 Chi-square (χ^2)

To determine the best-fitting isothermal model, the linear Chi-square values (χ^2) were used along with the linear regressions (R^2). The Chi-square test statistic is basically the sum of the squared errors of the differences between the experimental data and the data obtained by calculations with the models. Each square difference is divided by the corresponding data obtained by calculations using the models. If the values obtained using a model are similar to the experimental values, the value of χ^2 is very small and close to zero. High values of χ^2 imply a high discrepancy between the experiment and the model. Therefore, the analysis of the dataset of the Chi-square test may confirm the isotherm that best fits the adsorption system. The mathematical expression of the Chi-square test is explained below [29].

$$x^2 = \sum_{i=1}^{n} \frac{\left(q_{i,exp} - q_{i,cal}\right)^2}{q_{i,cal}} \tag{36}$$

If values $\chi^2 \leq 0.05$, the experimental data and the data obtained by calculations using the models have a statistically significant association.

3. Conclusions

The adsorption process of various compounds and elements is one of the most important environmental issues of the last decade due to its relevance, especially in emerging countries. To better understand how a solute can interact with the surface of a solid, it is useful to use the adsorption isotherm.

The effect of different variables on the adsorption process is of great importance for the development of treatment systems. The use of isotherms helps to determine these effects.

The analysis of equilibrium data is necessary to understand and interpret the adsorption mechanism and predict the removal of contaminants. They are necessary to know the mechanisms or transport phenomena that control the adsorption rate and thus be able to calculate, scale, and design an adsorption treatment system.

The obtained equilibrium data used to calculate isotherms can be used to scale up to a pilot plant experiment.

Among all the phenomena that determine the mobility of substances in porous aqueous media and in the aquatic environment, the transfer of substances from a mobile phase (liquid or gas) to a solid phase is the relevant phenomenon.

An isotherm is a curve describing the retention of a substance in a solid at different concentrations. It is an important tool for describing and predicting the mobility of this substance in the environment.

The adsorption process is a widely accepted treatment method because it is easy to apply and effective even at low concentrations. Adsorption has advantages over other treatment methods because it has low start-up and installation costs, is easy to operate, has low environmental risk, is resistant to toxic components, and offers significant potential for removing hazardous and unsafe contaminants.

The description of equilibrium data is valuable when using an equation that has a physical meaning, as it is then possible to correlate what is observed in one experiment with what is physically seen in a larger experiment.

Conflict of interest

The authors declare no conflict of interest.

Author details

Andres Abin-Bazaine[1]*, Alfredo Campos Trujillo[1] and Mario Olmos-Marquez[2]

1 Advanced Materials Research Center (CIMAV), Chihuahua, Mexico

2 Faculty of Animal Science and Ecology (FZYE), Chihuahua, Mexico

*Address all correspondence to: abinsdreamkennel@gmail.com

IntechOpen

References

[1] Misran E, Bani O, Situmeang EM, Purba AS. Banana stem based activated carbon as a low-cost adsorbent for methylene blue removal: Isotherm, kinetics, and reusability. Alexandria Engineering Journal. 2022;**61**:1946-1955. DOI: 10.1016/j.aej.2021.07.022

[2] Tonk S, Aradi LE, Kovács G, Turza A, Rápó E. Effectiveness and characterization of novel mineral clay in Cd^{2+} adsorption process: Linear and non-linear isotherm regression analysis. Water (Switzerland). 2022;**14**:1-25

[3] Demiral H, Güngör C. Adsorption of copper(II) from aqueous solutions on activated carbon prepared from grape bagasse. Journal of Cleaner Production. 2016;**124**:103-113

[4] Motsi T, Rowson NA, Simmons MJH. Adsorption of heavy metals from acid mine drainage by natural zeolite. International Journal of Mineral Processing. 2009;**92**:42-48. DOI: 10.1016/j.minpro.2009.02.005

[5] Shukla PR, Wang S, Ang HM, Tadé MO. Synthesis, characterisation, and adsorption evaluation of carbon-natural-zeolite composites. Advanced Powder Technology. 2009;**20**:245-250. DOI: 10.1016/j.apt.2009.02.006

[6] Ghasemi M, Javadian H, Ghasemi N, Agarwal S, Kumar V. Microporous nanocrystalline NaA zeolite prepared by microwave assisted hydrothermal method and determination of kinetic, isotherm and thermodynamic parameters of the batch sorption of Ni (II). Journal of Molecular Liquids. 2016; **215**:161-169. DOI: 10.1016/j. molliq.2015.12.038

[7] Park D, Yun Y-S, Park JM. The past, present, and future trends of

biosorption. Biotechnology and Bioprocess Engineering. 2010;**15**:86-102. Available from: http://link.springer.com/ 10.1007/s12257-009-0199-4

[8] Abdel Salam OE, Reiad NA, ElShafei MM. A study of the removal characteristics of heavy metals from wastewater by low-cost adsorbents. Journal of Advanced Research. 2011;**2**: 297-303. DOI: 10.1016/j.jare.2011.01.008

[9] Lin J, Zhan Y, Zhu Z. Adsorption characteristics of copper(II) ions from aqueous solution onto humic acid-immobilized surfactant-modified zeolite. Colloids and Surfaces A: Physicochemical and Engineering Aspects. 2011;**384**:9-16. DOI: 10.1016/j. colsurfa.2011.02.044

[10] Gupta VK, Mittal A, Mittal J. Batch and bulk removal of hazardous colouring agent rose Bengal by adsorption techniques using bottom ash as adsorbent. RSC Advances. 2012;**2**: 8381-8389

[11] Malamis S, Katsou E. A review on zinc and nickel adsorption on natural and modified zeolite, bentonite and vermiculite: Examination of process parameters, kinetics and isotherms. Journal of Hazardous Materials. 2013; **252–253**:428-461. DOI: 10.1016/j. jhazmat.2013.03.024

[12] Kim N, Park M, Park D. A new efficient forest biowaste as biosorbent for removal of cationic heavy metals. Bioresource Technology. 2015;**175**: 629-632. DOI: 10.1016/j. biortech.2014.10.092

[13] Yan L, Huang Y, Cui J, Jing C. Simultaneous As(III) and Cd removal from copper smelting wastewater using granular TiO_2 columns. Water Research.

2015;**68**:572-579. DOI: 10.1016/j. watres.2014.10.042

[14] Demiral H, Güngör C. Adsorption of copper(II) from aqueous solutions on activated carbon prepared from grape bagasse. Journal of Cleaner Production. 2016;**124**:103-113. Available from: https://www.sciencedirect.com/science/article/pii/S0959652616002651#fig 6 [cited 2018 Feb 19]

[15] Motsi T, Rowson NA, Simmons MJH. Adsorption of heavy metals from acid mine drainage by natural zeolite. International Journal of Mineral Processing. 2009;**92**:42-48. DOI: 10.1016/j.minpro.2009.02.005

[16] Shukla PR, Wang S, Ang HM, Tadé MO. Synthesis, characterisation, and adsorption evaluation of carbon-natural-zeolite composites. Advanced Powder Technology. 2009;**20**:245-250. DOI: 10.1016/j.apt.2009.02.006

[17] Li C, Zhong H, Wang S, Xue J, Zhang Z. A novel conversion process for waste residue: Synthesis of zeolite from electrolytic manganese residue and its application to the removal of heavy metals. Colloids and Surfaces A: Physicochemical and Engineering Aspects. 2015;**470**:258-267. DOI: 10.1016/j.colsurfa.2015.02.003

[18] Ayawei N, Ebelegi AN, Wankasi D. Modelling and interpretation of adsorption isotherms. Journal of Chemistry. 2017;**2017**:1-11

[19] Sadeek SA, Negm NA, Hefni HHH, Abdel Wahab MM. Metal adsorption by agricultural biosorbents: Adsorption isotherm, kinetic and biosorbents chemical structures. International Journal of Biological Macromolecules. 2015;**81**:400-409. DOI: 10.1016/j.ijbiomac.2015.08.031

[20] Li X, Zhou H, Wu W, Wei S, Xu Y, Kuang Y. Studies of heavy metal ion adsorption on chitosan/sulfydryl-functionalized graphene oxide composites. Journal of Colloid and Interface Science. 2015;**448**:389-397

[21] Malamis S, Katsou E. A review on zinc and nickel adsorption on natural and modified zeolite, bentonite and vermiculite: Examination of process parameters, kinetics and isotherms. Journal of Hazardous Materials. 2013; **252–253**:428-461. DOI: 10.1016/j.jhazmat.2013.03.024

[22] Humelnicu I, Băiceanu A, Ignat ME, Dulman V. The removal of basic blue 41 textile dye from aqueous solution by adsorption onto natural zeolitic tuff: Kinetics and thermodynamics. Process Safety and Environment Protection. 2017; **105**:274-287. Available from: https://www.sciencedirect.com/science/article/pii/S0957582016302865?via%3Dihub

[23] Zendelska A, Golomeova M, Blazev K, Krstev B, Golomeov B. Equilibrium studies of zinc ions removal from aqueous solutions by adsorption on natural zeolite. Journal of Materials Engineering. 2014;**4**:202-208. Available from: http://eprints.ugd.edu.mk/id/eprint/11369

[24] Li X, Zhou H, Wu W, Wei S, Xu Y, Kuang Y. Studies of heavy metal ion adsorption on chitosan/Sulfydryl-functionalized graphene oxide composites. Journal of Colloid and Interface Science. 2015;**448**:389-397. DOI: 10.1016/j.jcis.2015.02.039

[25] Arshadi M, Amiri MJ, Mousavi S. Kinetic, equilibrium and thermodynamic investigations of Ni(II), Cd(II), Cu(II) and Co(II) adsorption on barley straw ash. Water Resources and Industry. 2014;**6**:1-17. DOI: 10.1016/j.wri.2014.06.001

[26] Angin D. Utilization of activated carbon produced from fruit juice industry solid waste for the adsorption of yellow 18 from aqueous solutions. Bioresource Technology. 2014;**168**: 259-266. DOI: 10.1016/j. biortech.2014.02.100

[27] Ben-Ali S, Jaouali I, Souissi-Najar S, Ouederni A. Characterization and adsorption capacity of raw pomegranate peel biosorbent for copper removal. Journal of Cleaner Production. 2017;**142**: 3809-3821. DOI: 10.1016/j. jclepro.2016.10.081

[28] Rajabi M, Mirza B, Mahanpoor K, Mirjalili M, Najafi F, Moradi O, et al. Adsorption of malachite green from aqueous solution by carboxylate group functionalized multi-walled carbon nanotubes: Determination of equilibrium and kinetics parameters. Journal of Industrial and Engineering Chemistry. 2016;**34**:130-138. DOI: 10.1016/j. jiec.2015.11.001

[29] Tran HN, You SJ, Chao HP. Thermodynamic parameters of cadmium adsorption onto orange peel calculated from various methods: A comparison study. Journal of Environmental Chemical Engineering. 2016;**4**: 2671-2682. DOI: 10.1016/j. jece.2016.05.009

[30] Benzaoui T, Selatnia A, Djabali D. Adsorption of copper (II) ions from aqueous solution using bottom ash of expired drugs incineration. Adsorption Science and Technology. 2017;**1-2**:114-129. Available from: http:// journals.sagepub.com/doi/10.1177/ 0263617416685099

[31] Wu X, Zhou H, Zhao F, Zhao C. Adsorption of Zn^{2+} and Cd^{2+} ions on vermiculite in buffered and unbuffered. Adsorption Science & Technology. 2010; **3**:907-920

[32] Foo KY, Hameed BH. Insights into the modeling of adsorption isotherm systems. Chemical Engineering Journal. 2010;**156**:2-10. Available from: https:// www.sciencedirect.com/science/article/ pii/S1385894709006147

[33] Sogut EG, Caliskan N. Isotherm and kinetic studies of Pb(II) adsorption on raw and modified diatomite by using non-linear regression method. Fresenius Environmental Bulletin. 2017;**26**: 2721-2729. Available from: https://www. researchgate.net/publication/316276207

[34] Shahul Hameed K, Muthirulan P, Meenakshi SM. Adsorption of chromotrope dye onto activated carbons obtained from the seeds of various plants: Equilibrium and kinetics studies. Arabian Journal of Chemistry. 2017;**10**: S2225-S2233. DOI: 10.1016/j. arabjc.2013.07.058

[35] Vijayaraghavan K, Padmesh TVN, Palanivelu K, Velan M. Biosorption of nickel(II) ions onto *Sargassum wightii*: Application of two-parameter and three-parameter isotherm models. Journal of Hazardous Materials. 2006;**133**:304-308

[36] Dlugosz O, Banach M. Kinetic, isotherm and thermodynamic investigations of the adsorption of Ag^+ and Cu^{2+} on vermiculite. Journal of Molecular Liquids. 2018;**258**:295-309

[37] Hamdaoui O, Naffrechoux E. Modeling of adsorption isotherms of phenol and chlorophenols onto granular activated carbon. Part II. Models with more than two parameters. Journal of Hazardous Materials. 2007;**147**:401-411

[38] Nagy B, Mânzatu C, Măicăneanu A, Indolean C, Barbu-Tudoran L, Majdik C. Linear and nonlinear regression analysis for heavy metals removal using *Agaricus bisporus macrofungus*. Arabian Journal of Chemistry. 2017;**10**:S3569-S3579

[39] Ekebafe LO, Ogbeifun DE, Okieimen FE. Equilibrium, kinetic and thermodynamic studies of Lead(II) sorption on hydrolyzed starch graft copolymers. Journal of Polymers and the Environment. 2017;**26**:807-818. Available from: http://link.springer.com/10.1007/s10924-017-0949-x

[40] Amin MT, Alazba AA, Shafiq M. Adsorptive removal of reactive black 5 from wastewater using bentonite clay: Isotherms, kinetics and thermodynamics. Sustainability. 2015;7: 15302-15318

[41] Kaveeshwar AR, Ponnusamy SK, Revellame ED, Gang DD, Zappi ME, Subramaniam R. Pecan shell based activated carbon for removal of iron(II) from fracking wastewater: Adsorption kinetics, isotherm and thermodynamic studies. Process Safety and Environment Protection. 2018;**114**:107-122. DOI: 10.1016/j.psep.2017.12.007

[42] Farouq R, Yousef NS. Equilibrium and kinetics studies of adsorption of copper(II) ions on natural biosorbent. International Journal of Chemical Engineering and Applications. 2015;**6**: 319-324. Available from: http://www.ijcea.org/index.php?m=content&c=index&a=show&catid=66&id=854

[43] Rangabhashiyam S, Anu N, Giri Nandagopal MS, Selvaraju N. Relevance of isotherm models in biosorption of pollutants by agricultural byproducts. Journal of Environmental Chemical Engineering. 2014;**2**:398-414. DOI: 10.1016/j.jece.2014.01.014

[44] Ksakas A, Tanji K, El Bali B, Taleb M, Kherbeche A. Removal of Cu (II) ions from aqueous solution by adsorption using natural clays: Kinetic and thermodynamic studies. Journal of Materials Science and Environmental Science. 2018;**9**:1075-1085. Available

from: https://www.researchgate.net/profile/Adil_Ksakas/publication/323385122_Removal_of_Cu_II_Ions_from_Aqueous_Solution_by_Adsorption_Using_Natural_Clays_Kinetic_and_Thermodynamic_Studies/links/5a91d8f3a6fdccecff04016a/Removal-of-Cu-II-Ions-from-Aqueous-Soluti

Section 2

Wastewater Treatment Technologies

Chapter 3

Available Technologies for Wastewater Treatment

Ifeanyi Michael Smarte Anekwe, Jeremiah Adedeji, Stephen Okiemute Akpasi and Sammy Lewis Kiambi

Abstract

During the last three decades, environmental challenges related to the chemical and biological pollution of water have become significant as a subject of major concern for society, public agencies, and the industrial sector. Most home and industrial operations generate wastewater that contains harmful and undesirable pollutants. In this context, it is necessary to make continuous efforts to protect water supplies to ensure the availability of potable water. To eliminate insoluble particles and soluble pollutants from wastewaters, treatment technologies can be employed including physical, chemical, biological (bioremediation and anaerobic digestion), and membrane technologies. This chapter focuses on current and emerging technologies that demonstrate outstanding efficacy in removing contaminants from wastewater. The challenges of strengthening treatment procedures for effective wastewater treatment are identified, and future perspectives are presented.

Keywords: anaerobic digestion, bioremediation, coagulation, expanded granular sludge bed, ion exchange, membrane technology, microfiltration, nanofiltration

1. Introduction

Wastewater is produced as a result of human and industrial activities. Different kinds of firms are emerging because of ever-changing needs and demands, and as a response, numerous new pollutants are deposited in wastewater, necessitating the development of advanced treatment techniques. To manage ever-changing wastewater discharges, advanced methods are essential, and there is always a connection between water and energy. Although it is impossible to completely eliminate wastewater formation because no business is 100% efficient, however, it is feasible to develop novel and improve existing wastewater treatment and reuse methods to satisfy water demand. Moreover, water reuse has an enormous prospective for replenishing water resource portfolios that are already overburdened.

Since wastewater treatment and reuse are linked to public health, they are extremely important. The existence of pathogenic organisms and polluted substances in wastewater presents the possibility of harmful health effects where contact, inhalation, or ingestion of substance or microbiological elements of health concern occurs. The impact of several factors (such as pH, temperature, colour, and particle matter) and chemical components (cations, anions, and heavy metals) on human health have already been proven, and acceptable thresholds have been set. However, if industrial emission comprises a major portion of the wastewater, the influence of organic elements in treated water utilized for non-potable activities requires investigation [1]. Furthermore, while modern technologies can assist in reducing energy consumption and improving reliability, the difficulties in human understanding can be even more worrisome. Past and contemporary proof of disease carried by water (such as cholera, typhoid, malaria, dengue fever, and anaemia) has sparked public debate about the safety of reusing water [2]. On-line sensors, membranes, and enhanced oxidation mechanisms are examples of sophisticated technology that can aid to alleviate this impression. Nevertheless, a clearer knowledge of the processes of reuse and the qualities of reused water in comparison to freshwater resources will lead to a more favorable public opinion.

Wastewater treatment is an eco-friendly process because it protects the ecosystem by releasing less contamination; it employs sustainable resources; it offers the opportunity for unused products to be recycled, and it manages leftover wastes in a more biologically acceptable manner. The features and kinds of contaminants contained in the water, as well as the anticipated use of treated water, influence the choice of treatment technique. Activated sludge mechanisms and anaerobic digestion are century-old methods that continue to work well and have become the treatment of choice [3]. Emerging pollutants in wastewater and rising wastewater loads in water bodies necessitate immediate studies in this field to provide safe and clean water while also ensuring freshwater supplies. With this goal in mind, this chapter focuses on research into the present and emerging wastewater treatment and reuse technologies while highlighting their limitations and prospects [4].

2. Wastewater treatment technologies

Physical, chemical, biological, and combined technologies are commonly used in wastewater treatment facilities. Primary, secondary, and tertiary treatment procedures make up a conventional wastewater treatment plant (WWTP). Primary processes consist of screening, filtration, centrifugation, sedimentation, coagulation, and flotation. Biological treatment, which can be oxic or anoxic, is the most common secondary procedure while oxidation, precipitation, reverse osmosis, electrolysis, and electrodialysis are examples of tertiary treatment. Advanced oxidation processes (AOPs), ion exchange, ultra and nanofiltration, adsorption/biosorption, and advanced biological treatment combining algae, bacteria, and fungi are all emerging treatment methods that offer healthy and clean treated water [3].

2.1 Physical wastewater treatment technologies

Physical methods, in which physical forces are utilized to remove contaminants, were among the first wastewater treatment technologies used. They are still used in most wastewater treatment process flow systems. These methods are typically employed when water is heavily polluted. The most often used physical wastewater treatment methods are:

2.1.1 Screening, filtration, and centrifugal separation

The first phase in a wastewater treatment operation is screening. The purpose of screening is to eliminate solid waste from wastewater, and it is applied to remove items such as faecal solids, fibre, cork, hair, fabric, kitchen trash, wood, paper, cork, and so on. As a result, different-sized screens are utilized, the size of which is dictated by the requirement, i.e. the size of the particles in the wastewater.

In the filtering process, water is filtered in via a substance having fine holes. This is usually done with a set-up having pore diameters ranging from 0.1 to 0.5 mm. It is used to remove suspended particles, greases, oils, germs, and other contaminants. Membranes and cartridges are examples of filters that can be employed. Filtration can remove particles smaller than 100 mg l^{-1}, as well as oil smaller than 25 mg l^{-1}, reducing it by up to 99%. For water purification, the filtering process is used. Filtration water is utilized in ion exchange, adsorption, and membrane separation processes. Furthermore, filtration devices create potable water [5, 6].

To remove suspended noncolloidal particles, centrifugal separation is performed (size up to 1 mm). Solids (sludges) are separated and released after the wastewater is put to centrifugal devices and rotated at different speeds. Suspended solids segregate to a degree proportional to their densities. Furthermore, the centrifugal machine's speed is also important for the removal of suspended materials. Oil and grease separation, as well as source reduction, are examples of applications.

2.1.2 Sedimentation and gravity separation

This process removes suspended particles, grits, and silts by leaving water undisturbed/semi-disturbed in various types of tanks for varied time intervals. Under the pull of gravity, the suspended solids settle [5–8]. The size and density of the solids, as well as the velocity of the water if it is moving, determine the settling time. To speed up the sedimentation process, alums are occasionally utilized. Gravity separation alone can remove up to 60% of suspended particles. Sedimentation is normally carried out before the application of standard treatment methods. It's a cost-effective way to treat waste from the paper and refinery industries. Water is generated for membrane processes, ion exchange, industrial water supply, using this technology. Source reduction is another application of technology.

2.1.3 Coagulation

Non-settleable solids are allowed to settle when suspended solids do not settle down through sedimentation or gravity. Coagulation is the term for this process [5, 7]. It is possible to employ alum, starch, ferrous minerals, aluminum salts, and activated silica. Coagulants made of non-ionic polymers, anionic polymers, and synthetic cationic polymers are also effective, but they are usually more expensive than natural coagulants. The most essential governing parameters in the coagulation process are temperature, pH, and contact time. Specific coagulants are added to biological treatment units to remove bacteria and other organics that may be floating in the water. It's the most significant part of a wastewater treatment unit, and it's used for a variety of purposes, including wastewater treatment, recycling, and pollution removal.

2.1.4 Flotation

A conventional water treatment facility's flotation is a typical and necessary component. Flotation removes suspended particles, greases, oils, biological materials, and other contaminants by attaching them to air or gas [5, 9]. The solids bind to the gas or air and create agglomerates, which float to the water's surface and can be skimmed off easily. Alum, activated silica, and other substances enhance the flotation process. The flotation process is aided by compressed air flowing through the water. Electro-flotation (electro-flocculation) has been utilized for recycling and water treatment for a long time. This method may remove up to 75% of suspended particles while also eliminating up to 95% of grease and oil. It's a promising treatment method for paper and refinery sectors [5].

2.1.5 Membrane technologies of wastewater treatment

Over the last two decades, as an emerging wastewater treatment approach, membrane technology has evolved into a substantial separation technique. The water world has been looking for new solutions as regulatory limits and esthetic criteria for consumer water quality have continued to progress. Membrane technology is an example of a novel technology. Membranes are employed as filters in separation processes in a variety of applications in this technology. Adsorption, sand filters, and ion exchangers are just a few of the technologies they can replace. Water filtration (covering desalination) and purification (such as groundwater and wastewater) are major applications of this technology, as are sectors such as biotechnology and food & beverage [10, 11]. **Table 1** illustrates the pore size different membranes technologies ranges.

2.1.5.1 Ultrafiltration (UF)

Ultrafiltration has been utilized to remediate a wide range of waterways around the world. According to reports, surface waters, including lake waters, rivers, and reservoirs, have been employed in 50% of UF membrane plants. This technology has been used to treat municipal drinking water for over a decade [12]. UF pores are typically between 0.01 and 0.05 mm (roughly 0.01 mm) in diameter or less. Larger organic macromolecules can be retained by UF membranes. They used to be defined by a molecular weight cut-off (MWCO) rather than a definite pore size [13]. Since the osmotic pressure of the feed solution is low, hydrostatic pressures in UF are typically in the range of 2–10 bar. The operation of a pressure-driven UF process can be separated into three distinct pressure ranges based on the relationship of permeate flow on

Membrane process	Transmembrane pressure (kPa)	Pore size (nm)	Removable components
Microfiltration	100–200	100–1000	Suspended solids, bacteria
Ultrafiltration	200–1000	1–100	Macromolecules, viruses, proteins
Nanofiltration	1000–3000	0.5–5	Micropollutants, bivalent ions
Reverse-osmosis	3500–10,000	<1	Monovalent ions, hardness

Table 1.
Pressure-driven membrane process.

applied pressure (i) linearly increasing flux (sufficiently low), (ii) intermediate, (iii) and limiting flux (sufficiently high).

Even though its concentration polarization layer has not formed appreciably in the linearly increasing flux pressure range, the membrane is the only source of permeate flux resistance. Permeate flux in the limiting flux pressure range, on the other hand, is unaffected by the applied pressure. The process performance is primarily determined by these boundary layer phenomena, just as it is in MF [14]. Water and wastewater can be treated in a variety of ways using the UF process, including the manufacture of ultra-pure water for the electronics industry, COD levels are decreasing in maize starch plants, chemical treatment of groundwater combined with selective removal of dissolved hazardous metals, the dairy industry's whey treatment, wine, or fruit juice clarification.

The UF technology has several benefits such as perfect pore size range thus can be applied for the separation of most of the feed components, low energy usage owing to the unavailability of phase transition during separation, and simple and compact design makes it simple to use. In addition, for temperature-sensitive culinary, biological, and pharmaceutical applications, the most advanced membrane separation technology is UF. However, the application of this technology is faced with some drawbacks including an inability to desalinate saltwater because it cannot isolate dissolved salts or low molecular weight species. UF is ineffective at separating macromolecular mixtures; it can only be efficient if the species have a molecular weight difference of 10 times or more.

2.1.5.2 Microfiltration (MF)

Microfiltration is a pressure-driven membrane technology that can retain particles of molecular weight greater than 100 kDa and a diameter smaller than 1000 nm. The membrane pore size determines the separation or retention capacities. MF membrane pore size spans from 100 nm to 10,000 nm. Because the MF pore size is large, the separation pressure is low, ranging from 10 kPa to 300 kPa. Suspended particles, sediments, algae, protozoa, and bacteria are all separated with MF. Furthermore, the separation method is impractical since particles smaller than the pore size pass readily while larger particles are rejected. Darcy's law describes volume flow through MF membranes, where the applied pressure (ΔP) is directly proportional to the flux, J through the membrane:

$$J = A. \Delta P \qquad (1)$$

Where permeability is a constant A containing structural elements like pore size distribution and porosity. MF can be utilized in a variety of industrial settings, where particles with a diameter > 0.1 mm must be controlled in a suspension. The most fundamental operations still rely on cartridge-based dead-end filtering. However, crossflow filtration will gradually replace dead-end filtering in larger-scale applications. Clarification and sterilization of all types of drugs and beverages are two of the most common industrial applications. Ultrapure water in semiconductors, drinking water treatment, wine, beer, and fruit juice clarification, pre-treatment, and wastewater treatment are some of the other applications.

Microfiltration has shown to be viable due to its low energy consumption, operating pressure, and maintenance which result in low operating cost, fouling is not as bad as it could be because of two factors: larger pore sizes and low pressures. The application of this technology is limited due to its sensitivity to oxidizing agents, bacteria and suspended particles can only be eliminated, particles that are hard and sharp can disrupt the membrane, and cleaning pressures of more than 100 kPa can damage the membrane.

2.1.5.3 Nanofiltration (NF)

Nanofiltration is a filtration technology that separates different fluids or ions using membranes. Due to its broader membrane hole structure than the membranes used in RO, "Loose" RO is a term used to describe NF. More salt can pass through the membrane as a result of this. NF is employed in conditions where strong moderate inorganic removal and organic removal are sought since it can function at low pressures, typically 7–14 bars, and absorbs some inorganic salts. NF may concentrate proteins, sugars, bacteria, divalent ions, particles, colors, and other compounds with a molecular weight of more than 1,000 [15]. NF membranes are constructed of aromatic polyamide and cellulose acetate, displaying salt rejection rates ranging from 95% for divalent salts to 40% for monovalent salts and a molecular weight cut-off (MWCO) for organics of 300 [16]. Organics of low molecular weight, including methanol, are unaffected by NF.

Although NF membranes have strong molecular rejection properties for divalent cations such as magnesium and calcium and may be used instead of traditional chemical softening to effectively remove hardness, they can also be utilized to generate drinking water. Organics with a higher molecular weight that cause odor and taste, or that mix with chlorine to produce trihalomethanes or other particles, can be rejected by NF membranes, boosting the effectiveness of downstream disinfection treatments [17]. Rai and co-workers [18] reported using NF for tertiary treatment of distillery effluent, that the NF membrane had a very high separation efficiency for both inorganic and organic chemicals (around 85–95%, 98–99.5%, 96–99.5% removal of TDS, cooler, and COD, respectively). The advantage of nanofiltration is the lower operating pressure, which results in lower energy costs and potential pump and piping investment savings. The most important drawback of NF membranes is the difficulty in controlling membrane pore size and pore size distribution repeatability. Furthermore, NF membranes are prone to fouling, which could result in significant flow reduction.

2.1.5.4 Reverse osmosis (RO)

Reverse osmosis (RO), in general, is the reverse of the osmosis process. When a semi-permeable barrier is established between two solutions, a solvent flows from lower to higher solute concentrations. Reverse osmosis occurs when an external force causes a solvent to flow from a higher to lower solute concentration. The driving force in the typical osmosis process is a drop in the system's free energy, which diminishes as the system seeks to achieve equilibrium. When the system reaches equilibrium, the osmosis process comes to a stop. An external force larger than the osmotic pressure of the system drives the RO process. RO is like other pressure-driven membrane processes; however, other processes employ size exclusion or straining as the mode of separation and RO employs diffusion.

RO membranes are usually dense membranes having pore sizes less than 1 nm. They are generally a skin layer in the polymer matrix. The membrane material (polymer) forms a layer and a web-like structure. The water follows a tortuous path to get permeated through the membrane. RO membranes can reject the smallest entities from the feed. These include monovalent ions, dissolved organic content, and viruses, almost everything that other membrane processes are not capable of. RO membranes can also be used in both cross-flow and dead-end configurations, but on the other hand, crossflow is frequently favored due to its low energy usage and low fouling qualities. Spiral wound modules, in which the membrane is wound around the inner tube, are the most prevalent. RO has several applications, of which desalination is the most important and widely used. RO is also used in wastewater treatment, and dairy and food products.

Using RO technology, desalination of the sea and brackish water is possible when compared to other membrane processes where separation occurs without a phase change. In comparison to other desalting systems, it is compact and hence takes up less space while ensuring low maintenance and easy scalability. High-pressure requirements, energy-intensive process, lower flux, fouling, and the need to pre-treat feed before use are some of the shortcomings of RO.

2.1.5.5 Forward osmosis (FO)

The FO process is a designed osmotic process in which the treated water is on one side of a semi-permeable membrane and a draw solution (DS) is on the other. Even though FO is built on the osmosis principle, the word "forward osmosis" (FO) was most likely coined to differentiate it from "reverse osmosis," which has been the term for membrane desalination technology for decades. Forward osmosis (FO) employs a concentrated draw solution to create high osmotic pressure, which extracts water from the feed solution across a semi-permeable membrane [19]. As a result, the volume of the feed stream drops, the salt concentration rises, and the permeate flux to the draw solution side reduces [20]. The general equation characterizing water movement over the RO membrane, according to Lee et al. [21], is:

$$J_W = A(\sigma\Delta\pi - \Delta P) \tag{2}$$

where J_W is the water flux, A is the membrane's water permeability coefficient, $\sigma\Delta\pi$ the effective osmotic pressure difference in reverse osmosis, σ is the reflection coefficient, and ΔP is the applied pressure; for FO, $\Delta P = 0$; for RO, $\Delta P > \Delta\pi$ [21]. Since the parameter A and the reflection coefficient are calculated using the pressure applied to the brine, this equation is not suited for FO operations; also, the driving force employed is the difference between osmotic pressure and the applied hydraulic pressure (ΔP) [22, 23]. **Figure 1** displays the principles of osmotic processes.

The primary benefit of FO is how little energy is required to extract pure water from wastewater or recycled feed, with just the energy needed to recirculate the draw solution requiring additional energy [18]. The ultimate flux reduction of concentration polarization is a fundamental limiting element impacting the performance of FO systems [25, 26]. Since forward osmosis is gaining attention as a viable method for lowering the cost of wastewater treatment and generating freshwater, many potential applications for FO membranes have been investigated, including desalination, dilute industrial wastewater concentration, direct potable reuse for enhanced life support systems, food processing, landfill leachate concentration, pharmaceutical industry processes, and concentration of digested sludge liquids [26].

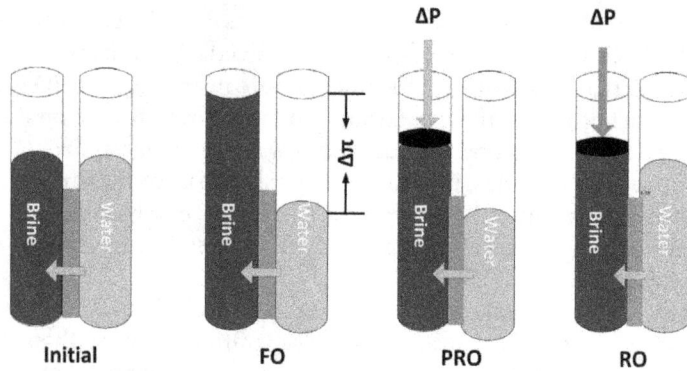

Figure 1.
Principles of osmotic processes: the initial state of the solutions, forward osmosis (FO), pressure retarded osmosis (PRO) and reverse osmosis (RO), adapted from Rao [24].

2.2 Chemical wastewater treatment technologies

Chemical methods employed in waste-water treatment are designed to create change through chemical reactions. They are always combined with physical and biological methods. Chemical methods, in comparison to physical ones, have an inherent disadvantage considering that they are additive processes. That is, the dissolved elements of wastewater usually increase. If the wastewater is to be reused, this is an important consideration. A brief description of chemical methods of wastewater treatment is given below.

2.2.1 Neutralization

The pH value of wastewater is adjusted through neutralization. Acids or alkalis are used to neutralize industrial wastewaters after operations such as precipitation and flocculation. Metal-containing acid wastewaters can be treated by adding an alkaline reagent to the acid waste, forming a precipitate, and collecting the precipitate. As a result, the pH of the input solution is adjusted to the optimal range for metal hydroxide precipitation. To meet the overall wastewater treatment objectives, the step is performed before the major phase of wastewater treatment [27].

2.2.2 Precipitation

By lowering their solubilities, dissolved contaminants become solid precipitates, which can be easily skimmed from the water's surface during precipitation [27]. While it effectively removes metal ions and organics, the accumulation of oil and grease may produce precipitation issues. Adding chemicals or reducing the temperature of the water reduces the solubility of dissolved pollutants. Adding organic solvents to the water could theoretically decrease the contaminant's solubility, however, this procedure is costly on a large scale. Precipitates form when these compounds react with soluble contaminants. The most used substances for this function include ferric chloride, lime, ferrous sulphate, sodium bicarbonates, and alum. The most critical moderating parameters for the precipitation process are temperature and pH. Precipitation can eliminate approximately 60% of pollutants [28]. This method can be used to recycle water and remediate wastewater from the chromium and nickel-plating industries. Among the applications are water softening and heavy metal removal and phosphate from water. The handling of the vast amount of sludge produced is the main issue related to precipitation [29, 30].

2.2.3 Ion exchange

An ion exchanger, a solid substance, exchanges hazardous ions in wastewater for non-toxic ions [31–35]. There are two types of ion exchangers: anion and cation exchangers, which can exchange anions and cations, respectively. Ion exchangers are resins with active sites on their surfaces, which might be natural or synthetic. The most used ion exchangers include metha-acrylic resins, zeolites, acrylic, polystyrene sulfonic acid, and sodium silicates. It is a reversible process that utilizes very little energy. Low amounts of inorganics and organics are removed using ion exchange (up to 250 mg l^{-1}). Concentrations of inorganic and organic compounds can be reduced by up to 95%. Potable water production, industrial water, pharmacy, fossil fuels, softening and other sectors are among the applications. It's also being utilized to cut down on pollution. If there is oil, grease, or large quantities of organics and inorganics in the water, it may be necessary to pre-treat it.

2.2.4 Oxidation/reduction

Redox reactions are commonly used in chemical wastewater treatment and potable water treatment. Chlorinated hydrocarbons and pesticides are effectively removed from drinking water using ozone and hydrogen peroxide oxidation methods. Oxidation techniques are utilized in wastewater treatment to remove problematic biodegradable chemicals. Photochemical purification, which uses UV light to create hydroxyl radicals from hydrogen peroxide or ozone, is very effective. These Advanced Oxidation Processes (AOP) destroy antibiotics, cytostatic medications, hormones, and other anthropogenic trace chemicals. Advanced Oxidation Processes (AOPs) are efficient methods to remove organic contamination not degradable through biological processes in water and wastewater. Ozone also helps with the oxidation of iron and manganese in well water. To convert heavy metal ions, for example, into easily dissolvable sulfides, reduction procedures are necessary [36].

2.2.5 Electrodialysis

Ion-selective semi-permeable membranes allow water-soluble ions to pass through them when an electric current passes through them [37, 38]. Ion-selective membranes are ion exchange materials that are selective. They can be anion or cation exchangers, allowing anion and cations to flow out of the system. The technique uses two electrodes to which a voltage is supplied in either a continuous or batch mode. The membranes are arranged in a series or parallel pattern, to obtain the required degree of demineralization [39, 40]. Factors such as pH, temperature, the type of contaminants, membrane selectivities, scaling and fouling of wastewater, the wastewater flow rate, and the volume and design of phases all affect dissolved solids removal. The creation of drinkable water from brackish water is one of the applications. Furthermore, this technology has been utilized to reduce water sources. Total dissolved solids (TDS) concentrations of up to 200 mg l^{-1} can be decreased by electrodialysis by up to 90% [41]. Membrane fouling happened in the same way that reverse osmosis does. Carbon nanotubes have been used in composite membranes to alleviate this problem and increase flow.

2.2.6 Disinfection

Disinfection in wastewater treatment aims to limit the number of microorganisms in the water that will be released back into the environment for later use as irrigation water, bathing water, drinking water, and so on. The quality of the treated water (pH, cloudiness, and other parameters), the type of disinfection used, the disinfectant dosage (time and concentration), and other external conditions all influence disinfection efficiency. Due to the obvious nature of wastewater, which contains several human enteric organisms linked to a variety of waterborne diseases, this technique is critical in waste-water treatment [42]. Physical agents such as heat and light, mechanical means such as screening, sedimentation, and filtration, radiation, primarily gamma rays, chemical agents such as chlorine and its compounds, bromine, iodine, ozone, phenol and phenolic compounds, alcohols, heavy metals, dyes, soaps, and synthetic detergents, quaternary ammonium compounds, hydrogen peroxide, and various alkali and acids are among the most used disinfection methods. Oxidizing chemicals are the most frequent chemical disinfectants, and chlorine is the most widely utilized of these.

2.3 Biological wastewater treatment technologies

Biological water treatment technologies are critical components of a wastewater treatment strategy since they are utilized to produce safe drinking water. Aerobic, anaerobic and bioremediation processes are the techniques employed for this. These operations are outlined below.

2.3.1 Aerobic processes

Aerobic and facultative bacteria cause biodegradable organic matter to break down aerobically when oxygen or air is freely accessible in wastewater in the dissolved form [43, 44]. Temperature, retention time, oxygen availability, and the biological activity of the bacteria all limit the extent of the process. Furthermore, the addition of specific compounds essential for bacterial development may increase the rate at which organic pollutants are biologically oxidized. This approach can remove phosphates, nitrates, volatile organics, dissolved and suspended organics, chemical oxygen demand (COD), biological oxygen demand (BOD), and other pollutants. It is possible to reduce the number of biodegradable organics in the environment by up to 90%. The method's downside is that it produces a huge number of bio-solids, which necessitates additional costly treatment and management. Oxidation ponds, aeration lagoons, and activated sludge processes are used to carry out the aerobic process [44]. The following Eq. (3) gives a simple depiction of aerobic decomposition.

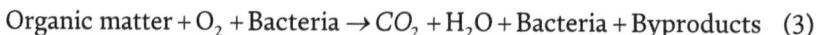

$$\text{Organic matter} + O_2 + \text{Bacteria} \rightarrow CO_2 + H_2O + \text{Bacteria} + \text{Byproducts} \quad (3)$$

2.3.1.1 Oxidation pond

Oxidation ponds are aerobic systems in which the heterotrophic microbes consume oxygen that is supplied by both the atmosphere and photosynthetic algae. In this process, algae utilize the inorganic substances (N, P, CO_2) generated by aerobic bacteria to fuel their growth, which is powered by sunlight. They discharge oxygen into the fluid, which the bacteria then use to complete the symbiotic cycle [44].

2.3.1.2 Aeration lagoon

Aeration lagoons are deeper than oxidation ponds, because aerators supply oxygen rather than algal photosynthetic activity, as in oxidation ponds. The aerators maintain the microbial biomass afloat and supply enough dissolved oxygen for the aerobic process to be maximized. Although there is no deposition or sludge return, this process relies on properly mixed liquor formation in the tank/lagoon. As a result, aeration lagoons are appropriate for effluent that is both strong and biodegradable, such as wastewater from the food industry [44].

2.3.1.3 Activated sludge

The activated sludge method works by suspending a substantial bacterial colony in wastewater under aerobic conditions. Greater levels of bacterial proliferation and respiration can be achieved with limitless nutrients and oxygen, resulting in the conversion of accessible organic compounds to oxidized end-products or the formation of new microbes. The activated sludge system is comprised of five interconnected components: bioreactor, activated sludge, aeration and mixing system, sedimentation tank, and returned sludge [44]. The biological mechanism employing activated sludge is a widely utilized technology for wastewater remediation that has low operating costs.

2.3.2 Anaerobic processes or anaerobic digestion

Anaerobic treatment of waste is a biological process in which microorganisms degrade organic pollutants without oxygen. When there is no free dissolved oxygen in the wastewater, anaerobic breakdown or putrefaction takes place where anaerobic and facultative bacteria break down complex organic substances into sulfur-based organic molecules, carbon, and nitrogen. This sequence of biochemical events produces biogas such as methane, hydrogen sulfide, ammonia, and nitrogen. This approach minimizes the number of bacteria in wastewater [45–47]. Anaerobic technologies are generally used before aerobic treatment for streams with high organic material (measured as high BOD, COD, or TSS). Anaerobic treatment is a tried-and-tested low-energy way of treating industrial effluent. The following Eq. (4) represents the anaerobic process.

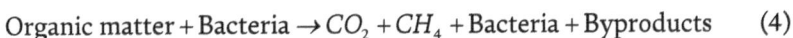

$$\text{Organic matter} + \text{Bacteria} \rightarrow CO_2 + CH_4 + \text{Bacteria} + \text{Byproducts} \qquad (4)$$

The anaerobic digestion (AD) approach is appealing because it treats wastewater, provides renewable energy, and generates byproducts that may be utilized as farm fertilizers, making it an environmentally benign process [48]. When compared to the aerobic wastewater treatment process, the AD process offers the following advantages: fewer nutrients required and the creation of less biological sludge, which requires simply drying as further treatment [49]. It also necessitates a small reactor capacity and no oxygen, reducing the power needed to deliver oxygen in the aerobic approach, and the organic loading on the system is not restricted to an oxygen supply. Thus, a higher loading rate can be used in AD, allowing for a faster response to substrate addition after long periods without feeding and semi-feed strategies for a few months. This benefits the system, making AD a viable option for seasonal industrial wastewater treatment and off-gas elimination that causes air pollution. Examples of anaerobic treatment systems are upflow anaerobic sludge bed (UASB) reactor, expanded granular sludge bed (EGSB), anaerobic baffled reactor (ABR), anaerobic filter reactors and anaerobic Lagoons

2.3.2.1 Upflow anaerobic sludge bed (UASB) reactor

The Upflow anaerobic sludge blanket (UASB) technology is particularly effective for treating wastewater with a high carbohydrate content. As a result, the UASB reactor has become one of the most common designs for treating wastewater from agro-industrial processing companies because it can endure fluctuations in effluent quality and complete reactor shut down during the season [50]. In addition, wastewater containing carbohydrates are readily degraded by bacteria and acts as a nutrient-rich precursor for the anaerobic process. Because of its minimal sludge production and low energy and space requirements, the UASB technique has become well-known for treating wastewater. However, the most significant benefit of this technology is that it can generate energy rather than consume it while treating wastewater [51].

The treated wastewater enters the reactor from the bottom and runs upward through a blanket of biologically activated sludge, typically in granular aggregates. The anaerobic bacteria digest (degrade) the wastewater as it moves upward through the blanket. Under realistic conditions, the blanket is held by the upward flow coupled with gravity's settling action with the support of flocculants and does not wash off, resulting in better treatment efficiency. Intrinsic mixing is facilitated by anaerobic gas production, which aids in the creation and enhancement of biological granules. However, because some of the gas created in the sludge blanket is connected to the granules, a gas-liquid-solid separator (GLSS) is added to the reactor's top for effective gas, liquid, and granule separation. In GLSS, gas-enclosed particles collide with the bottom of degassing baffles, fall back into the sludge blanket, and treated water exits the reactor [52].

2.3.2.2 Expanded granular sludge bed (EGSB)

An improved anaerobic treatment system based on an up-flow anaerobic sludge blanket is the expanded granular sludge bed (EGSB). The differentiating feature is that the wastewater passing through the sludge bed has a faster rate of upward flow velocity. In addition, the enhanced flux allows for partial expansion (fluidisation) of the granular sludge bed, boosting wastewater-sludge interaction and enhancing sludge bed segregation of small inactive, suspended particles

2.3.2.3 Anaerobic baffled reactor (ABR)

McCarty and colleagues created the anaerobic baffled reactor (ABR) at Stanford University in the early 1980s. It is a simple linear reactor with a simple operational design that has widespread use in wastewater treatment. The ABR primarily treats wastewater through sludge and scum retention as well as anaerobic degradation of particulate and dissolvable organic substances. As a result, any factors impacting these processes impact ABR treatment. Baffles guide the flow within the reactor in an ABR reactor under the force of the pressure head at the influent. There is no need for mechanical mixing because the flow directly touches the biomass as it is driven through the sludge bed. As a result, no electricity is required during regular operation for an underground ABR design, while ABR above ground design necessitates pumping energy. In ABR, byproduct sludge is recirculated, discharged, or used as manure.

According to Reynaud and Buckley [53], a long solid retention time is required for anaerobic treatment of low-strength wastewater, and the required reactor capacity is influenced by the hydraulic load instead of the organic load. The upflow velocity of the wastewater inside the reactor compartments containing sludge influences solid retention in the ABR design. Low-strength applications, on the other hand, have negligible solid flotation as well as carry-over due to gas production.

2.3.2.4 Anaerobic filter reactors

In 1969, Young and McCarty invented the upflow anaerobic filter. An anaerobic filter was the first high-rate bioreactor that excluded the separation and effluent recycling requirement. In addition, it offers the advantages of eliminating the mechanical mixing stage, having improved stability even at loading rates higher than 10 kg/m^3 day COD, enduring hazardous shock loads, and being inhibitor-resistant. Because the upflow anaerobic filter is loaded with inert support material such as gravel, pebbles, coke, or plastic media, it works similarly to an aerobic trickling filter. As a result, there is no need for biomass separation or sludge recycling in the system. The reactor's designation is to trap particles in the wastewater as it runs through it, while active biomass connected to the surface of the filter material degrades the organic matter [43]. The anaerobic filter reactor can be used as a downflow or upflow filter reactor, with an OLR range from 1 kg/m^3 to 15 kg/m^3 day COD and separation efficiencies ranging from 75 to 95%. The treatment temperature ranges from 20 to 35.8°C, with HRTs varying from 0.2 to 3 days. The main disadvantage of the upflow anaerobic filter is the possibility of blockage due to undegraded sewage sludge, mineral precipitates, or bacterial biomass [43].

2.3.2.5 Anaerobic lagoons

An anaerobic lagoon is a deep earthen basin with enough volume to allow sedimentation of sedimentable solids, digestion of residual sludge, and anaerobic reduction of some soluble organic substrate [54]. Anaerobic lagoons are typically designed to store and treat wastewater for 20–150 days. They're deep (normally 8–15 feet) and function similarly to septic tanks, where anaerobic microorganisms break down contaminants in the absence of oxygen. Solids in wastewater segregate and settle into strata inside an anaerobic lagoon. Grease, scum, and other floating debris make up the top layer. The layer of sludge that settles at the bottom of an anaerobic lagoon gradually accumulates and must be removed if septic tanks are not used first. The effluent from an anaerobic lagoon will need to be treated further [55].

2.3.3 Bioremediation

Bioremediation is a biological treatment process that uses biological resources to convert environmental pollutants into less hazardous forms. For example, the innate ability of microorganisms, plants, bacteria, algae or fungi to survive, adapt and thrive in unseemingly harsh conditions has been exploited to treat contaminated water bodies or soils. Like any other biological treatment process, bioremediation is preferred because it does not require chemicals or a lot of energy. This technology can be applied both in-situ (on-site) or ex-situ; for example, the wastewater can be treated on-site where the pollution takes place or transported to an external site for proper manipulation of the operating condition if it cannot be achieved at the contaminated site. Bioremediation can occur in either aerobic or anaerobic environments. Living organisms require ambient oxygen to thrive in aerobic environments. There is no oxygen in anaerobic situations. Microbes in this situation decompose chemical molecules or ions like sulfates in the wastewater to obtain the required energy [56].

Bioremediation is broadly classified into the following;

i. Microbial bioremediation—employs microorganisms as food sources to break down contaminants.

ii. Mycoremediation—breaks down contaminants using the digestive enzymes of fungi.

iii. Phytoremediation—employs plants to extract, break down and clean up contaminants.

Microbial remediation and mycoremediation can be classified further based on the strategy used as bioattenuation (natural attenuation), biostimulation (use of organic or inorganic nutrients for remediation), and bioaugmentation (use of genetically engineered microbe).

3. Limitations and prospects of wastewater treatment technologies

3.1 Physical and chemical technologies

Conventional wastewater treatment methods are currently beset by several issues, including increased chemical usage, sludge disposal, and increased energy and space needs. Furthermore, effective elimination of recalcitrant organic components, the inability to handle more wastewater than the limited design capacity, and a scarcity of experienced labour are all major operational issues in these systems. Because of all of these operational and technological limitations in traditional wastewater treatment methods, researchers are working to establish novel categories of advanced wastewater treatment techniques to address the aforementioned issues. Advanced wastewater techniques must integrate membrane technology, Advanced Oxidation Processes, Less sludge formation and if sludge is formed, how to use the sludge rather than disposing of it at the dumpsite, adsorption materials with a low cost, fewer chemical or bioflocculant usage, a new group of nanoparticles for wastewater treatment. Although there is a large body of study on the aforementioned topics, there are still areas that need improvement in the open literature to tackle the concerns of developments in wastewater treatment methods. The employment of modern wastewater technologies in conjunction with traditional methods may lead to more efficient wastewater treatment as well as increased reuse and recycling of treated water.

3.1.1 Membrane technologies

Membrane technology has several drawbacks, including greater energy consumption and fouling. Developing novel membrane materials, calculating hydrodynamics, incorporating modules, and exploring innovative modes of operation to reduce energy usage or application parameters to improve the treatment of water or wastewater are all examples of current advancements linked to membrane technology. All membrane processes have a minimal impact on the environment. There are no hazardous chemicals that must be disposed of, and no heat is generated in the operations. Future trends will include the recovery of valuable compounds, utilization of process waters, technological development including forwarding osmosis and pervaporation, real-time fouling monitoring, the advancement of existing fouling analysis techniques, the creation of custom-made novel membranes, and the development of membranes that can be applied in extreme circumstances. As these objectives are met, capacity, selectivity, and cost, as well as environmental effects including chemical consumption and concentrate handling should be addressed.

Membrane processes play an important role as well. As materials and membrane processes advance, new applications such as new MBRs (membrane bioreactor technologies), advanced osmosis, and pervaporation systems will be accessible. Anaerobic MBRs decompose organic compounds using anaerobic bacteria. In this configuration, biogas can replace the air in the submerged reactor. Due to their lower energy use, MBR systems outperform conventional systems. Since anaerobic MBR systems can retain high biomass concentrations, withstand high organic loadings, recover organic and energy acid, and generate little sludge, they are promising. Another promising technique is microbial fuel cells, a new form of MBR. Decentralized treatment systems can be utilized in wastewater systems to reduce costs and promote sanitation and reuse [57, 58].

3.2 Biological technologies

The biological treatment process is a well-known technique for dealing with problems associated with the treatment of industrial effluents and municipal wastewaters, where conventional technologies have proven to be prohibitively expensive, time-consuming, and ineffective. Though the aerobic technique has been successful in terms of industrial application, there are some drawbacks, such as greater capital costs for aeration facilities, increased operational costs (especially for energy for pumps or aerators), increased maintenance demands, and probably surveillance requirements for detecting the dissolved oxygen content in the liquid. While for the anaerobic treatment post-treatment of wastes generated because treated water does not meet standards, odor generation, fouling/clogging of the membrane, and a slower start-up time are some of the limitations. Bioremediation is only possible with biodegradable chemicals. Not all substances can be completely degraded in a short period. There are concerns that the biodegradation byproducts will be more persistent or dangerous than the main contaminant. Extrapolating some biological technologies from bench and pilot-scale to large scale operations is still challenging. Biological mechanisms are frequently very specialized. The availability of metabolically competent microbial communities, proper environmental growth parameters, and optimum quantities of nutrients and pollutants are all crucial site considerations.

Biological treatment technology is an innovative tool with significant future potential. As scientists understand more about its functionalities, it is possible to become one of the most effective methods for wastewater and environmental remediation. The tremendous improvement of molecular biological technologies has made it possible to analyze the organization of microbial communities without being influenced by cultivation. To achieve effective system operation with diverse functional microorganisms, careful management and modification of environmental parameters are required for system performance. The invention of innovative techniques and new concepts (e.g., new functional components and novel biological metabolism pathways) will facilitate the advancement of biological wastewater remediation systems. The best approach to achieving this goal is interdisciplinary collaboration.

4. Conclusion

The treatment of wastewater is crucial because of its effect on the environment. Due to increased urbanization and industrialization, wastewater generation and treatment have become a growing concern in the twenty-first century. Wastewater treatment ensures the long-term viability of the ecosystem. Many wastewater treatment options are employed to address the problem of growing environmental pollution, including physical, chemical, and biological (primary to tertiary treatment) technologies. The employment of some treatment strategies has the potential to produce secondary contaminants. The effective implementation of wastewater treatment options in water resource management necessitates planning, activity, design, storage, and operation. Advances in wastewater recycling have made it possible to produce water of virtually any quality. Water recovery systems incorporate a variety of safety precautions to reduce the environmental risks associated with various reuse applications. Continuous advancements have been made in the fundamental science of water treatment methods, as well as the innovation used in the process. However, based on the known treatment methods, attaining considerable wastewater treatment with a single treatment technology is difficult. Under the present conditions, improved or integrated wastewater treatment technologies are critically required to ensure high-quality water, reduce chemical and biological pollutants, and enhance industrial production operations. Integrated approaches, which may overcome the limits of single treatment techniques, seem to be viable options for efficient wastewater remediation. Regrettably, most viable treatment techniques are on the small scale and lack commercial application feasibility.

Author details

Ifeanyi Michael Smarte Anekwe[1]*, Jeremiah Adedeji[2], Stephen Okiemute Akpasi[3]
and Sammy Lewis Kiambi[3]

1 School of Chemical and Metallurgical Engineering, University of the
Witwatersrand, Johannesburg, South Africa

2 Department of Chemical Engineering, School of Engineering, University of
KwaZulu-Natal, Durban, South Africa

3 Department of Chemical Engineering, Durban University of Technology, Durban,
South Africa

*Address all correspondence to: anekwesmarte@gmail.com

IntechOpen

References

[1] Crook J, Surampalli RY. Water reclamation and reuse criteria in the US. Water Science and Technology. 1996;**33**(10-11):451-462

[2] Angelakis AN, Snyder SA. Wastewater treatment and reuse: Past, present, and future. Water. 2015;**7**(9):4887-4895

[3] Ding GKC. Wastewater treatment and reuse-The future source of water supply. Encyclopedia of Sustainable technologies. 2017;**2017**:43-52

[4] Krishnamoorthy S, Selvasembian R, Rajendran G, Raja S, Wintgens T. Emerging technologies for wastewater treatment and reuse. Water Science and Technology. 2019;**80**(11):3-4

[5] Tchobanoglous G, Burton FL, Stensel H. Wastewater engineering. Management. 1991;**7**:1-4

[6] Nemerow NL, Dasgupta A. Industrial and hazardous waste treatment. New Jersey: Noyes Publications; 1991

[7] Gupta VK, Ali I, Saleh TA, Nayak A, Agarwal S. Chemical treatment technologies for waste-water recycling—An overview. RSC Advances. 2012;**2**(16):6380-6388

[8] Cheremisinoff NP. Handbook of Water and Wastewater Treatment Technologies. Boston: Butterworth-Heinemann; 2001

[9] Sinev I, Sinev O, Linevich S. Apparatus of flotation treatment of natural waters and wastewater. Izobreteniya. 1997;**26**:369-370

[10] Kurt E, Koseoglu-Imer DY, Dizge N, Chellam S, Koyuncu I. Pilot-scale evaluation of nanofiltration and reverse osmosis for process reuse of segregated textile dyewash wastewater. Desalination. 2012;**302**:24-32

[11] Ozgun H, Ersahin ME, Erdem S, Atay B, Kose B, Kaya R, et al. Effects of the pre-treatment alternatives on the treatment of oil-gas field produced water by nanofiltration and reverse osmosis membranes. Journal of Chemical Technology & Biotechnology. 2013;**88**(8):1576-1583

[12] Kasim NO, Mahmoudi EB, Mohammad AW, Sheikh Abdullah SR. Study on the effect of applied pressure on iron and manganese rejection by polyamide and polypiperazine amide nanofiltration membranes. Solid State Phenomena. 2021;**317**:283-290

[13] Ryu H, Addor Y, Brinkman NE, Ware MW, Boczek L, Hoelle J, et al. Understanding microbial loads in wastewater treatment works as source water for water reuse. Water. 2021;**13**(11):1452

[14] Birrenbach O, Faust F, Ebrahimi M, Fan R, Czermak P. Recovery and purification of protein aggregates from cell lysates using ceramic membranes: Fouling analysis and modeling of ultrafiltration. Frontiers in Chemical Engineering. 2021;**3**:9

[15] Doménech NG, Purcell-Milton F, Gun'ko YK. Recent progress and prospects in development of advanced materials for nanofiltration. Materials Today Communications. 2020;**23**:100888

[16] Hao Y. Black liquor in pulp mill and its treatment. Jakobstad: University of Applied Sciences; 2021

[17] Noyes R. Unit operations in environmental engineering. New Jersey: Noyes Publications; 1994

[18] Choudhury RR, Gohil JM, Mohanty S, Nayak SK. Antifouling, fouling release and antimicrobial materials for surface modification of reverse osmosis and nanofiltration membranes. Journal of Materials Chemistry A. 2018;**6**(2):313-333

[19] Zhu X-Z, Wang L-F, Zhang F, Lee LW, Li J, Liu X-Y, et al. Combined fouling of forward osmosis membrane by alginate and TiO_2 nanoparticles and fouling mitigation mechanisms. Journal of Membrane Science. 2021;**622**:119003

[20] Guo B-B, Zhu C-Y, Xu Z-K. Surface and interface engineering for advanced nanofiltration membranes. Chinese Journal of Polymer Sciences. 2022;**40**:1-14

[21] Lee J, Kim B, Hong S. Fouling distribution in forward osmosis membrane process. Journal of Environmental Sciences. 2014;**26**(6):1348-1354

[22] Fareed H, Qasim GH, Jang J, Lee W, Han S, Kim IS. Brine desalination via pervaporation using kaolin-intercalated hydrolyzed polyacrylonitrile membranes. Separation and Purification Technology. 2022;**281**:119874

[23] Rai B, Shrivastav A. Chapter 26 - Removal of emerging contaminants in water treatment by nanofiltration and reverse osmosis. In: Shah M, Rodriguez-Couto S, Biswas J, editors. Development in Wastewater Treatment Research and Processes. Elsevier; 2022. pp. 605-628

[24] Rao AK, Li OR, Wrede L, Coan SM, Elias G, Cordoba S, et al. A framework for blue energy enabled energy storage in reverse osmosis processes. Desalination. 2021;**511**:115088

[25] Xiang Q, Nomura Y, Fukahori S, Mizuno T, Tanaka H, Fujiwara T. Innovative treatment of organic contaminants in reverse osmosis concentrate from water reuse: A mini review. Current Pollution Reports. 2019;**5**(4):294-307

[26] Bahoosh M, Kashi E, Shokrollahzadeh S. The effect of concentration polarization in the process of water desalination by forward osmosis method. Journal of Environmental Science and Technology. 2020;**22**(3):241-252

[27] Son M-K, Sung H-J, Lee J-K. Neutralization of synthetic alkaline wastewater with CO2 in a semi-batch jet loop reactor. Journal of the Korean Society of Combustion. 2013;**18**(2):17-22

[28] Lelieveld J, Berresheim H, Borrmann S, Crutzen P, Dentener F, Fischer H, et al. Global air pollution crossroads over the Mediterranean. Science. 2002;**298**(5594):794-799

[29] Zinkus GA, Byers WD, Doerr WW. Identify appropriate water reclamat'technologies. Chemical Engineering Progress. 1998;**94**(5):19-32

[30] Iftekhar MS, Blackmore L, Fogarty J. Non-residential demand for recycled water for outdoor use in a groundwater constrained environment. Resources, Conservation and Recycling. 2021;**164**:105168

[31] van der Bom FJ, Kopittke PM, Raymond NS, Sekine R, Lombi E, Mueller CW, et al. Methods for assessing laterally-resolved distribution, speciation and bioavailability of phosphorus in soils. Reviews in Environmental Science and Bio/Technology. 2022;**21**:1-22

[32] Cao R, Liu S, Yang X, Wang C, Wang Y, Wang W, et al. Enhanced remediation of Cr (VI)-contaminated groundwater by coupling electrokinetics with $ZVI/Fe_3O_4/$

AC-based permeable reactive barrier. Journal of Environmental Sciences. 2022;**112**:280-290

[33] Singh R, Mondal P, Purkait MK. pH-responsive membranes. Biomedical Applications. Boca Raton: CRC Press; 2022

[34] Yang C, Wang Y, Alfutimie A. Comparison of nature and synthetic zeolite for waste battery electrolyte treatment in fixed-bed adsorption column. Energies. 2022;**15**(1):347

[35] Srivastava N, Chattopadhyay J, Yashi A, Rathore T. Heavy metals removal techniques from industrial waste water. In: Advanced Industrial Wastewater Treatment and Reclamation of Water. Tunisia: Springer; 2022. pp. 87-101

[36] Tufail A, Price WE, Mohseni M, Pramanik BK, Hai FI. A critical review of advanced oxidation processes for emerging trace organic contaminant degradation: Mechanisms, factors, degradation products, and effluent toxicity. Journal of Water Process Engineering. 2021;**40**:101778

[37] Hussain S, Hussain A, Aziz MU, Song B, Zeb J, George D, et al. A review of zoonotic babesiosis as an emerging public health threat in Asia. Pathogens. 2022;**11**(1):23

[38] Abarkan A, Grimi N, Métayer H, Sqalli Houssaïni T, Legallais C. Electrodialysis can lower the environmental impact of hemodialysis. Membranes. 2022;**12**(1):45

[39] Zhang S, Meng Y, Pang L, Ding Q, Chen Z, Guo Y, et al. Understanding the direct relations between various structure-directing agents and low-temperature hydrothermal durability over Cu-SAPO-34 during NH_3-SCR reaction. Catalysis Science & Technology. 2022;**12**:579-595

[40] Zentner DL, Raabe JK, Cross TK, Jacobson PC. Machine learning applied to lentic habitat use by spawning walleye demonstrates the benefits of considering multiple spatial scales in aquatic research. Canadian Journal of Fisheries and Aquatic Sciences. 2022;**71**(1):120-130

[41] Adhikary S, Tipnis U, Harkare W, Govindan K. Defluoridation during desalination of brackish water by electrodialysis. Desalination. 1989;**71**(3):301-312

[42] Ganguli S, Karmakar R, Singh M, Ghosh MM. Metagenomics-guided assessment of water quality and predicting pathogenic load. In: Handbook of Research on Monitoring and Evaluating the Ecological Health of Wetlands. India: IGI Global; 2022. pp. 71-91

[43] Goli A, Shamiri A, Khosroyar S, Talaiekhozani A, Sanaye R, Azizi K. A review on different aerobic and anaerobic treatment methods in dairy industry wastewater. Journal of Environmental Treatment Techniques. 2019;**6**(1):113-141

[44] Samer M. Biological and chemical wastewater treatment processes. Wastewater Treatment Engineering. 2015;**14**:150

[45] Jin Z, Zhao Z, Liang L, Zhang Y. Effects of ferroferric oxide on azo dye degradation in a sulfate-containing anaerobic reactor: From electron transfer capacity and microbial community. Chemosphere. 2022;**286**:131779

[46] Khan MA, Ngo HH, Guo W, Liu Y, Zhang X, Guo J, et al. Biohydrogen production from anaerobic digestion and its potential as renewable energy. Renewable Energy. 2018;**129**:754-768

[47] Chan YJ, Chong MF, Law CL, Hassell DG. A review on anaerobic–aerobic treatment of industrial and

municipal wastewater. Chemical Engineering Journal. 2009;**155**(1-2):1-8

[48] Ruiz B, Flotats X. Citrus essential oils and their influence on the anaerobic digestion process: An overview. Waste Management. 2014;**34**(11):2063-2079

[49] Buitron G, Kumar G, Martinez-ane production via a two-stage processes (H2-SBR+ CH4-UASB) using tequila vinasses. International Journal of Hydrogen Energy. 2014;**39**(33):19249-19255

[50] Daud MK, Rizvi H, Farhan Akram M, Ali S, Rizwan M, Nafees M, et al. Review of upflow anaerobic sludge blanket reactor technology: Effect of different parameters and developments for domestic wastewater treatment. Journal of Chemistry. 2018;**2018**:1-13. DOI: 10.1155/ 2018/1596319

[51] Sivaram NM, Barik D. Toxic Waste from Leather Industries. Energy from Toxic Org Waste Heat Power Generation. Cambridge: Woodhead Publishing; 2019. pp. 55-67

[52] Mainardis M, Buttazzoni M, Goi D. Up-flow anaerobic sludge blanket (UASB) technology for energy recovery: A review on state-of-the-art and recent technological advances. Bioengineering. 2020;**7**(2):43

[53] Reynaud N, Buckley CA. The anaerobic baffled reactor (ABR) treating communal wastewater under mesophilic conditions: A review. Water Science and Technology. 2016;**73**(3):463-478

[54] Stronach SM, Rudd T, Lester JN. Anaerobic Digestion Processes in Industrial Wastewater Treatment. Heidelberg: Springer, Science & Business Media; 2012

[55] Pal P. Biological treatment technology. Industrial Water

Treatment Processing and Technology. 2017;**1**:65-144

[56] Anekwe IMS, Isa YM. Comparative evaluation of wastewater and bioventing system for the treatment of acid mine drainage contaminated soils. Water-Energy Nexus. 2021;**4**:134-140

[57] Veress M, Bartik A, Benedikt F, Hammerschmid M, Fuchs J, Müller S, et al. Development and techno-economic evaluation of an optimized concept for industrial bio-SNG production from sewage sludge. In: Proceedings of the 28th European Biomass Conference. 2020

[58] Guo H, Li X, Yang W, Yao Z, Mei Y, Peng LE, et al. Nanofiltration for drinking water treatment: A review. Frontiers of Chemical Science and Engineering. 2021;**2021**:1-18

Chapter 4

Sustainable Treatment of Acidic and Alkaline Leachates from Mining and Industrial Activities: Current Practice and Future Perspectives

Thomas F. O'Dwyer, Bashir Ghanim, Ronan Courtney,
Ashlene Hudson, J. Tony Pembroke and John G. Murnane

Abstract

Water resources are under continued pressure from anthropogenic sources, including acidic waste from abandoned mine sites and alkaline waste from a variety of industrial activities. Large quantities of mine and industrial wastes are typically stored in tailings facilities which can generate significant quantities of leachates due to weathering. If released untreated to the aquatic environment these have the potential to contaminate surface and ground waters. In addition, generation of leachates from abandoned or closed sites presents a major long-term environmental challenge where the generation of leachates is expected to continue for decades if not centuries post closure. An overview of leachate production and associated treatment technologies are described, with an emphasis on passive and potentially sustainable technologies. Measures to prevent the formation of acidic leachates and the potential for resource recovery from acidic and alkaline wastes and leachates are also discussed. Finally, technologies that require further development for long term and sustainable treatment are highlighted.

Keywords: mine and industrial wastes, acid mine drainage, alkaline leachates, passive treatment, resource recovery, sustainability, circular economy

1. Introduction

The mining industry generates in excess of 6 billion tonnes of waste annually [1] with significant growth expected in the future. In the EU for example, annual mine waste comprised 26.3% (615 Mt) of the total waste generated in 2018 [2]. Mine waste is generally categorized as i) non-mineralized overburden (typically 2–20 cm diameter), which is removed to access valuable mine ores and stored in spoil heaps, and ii) tailings and process wastewater, arising from the extraction and processing of ores. Tailings from metal mining are enriched with heavy metals (metals with a density > 6 g cm^{-3}),

IntechOpen

whose extraction is no longer economically viable and are typically deposited indefinitely in storage lagoons, often referred to as tailings storage facilities. These storage facilities generate vast quantities of metal rich leachates, which if released to the aquatic environment, can result in elevated bioavailable metal concentrations and sediment loading leading to the stress and death of aquatic organisms and human health.

Leachates are often classified as acidic, alkaline or neutral depending on the geochemistry of the mine tailings and the processing steps utilized in the mining process. Acidic leachates and acid mine drainage (AMD) have a low pH (typically pH < 6) and are generated when sulfidic ores, most commonly pyrite ores (FeS_2, often referred to as 'fool's gold'), which are normally stable in anaerobic underground conditions are exposed to oxygenated environments during mining operations causing the sulfides to oxidize. This process results in acidic conditions (sulfuric acid generated) with associated elevated levels of sulfate, heavy metals and metalloids (semi metals having metallic and non-metallic properties), which if released to the environment result in significant and long-term pollution. The most abundant and common metal in AMD is Fe(II) which reacts with dissolved oxygen to produce iron oxide precipitates. Alkaline leachates on the other hand have a high pH (typically pH > 10) and are generated at disposal sites of industrial by-products such as steel slag, coal ash, municipal waste incinerator ash and bauxite residue from the alumina processing industry. The high alkalinity of these leachates is typically generated from reagents used in industrial processes, such as sodium hydroxide (NaOH) and lime (CaO), and are often enriched with trace metals such as chromium (Cr), vanadium (V), molybdenum (Mo) and gallium (Ga). Neutral leachates (typically pH 6–10) are normally generated from mine wastes low in sulfides or when the oxidation of sulfides is weak or when waste is neutralized by carbonate content in the material [3]. Although neutral, these leachates can contain potentially toxic elements such as water-soluble forms of nickel (Ni), zinc (Zn), cobalt (Co), arsenic (As) and antimony (Sb).

An estimated 3.5 billion tonnes of bauxite residue, a byproduct of alumina refining and more commonly known as red mud, are deposited globally and this amount is increasing at rates of between 120 and 150 Mt. per annum. While it is difficult to accurately predict alkaline leachate quantities generated from these deposits, a global estimate is in the region of 150 million m^3 per annum [4]. These leachates can contain elevated concentrations of metals such as aluminum (Al), potassium (K), sodium (Na), V, Mo, Ga and Ni, which are potentially toxic if released untreated to the aquatic environment. In addition to this, an estimated 30–40 Mt. of incinerated bottom ash and 2–6 Mt. of fly ash are generated annually from incineration of municipal solid waste (MSW). Most of these residues are generated in the EU (33%), China (29%), Japan (20%) and the USA (16%). Despite the varied, significant and potentially valuable metal content of these ashes, most are deposited to landfill with associated generation of metal rich alkaline leachate production [4].

As well as posing a serious threat to the environment and to human health, billions of euros worth of valuable metals contained in industrial and domestic wastes are disposed of in hazardous waste disposal sites [4]. However, metal recovery from these wastes and associated leachates is for the most part technically difficult, uneconomical and unsustainable, primarily because the metals tend to be present in low concentrations and in complex matrices. Nevertheless, resource recovery must remain a priority, particularly in an age where technological advancement is a key driver for global sustainability.

The following sections will examine current treatment options with a focus on passive treatment of acid mine leachates and alkaline industrial leachates. An overview of recent attempts at resource recovery from these leachates will also be discussed before considering future requirements for treatment of acidic and alkaline leachates.

2. Treatment of acid mine leachates

Source control techniques to prevent the formation of AMD is an ideal scenario for the mining industry, which would significantly reduce the environmental burden from mining operations. Source control operates by limiting the exposure of sulfidic waste to air, water or oxidizing bacteria (e.g. sulfide reducing or iron oxidizing bacteria) thereby preventing or reducing its acidification. The most common source control treatments include underwater storage of mine tailings or dry covering with non-reactive materials (oxygen barrier), co-disposal with acid consuming or alkaline producing materials, microencapsulation and passivation. Microencapsulation involves forming an iron hydroxide coating on the surface of the pyrite to inhibit pyrite oxidation and reduce the formation of AMD while addition of a passivation agent facilitates a series of reactions to form a dense inert layer on the surface of the metal sulfide materials which reduces contact with oxygen, water, microorganisms and metal sulfide materials and therefore reduces AMD formation. However, such methods do not always successfully prevent the formation of AMD and are difficult to implement in practice. While source control approaches are a focus for future research [5], a more realistic and common approach is to treat the generated AMD and leachates. Generation of AMD and leachate from tailings storage facilities at both active and historic mine sites is predicted to continue over a multi decadal time span and will therefore need corresponding long-term treatment. However, long-term treatment presents a difficulty, particularly where active treatment processes **Table 1** [6] require indefinite operational and maintenance inputs, which incur large costs, including long-term energy usage and treatment of metal rich sludges [7].

Given the costly, unsustainable and largely unknown operational timescale for active treatment processes for acid mine leachates, there has been a focus in the past 20+ years on passive treatment technologies, which tend to have lower capital

Active treatment process	Summary description
Chemical precipitation	Precipitates are formed by addition of chemicals such as metal hydroxides and are separated from the water by sedimentation and/or filtration
Ion exchange	Synthetic or natural resins are used to exchange cations with soluble metals in the wastewater
Membrane filtration	Technology which uses different types of membrane filtration methods such as ultrafiltration, reverse osmosis, nanofiltration and electrodialysis to separate solutes from the water across semi-permeable membranes
Coagulation and flocculation	Colloidal particles are destabilized by charge neutralization so that they agglomerate into larger flocs which settle more readily as a metal rich sludge

Table 1.
Summary descriptions of some active treatment processes for removal of heavy metals from acid mine leachates.

construction costs and generally rely on gravity rather than pumped flow. They also require much lower operational and supervision inputs, although some level of maintenance will be required to ensure effective removal efficiency. Passive treatment technologies also do not require continuous chemical inputs and are therefore more sustainable than active treatment processes; however their ability to effectively treat mine waste streams in the long-term is largely unknown [7]. The key characteristics of passive treatment systems are their ability to produce alkalinity and to efficiently remove metals from the leachates. Some of the more promising passive treatment technologies are assessed below.

2.1 Neutralization

Given that pH is an important influence on trace metal solubility, passive treatment by neutralization is sometimes used to remove metals from acid mine leachates. One such method is the installation of oxic or anoxic limestone drains where acidic mine leachate is directed through the bedding material and neutralized to a pH $\approx$ 6 by dissolution of the limestone. The alkalinity production and neutralization rates are important criteria when selecting the limestone, as a high carbonate content induces quicker neutralization rates. An operational drawback with limestone drains however is that long-term metal hydroxide precipitation tends to clog the limestone and reduce their flow capacity. This can be overcome somewhat by use of anoxic drains, which inhibit the formation of these precipitates; however, accumulation of other particulate material within the drains also contributes to reduced permeability of the drains over time [8]. In addition, removal of some metals, such as Zn and manganese (Mn), require a pH < 6, which is lower than the pH that can be naturally provided by passive limestone drains. In such cases, alternative or additional treatments such as Dispersed Alkaline Substrate (DAS) systems may be used. These may include application of fine-grained alkaline reagents to provide high neutralizing capacity, such as magnesium oxide (MgO) powder or limestone sand mixed with high porosity inert materials to ensure continuous flow through the medium [9, 10].

Other alkaline waste byproducts, such as fly ash (FA) and bottom ash (BA), flue gas desulphurization material, recycled concrete aggregate (RCA) and alkaline industrial byproducts have also been investigated as potential alternatives to traditional treatment materials. A leach test study to evaluate and compare the efficacy of RCA and FA in remediating AMD found that while RCA's were effective in neutralizing AMD and reduced concentrations of iron (Fe), Cr, copper (Cu) and Zn, FA's actually increased Fe, Cu and Mn concentrations. In addition, RCA with a higher calcium oxide content and finer grained particles had greater efficacy in increasing the pH and reducing concentrations of magnesium (Mg), Mn and Zn in the AMD [11]. In a separate laboratory based filtration study to investigate the effectiveness of alkaline industrial byproducts, namely drinking water treatment residuals (WTR's), to neutralize and remove metals from AMD, the authors reported irreversible removal of more than 99% of Fe, Al, Zn, lead (Pb), As, Mn and 44% of sulfate (SO_4^{2-}) [12].

Natural neutralization of AMD has also been reported in two abandoned alum shale pit lakes, which originally contained acidic waters (pH < 4) with elevated levels of Na, K, Mg, calcium (Ca), Al, Mn, Fe, and sulfate. Inflow of leachates from an adjacent alkaline waste deposit gradually increased the pH from <4 to 8, which resulted in decreased concentrations of Fe, Al, Co, Ni and Zn in the lakes. However, accumulation of metal laden sediments in the lakes pose a long-term threat in the event that the lakes become re-acidified over time, leading to desorption/dissolution of trace elements. Thus, the long-term effects of changing lake chemistry in natural neutralization processes may alter the distribution and concentrations of trace elements with time in the lake outflows and this needs to be considered in the context of a sustainable solution [13].

2.2 Adsorption/biosorption

Although considered by many as an active treatment process, adsorption, and in particular biosorption, is considered to be an efficient treatment method which uses abundantly available waste organic material and biomass to adsorb toxic contaminants, such as heavy metals, from wastewaters. Depending on the biomass used, biosorption of heavy metals occurs via electrostatic interactions or hydrogen bonding due to the formation of carboxyl groups on the adsorbents for binding cationic metals and amine groups for binding of either cationic or anionic metals [14]. Biosorption also offers the possibility of adsorbent regeneration and metal recovery while producing a minimal amount of chemical sludge. While many biosorption studies have examined the uptake of single metals with a variety of results (**Table 2**), adsorption of multiple metals is more difficult where competing ions reduce the capacity of the adsorbent to remove target metals [23]. For example, in a batch study to investigate the effectiveness of waste digested activated sludge (WDAS) as a biosorbent to remove and recover metals from AMD, the authors reported high (>70%) removal of V and Cu, and slightly lower (40–70%) removal of uranium (U), thorium (Th) and Cr; however the removal

Metal	Biosorbent	Solution pH	Maximum Uptake (mg g−1)	Reference
Pb(II)	Sewage sludge	5	98.5	[15]
Cd(II)		5	67.3	
Cu(II)		5	48.7	
Cr(III)	Garden grass	4	19.4	[16]
Cr(III)	Fugal biomass of *Termitomyces clypeatus*	4	24.8	[17]
Ni(II)	Activated carbon (peanut shells)	4.8	26.4	[18]
Mn(II)	Activated carbon (bone char)	5.7	22	[19]
Cu(II)	Rice straw	6	12.3	[20]
	Rice husk	6	8.9	
Cd(II)	Rice straw	6	9.1	
	Rice husk	6	1.6	
Cu(II)	Algal biomass (*Cystoseira crinitophylla*)	4.5	160	[21]
Fe (as $FeSO_4.7H_2O$)	Shrimp shells	2.8	17.4	[22]
Mn (as $MnSO_4.H_2O$)		2.8	3.9	

Table 2.
Metal uptakes from acid mine drainage by a variety of natural biosorbents.

rates were dependent on WDAS concentrations. The authors also noted that that there was no removal of Mn, Ni, Zn and yttrium (Y) at any WDAS concentration [24].

While many laboratory scale biosorption studies have been carried out for the removal of heavy metals from AMD, the development of full-scale biosorption treatment systems is at an early stage. Like many filtration systems, operational issues such as clogging of the adsorbent pore spaces and the need to recycle spent adsorbents are difficult issues to overcome and ultimately lead to increased maintenance. However, the possibility of resource recovery from such systems may eventually be a consideration for their full-scale development.

2.3 Constructed wetlands

Constructed wetlands (CW's) are a passive wastewater treatment technology that combine biogeochemical and physical interactions between the wetland's soil matrix, vegetation and microbial communities. Constructed wetlands may be categorized in terms of their hydrology (whether surface or subsurface flow), their flow path (whether horizontal or vertical) and the type of macrophytic growth (whether free floating, submerged or emergent plant growth) [25]. When treating AMD, some or all of these components can be adjusted to suit the local and environmental conditions making CW's very flexible and efficient treatment systems. The key components and operating parameters for effective and efficient operation of a CW include: number of cells within the CW, substrate type and composition, plant type and planting density, hydraulic flow paths, hydraulic loading rate and hydraulic retention time. Wastewater pH is a key treatment indicator for AMD as it affects metal removal efficiencies. In a bench scale study to evaluate the performance of a CW using a mixed substrate of 75% soil, 20% powdered goat manure and 5% wood shavings, the pH of the AMD increased from 2.93 to 7.22 within 24 hours with corresponding enhanced removal rates for Fe (95%), Cu (90%), Zn (77%), Pb (89%), Co (70%), Ni (47%) and Mn (56%). In addition the sulfate content of the AMD decreased by an average 25% with an increase in alkalinity from 0 to 204 mg $CaCO_3$ L^{-1} [26]. The authors attributed the sulfate reduction to the addition of biodegradable organic substrate to the soil (in the form of goat manure and wood shavings) which provided a carbon source for the anaerobic microbes to generate alkalinity, leading to sulfate reduction and associated metal removal.

Leachate metal removal by CW's include physical, chemical and biological processes which are both complex and interactive. The metal removing mechanisms include sedimentation, sorption, precipitation, cation exchange, photodegradation, phytoaccumulation, biodegradation, microbial activity and plant uptake [27]. During CW treatment of acidic and alkaline leachates, many metals are precipitated from solution, because of a change towards circumneutral pH (**Figure 1**). Once this happens, they settle through the liquid and into the substrate /sediment of the CW, provided flow conditions are sufficiently acquiescent. For example, Fe, Al and Mn can form hydroxides through hydrolysis and/or oxidation, which deposit in the substrate. The rate of change in pH varies as the effluent moves through the CW and this determines how quickly precipitation will occur and also the locations where most sedimentation occurs. If pH changes quickly it can be expected that metal accumulation in the sediment will occur at the inlet end of the CW and conversely if pH changes are slow then metal accumulation will be more dispersed. The retention time of CW's are therefore important design considerations. While settled

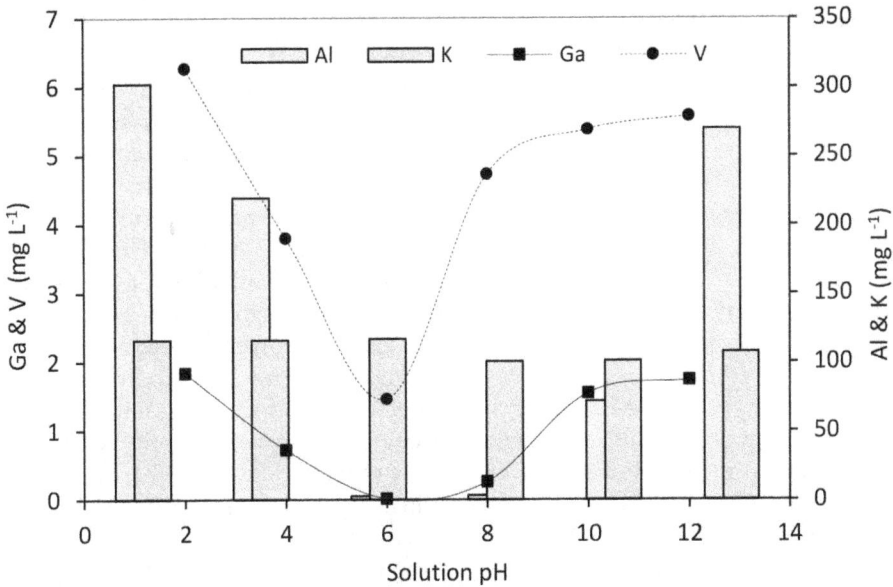

Figure 1.
Metal solubility versus solution pH for a selection of synthetic amphoteric metal solutions [unpublished data].

metals will ideally remain in the sediment indefinitely, there is a risk that long term changes in the pH may result in resolubilisation and emission of high metals concentrations in the treated effluent. Such a risk is related to the composition of the substrate/sediment and the amount of organic matter it contains [26]. As well as precipitation and sedimentation, adsorption of the metals by the soil is probably one of the more significant metal removal processes in CW treatment. Metal adsorption occurs either by reversible cation exchange or by irreversible chemisorption. Adsorption to humic or clay colloids is more permanent than adsorption to soil organic matter which ultimately decomposes and releases the adsorbed metals back into solution. The role of biological processes for metal removal in CW's is an important one, and these are normally centered around the wetland plants. As well as providing direct uptake of metals from the wastewater, wetland plants generate organic particulate matter that contributes to sedimentation processes and to symbiotic bacterial processes. Plant species, particularly emergent macrophytes such as *Phragmites* or *Typha* species, also influence the rhizosphere. Root exudates and oxygen gradients within the sediment/substrate can facilitate diverse microbial communities that can influence the oxidation state in the sediment and partial pressures of CO_2 or O_2 in solution, influencing metal removal [28]. The rate of metal uptake by plants varies significantly, depending on plant type (emergent, surface floating, or submerged), species, density, and growth rates with maximum uptake observed in the roots [29].

The long-term performance of CW's to treat AMD is variable with effectiveness determined by variables such as metal types and concentrations in the influent and the quality and quantity of the wastewater. Typically, removal of the contaminants Fe, Al and Zn is highly effective in both the short (1–3 years) and longer term (c.10 years). However, other metals, such as Mn, are more problematic and lower removal rates have been reported [7].

3. Treatment of alkaline leachates

Many of the processes used to treat AMD described in the previous sections are similarly used to treat alkaline leachates generated at disposal sites of industrial by-products. Neutralization or partial neutralization of alkaline leachates prior to discharge to a pH of approximately 8.5–9 using acids is a commonly used treatment process. Recent attempts have been made to examine passive neutralization techniques such as passive in-gassing of atmospheric of CO_2 [30], seawater and industrial waste brines [31]. Apart from proximity to an available supply, one of the main disadvantages of seawater neutralization is the high amount of seawater needed and so use of concentrated brines have also been investigated, mostly limited to laboratory studies [31]. Use of CO_2 may ultimately provide an attractive neutralization option for alkaline leachates, particularly if CO_2 emissions from the processing plant or from adjacent industries could be recycled, thereby reducing their carbon footprints. However, while passive remediation using CO_2 has been examined at a laboratory scale [30], further investigation is required at a pilot scale to evaluate issues such as pH rebound. To overcome this, a hybrid process combining CO_2 and brine treatments might be considered where the CO_2 converts hydroxides to insoluble carbonates and bicarbonates leading to a more stable leachate with a pH < 8.5 [31].

Biosorption of alkaline leachates has also been investigated with varying degrees of success. There are many factors which influence the metal adsorption capacity of biosorbents, including biosorbent characteristics such as pore volume and specific surface area, ionic strength, contact time, adsorbent dosage, solution temperature, initial metal concentration, and solution pH. Solution pH is one of the more important factors, particularly where electrostatic interactions play a key role in the adsorption process such as metal removal from alkaline leachates. The pH at which the net surface charge of the adsorbent is zero is termed the zero point charge (pH_{zpc}). When the pH is less than the pH_{zpc}, the adsorbent surface becomes positively charged and therefore has a high affinity towards negatively charged or anionic metal species. Conversely, when the pH is greater than pH_{zpc}, the surface becomes negatively charged and has an affinity towards positively charged or cationic metal species. In general, optimum metal adsorption is more common at acidic rather than basic pH values, given that alkali metals form cations, which are not attracted to protonated functional groups on the adsorbent surfaces. For example, in a batch study to investigate the efficacy of red mud modified sawdust biochar to adsorb V from aqueous solution, the authors reported a maximum uptake of 16.5 mg g^{-1} at a solution pH in the range of 3.5–5.5. They attributed the higher uptake levels at low pH to the positively charged adsorbent surface, which enhanced its binding capacity for anionic metal species in solution [32]. However if the pH_{zpc} of the adsorbent is relatively high, then this may promote the removal of heavy metals in alkaline solution. In a batch study to measure the uptake of Cu(II) and Cd(II) from aqueous solutions using ferromanganese binary oxide-biochar composites with a pH_{zpc} = 9.2, the authors reported maximum uptakes of 64.9 and 101 mg g^{-1} respectively, which increased as the pH increased from 3 to 6. Other adsorption studies at high pH values have also been reported (**Table 3**).

The use of CW's to treat alkaline leachates has also proved successful and as for AMD, the treatment processes rely on metal precipitation, sedimentation, sorption, biological activity and vegetation. The ability of CW's to buffer pH is a key treatment mechanism and recent studies have shown that CW's are effective in quickly reducing

Metal	Biosorbent	Solution pH	Maximum Uptake (mg g−1)	Reference
Cr (VI)	Ground nut shell	8	3.8	[33]
Ti (I)	Manganese dioxide coated magnetic pyrite cinder	12	320	[34]
Cd (II)	Cashew nut shell resin bonded with magnetic Fe₃O₄ nanoparticles	10	54.6	[35]

Table 3.
Selection of metal adsorption studies at high pH values.

pH from pH ≈ 11–13 to pH ≈ 7–10 with associated reductions in concentrations of Ca, Al, Ba, Cr, Ga, Ni, Zn and V; however a pH < 10 was needed for effective V reduction [35, 36]. Similar to AMD treatment, the use of CW's to treat alkaline leachates is an attractive long term passive treatment option; however while their longevity in the short term (ca 5 years) has shown to be effective, there is a lack of data to assess their long term performance, particularly with regard to metal concentrations and metal forms in the sediments [7]. While there is no evidence to date of metal accumulation in the CW vegetation treating alkaline leachates [35, 36], the long-term risks of metal saturation in the sediment and metal resolubilisation due to pH changes with concomitant increase in treated effluent concentrations needs to be assessed over a sustained period and under varying operating conditions to establish the long term viability of CW's.

4. Resource recovery from acidic and alkaline solid wastes and leachates

Large quantities of acidic and alkaline wastes are disposed of in storage facilities generating large quantities of metal rich leachates, which are potentially valuable but also toxic to the environment. Several attempts have been made to recycle these wastes but currently waste production far outstrips demand for their reuse. Potential uses for these waste materials are discussed below, many of which focus on the construction industry as an outlet.

4.1 Recycling of mine wastes

Mine tailings have been trialed for use as additives for the production of cement, building bricks and road construction materials, mainly aggregates and asphalt. The use of electric arc furnace steel slag and copper mine tailings were investigated as suitable substitutes for granite aggregates in road asphalt mixtures and were found to improve their performance when compared with conventional aggregates [37]. Similarly, magnetite tailings were used as a substitute for limestone aggregate in asphalt mixtures and were found to improve their high temperature properties and slightly decrease their splitting strength at low temperature [38]. The impact of copper mine tailings blended with cement mortars was also investigated and results indicated that their addition enhanced the mechanical strength of the mortar as well as increasing their resistance to chloride and acid attack [39]. Other studies however have noted that use of mine tailings for cement production involves increased energy consumption, increased dust generation and large emissions of CO_2, and have instead advocated the use of geopolymerization as a sustainable process. Geopolymerization chemically binds natural occurring silico-aluminates to form a stable material (geopolymer) with an amorphous polymeric structure. It has an advantage over other recycling processes in that it reduces the leaching potential of the waste, locking about 90% of the metal content into the geopolymeric matrix [39]. This process therefore requires mine tailings, which are rich in silicon (Si) and Al, both of which are essential materials for geopolymerization. Although an emerging technology, potential applications and properties of geopolymers are their high mechanical strength, good durability, good fire resistance (up to 1000–1200°C without loss of function) and are fast setting making them suitable for use as construction materials such as geopolymer concrete. They also have low energy consumption and generate low waste gas during manufacture with associated reduced CO_2 emissions (≈80% reduction) when compared with production of Ordinary Portland Cement [39]. However, as with many innovative emerging technologies, the long-term release of toxic metals from geopolymers requires further research in relation to its eventual leachability during weathering.

Sludge produced from AMD has also been used as an adsorbent for treatment of agricultural waste. In one such column study, the authors reported that AMD sludge was potentially an effective low cost adsorbent for the removal of phosphate from dairy wastewater [40]. It is worth remembering however, that pollution swapping must be considered when applying new recycling technologies and care should be taken not to increase one pollutant as a result of introducing a measure to reduce a different pollutant. For example while one pollutant (phosphorus) might be reduced, metals associated with the adsorbent may be released in the long term and this need to be assessed as part of a life cycle assessment (LCA) to evaluate the overall environmental impacts.

Metal recovery from AMD (as opposed to acid mine waste) remains a technically difficult process where selective precipitation remains the most common treatment method. Separation of a particular metal from a matrix of other metals in solution is a difficult one and typically involves an integrated process, particularly at low metal concentrations. The viability of such processes depends on the economic value of the target metal to be recovered as well as its relative concentration. Treatment technologies such as microbial fuel cells, biological sulfide precipitation, sulfate reducing bacteria, membrane separation and adsorption remain the most promising recovery methods in conjunction with coagulation and precipitation processes and the use of aeration and oxidation to improve efficiencies. While these technologies present opportunities for metal recovery, there are also challenges with their development, not least their economic and environmental viability. It is worth noting that, in addition to trace metals, water, rare earth metals and sulfuric acid are also valuable resources contained within AMD discharges.

4.2 Recycling of alkaline wastes

4.2.1 Municipal solid waste incineration byproducts

The final residue after MSW incineration is generally <10% of the original volume and < 30% of the original mass and typically comprises (i) bottom ash, made up of non-combustible organic matter and inert materials such as glass, ceramics and metals; (ii) grate siftings, fine materials which pass through the grate and are collected at the base of the combustion chamber; (iii) boiler and economizer ash, coarse particulate matter contained in flue gases and usually collected at the heat recovery section; (iv) fly ash, fine particulate matter downstream of the heat recovery section and (v) air pollution control (APC) residues in the form of particulate material captured prior to gas emissions to the atmosphere. The amounts of residue generated depend on the composition of the MSW, and the type and efficiency of the incineration process; however typical quantities, expressed as a percentage of the original waste mass on a wet basis, are 20–30% for bottom ash and grate siftings, 10% for boiler and economizer ash, 1–3% for fly ash, and 2–5% for APC residues.

Treatment and disposal of incineration ash residue varies considerably between countries. In China for example most of the bottom ash, an estimated 11 Mt. annum^{-1}, is disposed directly to landfill without pretreatment while in France, Denmark and the Netherlands 80, >90 and 100% respectively of bottom ash is used for road construction and embankments with the balance sent to landfill [41]. Similarly, in countries such as Belgium, Germany, Norway, Spain and Sweden reuse of bottom ash is incentivized resulting in ongoing efforts to establish new outlets for its reuse [42]; however, in the USA almost all incinerator ash is sent to landfill [43].

Typically, scrap iron and other metals are recovered from bottom ash before being landfilled or reused in the manufacture of different types of construction materials. Such applications may be commercially viable but have limitations, for example if ashes with high salt concentrations are used in the manufacture of cement, this may lead to accelerated corrosion of steel reinforcement. Bottom ash is most frequently used as a road granular sub-base material but is more susceptible to leaching in unbound aggregate than in cement bound or ceramic materials which lock in the heavy metals thereby restricting their leachability [44].

Recycling of fly ash, boiler and economizer ash, and APC residues on the other hand is at very low levels with almost all of these being landfilled [41], although other treatments such as thermal processing (melting technology) are also used [45]. The main reason for high disposal rates of fly ash to landfill is that fly ash typically contains high concentrations of heavy metals (for example Zn concentrations can be as high as 60,000 mg kg^{-1}), salts and organic micro pollutants due to their volatization and subsequent condensation during the incineration process. For this reason, fly ashes have a low reuse potential, for example in the cement industry, compared to other secondary raw materials. They are therefore classified as hazardous waste in many countries, which is disposed to either hazardous waste landfills or cement stabilized prior to disposal to non-hazardous waste landfill sites. Similar to bottom ash, one of the main environmental difficulties with recycling fly ash is its leaching potential and consequently there is an emphasis on improving its quality so that it can be used in more sustainable applications.

4.2.2 Industrial waste incineration byproducts

Coal fired power plants are one of the main global energy sources and currently contribute over 40% of power generation. Consequently coal combustion ashes are a major source of economic and environmental concern with >750 Mt. coal ash generated annually and < 50% reused with the remainder generally disposed to landfill or impounded. In Germany for example, approximately 10 million tonnes of stabilized ash is produced annually from lignite combustion power stations [46]. Globally, approximately 25% of coal fly ash is reused with the remainder disposed as waste to landfill. Coal fly ash is an alkaline residue with a variety of trace metals including barium (Ba), boron (B), cadmium (Cd), Co, Cu, Cr, mercury (Hg), Ni, Pb, Mn, tin (Sn), strontium (Sr) V and Zn [47]. While some of these metals are attenuated, at least in the short term, by the alkalinity of the fly ash, other oxyanionic species are released with consequent adverse environmental impacts. Recent research has focused on their recovery as critical and rare earth elements. Similar efforts are being made with regard to other industries including the steel industry and hazardous waste incinerators. In Sweden for example, trials have indicated recovery of >95% Fe and Mn, and 40% Zn from bottom ash from a crushed alkaline battery incinerator. These represented the metals with the highest concentrations in the bottom ash (Fe, 143,800 mg kg^{-1}; Mn, 154,600 mg kg^{-1} and Zn, 65,810 mg kg^{-1}). In a separate hazardous waste incinerator trial, valuable metals such as Ni, Sb, Mo, Zn, Cr, and Cu were recovered along with significant quantities of soluble salts, which can be subsequently used as deicing agents on motorways [unpublished data].

4.2.3 Bauxite residue reuse

Bauxite residue is characterized by extreme alkalinity (pH $\approx$ 10.5–13.5), its red color due to high Fe$_3$O$_3$ content ($\approx$10–50%) and its similarity to clay in terms of its mechanical and physical properties. It also has high concentrations of aluminum oxide (Al$_2$O$_3$, < $\approx$10–20%). The extreme alkalinity and leachability potential are the main barriers to its reusability, which is considered to be mainly in geotechnical engineering applications [48]. However, given its low strength, poor hydraulic conductivity and relatively poor compactability, additives may be required to render it suitable for many applications including as a road construction material [48]. Changes (reductions) in pH over time may also contribute to long-term leaching, resulting in potentially toxic metals being released to the environment and further research is needed to assess for example the application of pozzolanic materials as a low cost stabilization method.

The application of bauxite residue as an additive to masonry materials has also been investigated. For example, in a study to evaluate the use of bauxite residue co-mixed with agricultural residues as an additive to replace clay in the production of ceramic bricks, the authors concluded that samples produced with an additive of 10% hazelnut shells and 30% bauxite residue resulted in acceptable thermal conductivity and compressive strength values (0.45 W/mK and 9 MPa respectively). Importantly the authors reported that leaching toxicity values were within acceptable Environmental Protection Agency limits [49].

Similar to AMD, metal recovery from alkaline leachates is a technically difficult process, which tends to rely on selective metal precipitation. In a study to investigate V adsorption from aqueous solution by potassium hydroxide (KOH) modified seaweed hydrochar, the authors assessed the reusability of the adsorbent and found that while the adsorption levels remained consistent over three cycles, the physical condition of the adsorbent was the limiting factor in terms of recycling [50]. Thus, further investigation of low cost organic biosorbents in terms of mechanical and physical parameters such as particle size, hydraulic conductivity and porosity in a continuous flow system, as well as life cycle assessment are needed to develop the technology to a higher level.

5. Conclusions

Production of acidic and alkaline mine wastes is expected to continue into the future, with ever increasing amounts of acid mine drainage and alkali leachates being generated over a multi decade timescale. The detrimental impacts of these leachates on the aquatic environment is evident with in excess of 18,000 km of streams polluted or projected to be polluted from the coal mining industry alone in north America. Global treatment and remediation costs for existing and abandoned mines is significant, estimated in the range \$32–72 billion, while the remediation costs of treating AMD at abandoned mine sites is estimated to be higher than at operational sites. The difficulty with treatment of acidic or alkaline leachates is that metals tend to exist in low concentrations and in complex matrices. Current active treatment processes such as chemical precipitation, ion exchange, membrane filtration, and coagulation and flocculation processes require ongoing chemical and maintenance inputs, energy usage and treatment of metal rich sludges. To overcome these disadvantages, there has been a recent emphasis on developing passive and sustainable treatment solutions, which do not require continuous chemical and energy inputs. Passive treatment methods such as neutralization, adsorption/biosorption and constructed wetlands are considered to be some of the more promising techniques; however they are not yet fully proven and their ability to effectively treat AMD and alkaline leachates in the long-term is largely unknown. These and other technologies, including hybrid solutions, require further research for long term and sustainable treatment.

Many attempts have been made to reuse disposed acidic and alkaline wastes; however their production far outstrips their demand for reuse at present. The construction industry is a key outlet for mine and industrial waste reuse in products such as aggregates for road construction, cement manufacture and masonry materials. For many applications however, the long term performance of recycled wastes is uncertain and in many cases their use may require increased energy inputs resulting in higher CO_2 emissions when compared with traditional materials. To overcome these disadvantages, recent developments of geopolymer based products, formed from Si and Al rich mine tailings, are regarded as a promising emerging technology. Geopolymer based products have good mechanical and durability properties, which potentially make them suitable for use in a wide variety of construction materials. Additionally, the geopolymerization process binds in metals thus reducing their potential for long-term leaching.

With the identification of some metals as 'critical' for modern technology and their availability unpredictable, there has been a recent interest in examining routes to recover such valuable and sometimes scarce metals from mine and mineral processed waste. Metal recovery from the large volumes of leachates generated from acidic and alkaline wastes have had limited success, predominantly due to the complex nature of the metals which tend to exist in low concentrations. However, recent studies have reported efforts to enhance the metal adsorption properties of abundantly available biowaste materials which are sourced from other industries (e.g. agriculture/aquaculture industries) to facilitate selected metal recovery from mine and industrial leachates. This type of research fits well with the circular economy model of production and consumption, and reinforces the idea of using biowaste from one industry as a raw material to recover valuable resources from another. Although promising, these and other research developments need further technological and life cycle assessments to enhance their technology readiness levels, prior to implementation at an industrial scale.

Acknowledgements

The authors acknowledge support from the Geological Survey of Ireland (GSI, project no. 2018-ERAMIN2-002), the Irish Environmental Protection Agency (EPA) and an EU ERA-MIN2 award to the EU Biomimic Consortium (ID 86).

Conflict of interest

The authors declare no conflict of interest.

Author details

Thomas F. O'Dwyer[1,2], Bashir Ghanim[1,2], Ronan Courtney[2,3], Ashlene Hudson[2,3], J. Tony Pembroke[1,2] and John G. Murnane[4*]

1 Department of Chemical Sciences, School of Natural Sciences, University of Limerick, Limerick, Ireland

2 Bernal Institute, University of Limerick, Limerick, Ireland

3 Department of Biological Sciences, School of Natural Sciences, University of Limerick, Limerick, Ireland

4 School of Engineering, University of Limerick, Limerick, Ireland

*Address all correspondence to: john.murnane@ul.ie

IntechOpen

References

[1] Mudd GM, Boger DV. The ever growing case for paste and thickened tailings – Towards more sustainable mine waste management. J. Aust. Inst. Min. Metall. 2013;**2**:56-59

[2] Eurostat. Energy, transport and environment statistics. 2020. Available from: https://ec.europa.eu/eurostat/documents/3217494/11478276/KS-DK-20-001-EN-N.pdf/06ddaf8d-1745-76b5-838e-013524781340?t=1605526083000 [Accessed: 20 January 2022]

[3] Heikkinen PM, Räisänen ML, Johnson RH. Geochemical characterisation of seepage and drainage water quality from two sulphide mine tailings impoundments: Acid mine drainage versus neutral mine drainage. Mine Water and the Environment. 2009;**28**(1):30-49

[4] Murnane, J.G., Ghanim, B.M., Courtney, R., Pembroke, J.T. and O'Dwyer, T.F., 2021. Quantification and characterization of metals in alkaline leachates and the potential for vanadium adsorption using biochar and hydrochar. In AIP Conference Proceedings (Vol. 2441, No. 1, p. 020003). American Institute of Physics Publishing LLC. doi:10.1063/5.0073172

[5] INAP - The International Network for Acid Prevention. 2013. Last modified 18th December 2018. Available from: http://gardguide.com/index.php?title=Main_Page [Accessed: 20 January 2022]

[6] Fu F, Wang Q. Removal of heavy metal ions from wastewaters: A review. Journal of Environmental Management. 2011;**92**:407-418. DOI: 10.1016/j.jenvman.2010.11.011

[7] Hudson A, Murnane J, Courtney R. EPA Research Report 400: Use of Constructed Wetlands for Treating Mine Waste Leachates: Assessment of Longevity and Management Implications (2018-W-DS-32). 2021. Available from: https://www.epa.ie/publications/research/epa-research-2030-reports/research-400-use-of-constructed-wetlands-for-treating-mine-waste-leachates-assessment-of-longevity-and-management-implications.php [Accessed: 20 January 2022]

[8] Cravotta CA III. Applied geochemistry laboratory and field evaluation of a flushable oxic limestone drain for treatment of net-acidic drainage from a flooded anthracite mine, Pennsylvania, USA. Applied Geochemistry. 2008;**23**:3404-3422. DOI: 10.1016/j.apgeochem.2008.07.015

[9] Kefeni KK, Msagati TA, Mamba BB. Acid mine drainage: Prevention, treatment options, and resource recovery: A review. Journal of Cleaner Production. 2017;**151**:475-493. DOI: 10.1016/j.jclepro.2017.03.082

[10] Macías F, Caraballo MA, Nieto JM. Environmental assessment and management of metal-rich wastes generated in acid mine drainage passive remediation systems. Journal of Hazardous Materials. 2012;**229-230**:107-114. DOI: 10.1016/j.jhazmat.2012.05.080

[11] Mahedi M, Dayioglu AY, Cetin B, Jones S. Remediation of acid mine drainage with recycled concrete aggregates and fly ash. Environmental Geotechnics. 2020;**19**:1-4. DOI: 10.1680/jenge.19.00150

[12] RoyChowdhury A, Sarkar D, Datta R. Removal of acidity and metals from acid mine drainage-impacted water using industrial byproducts. Environmental Management. 2019;**63**(1):148-158. DOI: 10.1007/s00267-018-1112-8

[13] Åhlgren K, Sjöberg V, Grawunder A, Allard B, Bäckström M. Chemistry of acidic and neutralized alum shale pit lakes 50 years after mine closure, Kvarntorp, Sweden. Mine Water and the Environment. 2020;**39**(3):481-497. DOI: 10.1007/s10230-020-00665-y

[14] Kim N, Park D. Biosorptive treatment of acid mine drainage: A review. International journal of Environmental Science and Technology. 2021;**4**:1-14. DOI: 10.1007/s13762-021-03631-5

[15] Seo JH, Kim N, Park M, Lee S, Yeon S, Park D. Evaluation of metal removal performance of rod-type biosorbent prepared from sewage-sludge. Environmental Engineering Research. 2020;**25**(5):700-706. DOI: 10.4491/eer.2019.201

[16] Sulaymon AH, Mohammed AA, Al-Musawi TJ. Comparative study of removal of cadmium (II) and chromium (III) ions from aqueous solution using low-cost biosorbent. International Journal of Chemical Reactor Engineering. 2014;**12**(1):477-486. DOI: 10.1515/ijcre-2014-0024

[17] Fathima A, Aravindhan R, Rao JR, Nair BU. Biomass of Termitomyces clypeatus for chromium (III) removal from chrome tanning wastewater. Clean Technologies and Environmental Policy. 2015;**17**(2):541-547. DOI: 10.1007/s10098-014-0799-3

[18] Wilson K, Yang H, Seo CW, Marshall WE. Select metal adsorption by activated carbon made from peanut shells. Bioresource Technology. 2006;**97**(18):2266-2270. DOI: 10.1016/j.biortech.2005.10.043

[19] Sicupira DC, Silva TT, Leão VA, Mansur MB. Batch removal of manganese from acid mine drainage using bone char. Brazilian Journal of Chemical Engineering. 2014;**31**:195-204

[20] Li WC, Law FY, Chan YHM. Biosorption studies on copper (II) and cadmium (II) using pretreated rice straw and rice husk. Environmental Science and Pollution Research. 2017;**24**(10):8903-8915. DOI: 10.1007/s11356-015-5081-7

[21] Christoforidis AK, Orfanidis S, Papageorgiou SK, Lazaridou AN, Favvas EP, Mitropoulos AC. Study of Cu (II) removal by Cystoseira crinitophylla biomass in batch and continuous flow biosorption. Chemical Engineering Journal. 2015;**277**:334-340. DOI: 10.1016/j.cej.2015.04.138

[22] Nunez-Gomez D, Rodrigues C, Lapolli FR, Lobo-Recio MA. Adsorption of heavy metals from coal acid mine drainage by shrimp shell waste: Isotherm and continuous-flow studies. Journal of Environmental Chemical Engineering. 2019;**7**(1):102787. DOI: 10.1016/j.jece.2018.11.032

[23] Zhang M, Wang H, McDonald LM, Hu Z. Competitive biosorption of Pb (II), Cu (II), Cd (II) and Zn (II) using composted livestock waste in batch and column experiments. Environmental Engineering & Management Journal (EEMJ). 2017;**16**(2):431-438. DOI: 10.30638/eemj.2017.043

[24] Barthen R, Sulonen ML, Peräniemi S, Jain R, Lakaniemi AM. Removal and recovery of metal ions from acidic multi-metal mine water using waste digested activated sludge as biosorbent. Hydrometallurgy. 2022;**207**:105770. DOI: 10.1016/j.hydromet.2021.105770

[25] Murnane JG, Ghanim B, O'Donoghue L, Courtney R, O'Dwyer T, Pembroke JT. Advances in metal recovery from wastewaters using selected biosorbent materials and constructed wetlands. In: Eyvaz M, editor. Water and Wastewater Treatment. London:

InTech Press; 2019. DOI: 10.5772/intechopen.84335

[26] Sheoran AS. Management of acidic mine waste water by constructed wetland treatment systems: A bench scale study. European Journal of Sustainable Development. 2017;**6**(2):245-245. DOI: 10.14207/ejsd.2017.v6n2p245

[27] Sheoran AS, Sheoran V. Heavy metal removal mechanism of acid mine drainage in wetlands: A critical review. Minerals Engineering. 2006;**19**(2):105-116. DOI: 10.1016/j.mineng.2005.08.006

[28] Marchand L, Mench M, Jacob DL, Otte ML. Metal and metalloid removal in constructed wetlands, with emphasis on the importance of plants and standardized measurements: A review. Environmental Pollution. 2010;**158**(12):3447-3461. DOI: 10.1016/j.envpol.2010.08.018

[29] Khan S, Ahmad MT, Shah S, Rehman A, Khaliq A. Use of constructed wetland for the removal of heavy metals from industrial wastewater. Journal of Environmental Management. 2009;**90**:3451-3457. DOI: 10.1016/j.jenvman.2009.05.026

[30] Gomes HI, Rogerson M, Burke IT, Stewart DI, Mayes WM. Hydraulic and biotic impacts on neutralisation of high-pH waters. Science of the Total Environment. 2017;**601**:1271-1279. DOI: 10.1016/j.scitotenv.2017.05.248

[31] Kannan P, Banat F, Hasan SW, Haija MA. Neutralization of Bayer bauxite residue (red mud) by various brines: A review of chemistry and engineering processes. Hydrometallurgy. 2021;**206**:105758. DOI: 10.1016/j.hydromet.2021.105758

[32] Ghanim B, Murnane JG, O'Donoghue L, Courtney R, Pembroke JT,

O'Dwyer TF. Removal of vanadium from aqueous solution using a red mud modified saw dust biochar. Journal of Water Process Engineering. 2020;**33**:101076. DOI: 10.1016/j.jwpe.2019.101076

[33] Bayuo J, Pelig-Ba KB, Abukari MA. Adsorptive removal of chromium (VI) from aqueous solution unto groundnut shell. Applied Water Science. 2019;**9**(4):1-11. DOI: 10.1007/s13201-019-0987-8

[34] Li H, Li X, Xiao T, Chen Y, Long J, Zhang G, et al. Efficient removal of thallium (I) from wastewater using flower-like manganese dioxide coated magnetic pyrite cinder. Chemical Engineering Journal. 2018;**353**:867-877. DOI: 10.1016/j.cej.2018.07.169

[35] Gomes HI, Mayes WM, Whitby P, Rogerson M. Constructed wetlands for steel slag leachate management: Partitioning of arsenic, chromium, and vanadium in waters, sediments, and plants. Journal of Environmental Management. 2019;**243**:30-38. DOI: 10.1016/j.jenvman.2019.04.127

[36] O'Connor G, Courtney R. Constructed wetlands for the treatment of bauxite residue leachate: Long-term field evidence and implications for management. Ecological Engineering. 2020;**158**:106076. DOI: 10.1016/j.ecoleng.2020.106076

[37] Oluwasola EA, Hainin MR, Aziz MMA. Evaluation of asphalt mixtures incorporating electric arc furnace steel slag and copper mine tailings for road construction. Transportation Geotechnics. 2015;**2**:47-55. DOI: 10.1016/j.trgeo.2014.09.004

[38] Wang Z, Xu C, Wang S, Gao J, Ai T. Utilization of magnetite tailings as aggregates in asphalt mixtures.

Construction and Building Materials. 2016;**114**:392-399. DOI: 10.1016/j.conbuildmat.2016.03.139

[39] Onuaguluchi O, Eren Ö. Recycling of copper tailings as an additive in cement mortars. Construction and Building Materials. 2012;**37**:723-727. DOI: 10.1016/j.conbuildmat.2012.08.009

[40] Wang YR, Tsang DC, Olds WE, Weber PA. Utilizing acid mine drainage sludge and coal fly ash for phosphate removal from dairy wastewater. Environmental Technology. 2013;**34**(24): 3177-3182. DOI: 10.1080/09593330.2013.808243

[41] Beylot A, Muller S, Descat M, Ménard Y, Villeneuve J. Life cycle assessment of the French municipal solid waste incineration sector. Waste Management. 2018;**80**:144-153. DOI: 10.1016/j.wasman.2018.08.037

[42] Silva RV, De Brito J, Lynn CJ, Dhir RK. Use of municipal solid waste incineration bottom ashes in alkali-activated materials, ceramics and granular applications: A review. Waste Management. 2017;**68**:207-220. DOI: 10.1016/j.wasman.2017.06.043

[43] U.S. Environmental Protection Agency (US EPA). Municipal solid waste generation. 2016. Available from: https://archive.epa.gov/epawaste/nonhaz/municipal/web/html/basic.html [Accessed: February 2022]

[44] Verbinnen B, Billen P, Van Caneghem J, Vandecasteele C. Recycling of MSWI bottom ash: A review of chemical barriers, engineering applications and treatment technologies. Waste and Biomass Valorization. 2017;**8**(5):1453-1466. DOI: 10.1007/s12649-016-9704-0

[45] Sakai SI, Hiraoka M. Municipal solid waste incinerator residue recycling by thermal processes. Waste Management. 2000;**20**(2-3):249-258. DOI: 10.1016/S0956-053X(99)00315-3

[46] Kermer R, Hedrich S, Bellenberg S, Brett B, Schrader D, Schoenherr P, et al. Lignite ash: Waste material or potential resource-investigation of metal recovery and utilization options. Hydrometallurgy. 2017;**168**:141-152. DOI: 10.1016/j.hydromet.2016.07.002

[47] Gomes HI, Mayes WM, Rogerson M, Stewart DI, Burke IT. Alkaline residues and the environment: A review of impacts, management practices and opportunities. Journal of Cleaner Production. 2016;**112**:3571-3582. DOI: 10.1016/j.jclepro.2015.09.111

[48] Reddy PS, Reddy NG, Serjun VZ, Mohanty B, Das SK, Reddy KR, et al. Properties and assessment of applications of red mud (bauxite residue): Current status and research needs. Waste and Biomass Valorization. 2021;**12**(3):1185-1217. DOI: 10.1007/s12649-020-01089-z

[49] Atan E, Sutcu M, Cam AS. Combined effects of bayer process bauxite waste (red mud) and agricultural waste on technological properties of fired clay bricks. Journal of Building Engineering. 2021;**43**:103194. DOI: 10.1016/j.jobe.2021.103194

[50] Ghanim B, O'Dwyer TF, Leahy JJ, Willquist K, Courtney R, Pembroke JT, et al. Application of KOH modified seaweed hydrochar as a biosorbent of vanadium from aqueous solution: Characterisations, mechanisms and regeneration capacity. Journal of Environmental Chemical Engineering. 2020;**8**(5):104176. DOI: 10.1016/j.jece.2020.104176

Chapter 5

Adsorption Technique an Alternative Treatment for Polycyclic Aromatic Hydrocarbon (PAHs) and Pharmaceutical Active Compounds (PhACs)

Opololaoluwa Oladimarun Ogunlowo

Abstract

Water is essential to human consumption; however, its pollution is caused by populace activities from both organic and inorganic compounds sources that require serious attention, to provide clean water. Organic contaminants are known as persistent organic pollutants (POP). They are accumulated in the fat tissues of wildlife and human beings and are toxic to their organs. Degradations of POP are very difficult since they are persistent and also termed as semi-volatile, for example, polycyclic aromatic hydrocarbons (PAHs). Apart from POPs, others toxic organic contaminants with subtle ecological effects are the emerging organic contaminants (EOCs), like pharmaceutical actives contaminants (PhACs). They penetrate the aquatic environment and alter the natural quality. To obtain future discharge requirements, new technologies with granular activated carbon were developed using *Oxytenanthera abyssinica* and *Bambusa vulgaris* in remediating PhACs and PAHs. The activated carbon with KCl had removal efficiency of 73.3, 78.1, and 86.2%, which indicated the highest efficiency for PhACs removal, while adsorbent activated with H_3PO_4 gave 63.9, 66.7, and 82.2% for paracetamol, salbutamol, and chlorpheniramine, respectively. Removal efficiency of 42.5–81.2% and 8.9–65.5% ranges of PAHs were obtained for CBV and COA, respectively. The alternative adsorption treatment techniques are detailed in the chapter.

Keywords: wastewater, adsorption technique, organic pollutant, polycyclic aromatic hydrocarbon, pharmaceutical active compounds

1. Introduction

The earth's surface is covered by 70% of the water of which 97.5% are from seas and oceans which are salty for consumption, out of the remaining 2.5% water, 1.73% are in form of glaciers and ice-caps, left only with 0.77% available for freshwater supply. The amount of water available on the earth that can be renewable is only

IntechOpen

0.0008% in the rivers and lakes for humans and agricultural use [1]. Fundamentally water is needed by all living creations of God, hence it must be provided in the rest state for their consumption, therefore clean water becomes a critical issue as the world population increases [2]. The populace activities are a factor that causes an increase in pollution from both organic and inorganic compounds sources that require serious attention to ensure clean water that the same growing world needs to consume.

Clean water becomes a critical issue as the world population increases. It has been estimated that by the year 2025, there would be an additional 2.5 billion people on the earth that will live in a region already lacking sufficient clean water [1]. Similarly, scholars have indicated that the recent problems in water treatment originate primarily from the increasing pollution of water by an organic compound that is difficult to decompose biologically because these substances resist the self-purification capabilities of the rivers as well as decomposition in conventional wastewater treatment plants [3, 4]. Further observations state that the conventional mechanical-biological purification is no longer sufficient and must be supplemented by an additional stage of processing [5].

Adsorption is the capacity of the adsorbate to form a bond with the adsorbent [6]. It is also defined as a physical and chemical process in which substances are accumulated at the interface between the faces which may be liquid-liquid, liquid-solid, or gas-liquid [7]. Adsorption differs from absorption in that it is the process by which the surface concentrates fluid molecules by chemical or physical force while absorption is the partial chemical bonds formed between adsorbed species or when the absorbate gets into the channels of the solids [8]. In other words, fluid molecules are taken up by a liquid or solid and distributed throughout the liquid or solid. Adsorbate is the substance that is removed from the wastewater or the amount of contaminant adhering to the surface of the adsorbent, while the adsorbent is the solid phase that accumulates the pollutant. This may be activated carbon or other biosorption materials [7]. For adsorption to take place, the adsorbate must have less free energy on the surface of the adsorbent in solution.

Organic contaminants are occasionally termed persistent organic pollutants (POP), their occurrences in the environment are frequent and possess the ability to move fast across the water and settle from where they are sources. Accumulated in the fat tissues of wildlife and human beings and are very toxic to their organics. Degradations of POP are very difficult since they are persistent and also termed semi-volatile for example PAHs. Apart from POPs other toxic organic contaminants which can create subtle ecological effects are the Emerging Organic Contaminants (EOCs), the extent to which the environment can be adversely affected by EOCs is still under study. One of such is the (PhACs) which are products of synthetic chemicals, natural organic chemicals, or microorganisms not controlled. They possess the ability to penetrate the aquatic environment and alter the natural quality leading to adverse health issues in human and ecological disorders [9, 10].

The Petroleum and Pharmaceutical industries are seen among others as major contributors of organic contaminants because of continuous usage and pollutant from them are emerging and steady in the environment [10].

Produced water and crude oil spills are the major sources of pollutants generated by the petroleum industry. Produced water is the largest by-product of wastewater attributed to the petroleum industry and it is a mixture of salt, organic and inorganic compounds.

Among the organic constituent of crude oil is a group of hydrocarbons called PAHs [11]. These are large groups of organic contaminants, which are characterized by the presence of at least two fused aromatic rings and are seen by the United States Environmental Agency (USEPA) as priority organic pollutants [12]. PAHs are highly lipophilic contaminants that are ubiquitously present in the environment [13] because of their low biodegradation and bioaccumulation in the adipose tissues of organisms and biomagnifications through the food chain, they are considered persistent organic pollutants POPs [14].

The pharmaceutical industries on the other hand have to do with the well-being of living organisms. Their products refer to a group of chemicals used for the diagnosis, treatment, or prevention of health conditions. Most of the chemicals or ingredients used in production can be active, inactive, additive, or preservative. When most of these ingredients are no longer used for the intended purpose and if the pharmaceutical product is designated for discarding, it is then classified as pharmaceutical waste. Active chemicals like paracetamol (acetaminophen), salbutamol, amoxicillin, ibuprofen, chloramphenicol, etc. can be referred to as pharmaceutically active compounds (PhACs), and preservatives such as parabens, e.g., ethyl, propyl, etc. are called excipient [15]. Pharmaceutical wastes are EOCs of concern and are mostly unregulated contaminants that need future regulation [16].

Like any EOCs, they do not need to persist in the environment to cause negative effects because they are continually being released into the environment mainly from manufacturing processes, disposal of unused products, and excreta [17]. At the 2005 Burger AEC programme on EDCs, it was reported that most EOCs can disrupt the endocrine system-a health condition called endocrine disruptors [18]. The WHO defines an endocrine-disrupting substance as an exogenous substance that alters the function of the endocrine system and consequently causes adverse health effects in an organism or its progeny or subpopulations [19].

Petroleum and pharmaceutical Industries have been seen as major generating sources of organic contaminants that create adverse effects on surface water which are the primary source of livelihood. For water to be available in its pure state, identifying and remediating processes of those contaminants is key, if clean water is a necessity [10].

Most research works had employed analytical techniques like gas and liquid chromatography, UV-spectrophotometers and gravimetric, etc. to identify organic contaminants, gas and liquid chromatography with mass spectrophotometer followed by a cleanup method such as solid-phase extraction (SPE) and solid-phase micro-extraction (SPME) [20] are seen as most effective techniques in determinations of organic pollutants in trace amount (or micro-pollutant).

Induced Gas Flotation (IGF) or the Induced Air Flotation cells are usually used as conventional treatment methods by petroleum industries to separate produced water from crude oil with Enviro-cell as the newest technology that uses the principle of gravity with differences in density between the oil and water [20].

The hydrocarbon content in either the produced water or water polluted with crude spills can be classified as free, dispersed, and dissolved oil [21, 22]. The conventional method is seen to be effective in the removal of dispersed oil and grease but cannot be used in the removal of the dissolved hydrocarbon which includes the PAHs. Different literature had reviewed that conventional wastewater treatment plants (WWTPs) are not the best in the removal of PAHs pollutants from wastewater, additional methods that had been researched and are still being researched is the adsorption mechanism. It had been suggested that carbon and membrane filtration with reverse osmosis are very effective in the removal of dissolved and emulsified oils [23, 24]. Many materials in their raw or waste form had been developed or modified into adsorbent in adsorption of pollutants which could be agricultural materials, clay, zeolite, vibratory share enhanced process, etc.

The commonest adsorbent used by most industries for the removal and recovery of inorganic and organic substances from gaseous and liquid streams is activated carbon [25]. Because of its high internal surface area and porosity formed during the carbonization process, the adsorbent is said to have a high adsorption capacity. Similarly, the use of activating agents and heat during carbonization will influence the development of pore structure but its uses are limited to high cost hence the use of agricultural products or materials have been observed to be potential precursors in activated carbon production because of the abundant supply and low cost.

Most of the research conducted lately made use of agricultural product such as adsorbent in the removal of heavy metals from water and wastewater, such agricultural products are coconut shell and rice husk [26, 27], palm kernel shell, and oil palm fruit fiber [28], bamboo [3, 29], maize cob [6]. Other works had been done on the identification and remediation of organic contaminants in petroleum [30–31] and pharmaceutical wastes [32–33] but few works have been done on the use of adsorbents in remediating organic pollutants (PAHs and EOCs).

Since water is the prime necessity of life and very essential for the survival of all living organisms it is imperative to improve the quality of available water. The presence of pharmaceutical residues (PhACs) and PAHs as newly recognized contaminants in aquatic systems is one of the current environmental issues [34]. It should be noted that organic contaminants usually occur in multi-component in aquatic environments. Thus, it is expected that there will be interspecies interaction among these pollutants which will cause chemical reactions that can generate other metabolites compared to when the single contaminant is present [35].

Adsorption method of bioremediation had proven to be the chosen treatment option for PAHs and PhACs and other micro-pollutant in aqueous or any environmental media because it is easier to understand and has obvious advantages of convenience, easy operation, efficiency, effective, and very simple to design as compared to another kind of treatment. Apart from the identified attributes, it does not add harmful degradation metabolites or undesirable by-products [36].

Adsorption is better than any other wastewater treatment method due to its insensitivity to toxic substances and is economically based on types of materials employed as adsorbents [37]. But, the complete use of adsorption processes in purifying water is impeded by the insufficiencies of the commercial adsorbents like activated carbon and synthetic polymer resins, synthetic Nanomaterial. Hence, there is a need to develop a low-cost adsorbent for environmental research. So, the adsorbent of agricultural products is becoming the popular alternative for commercial and synthetic adsorbents due to the hydrophobic-oleophilic potential that is needed for bioremediation processes [38]. With this new trend in mind, this chapter will seek to explain courses of organic pollutants with a special interest in industrial wastewater and adsorption techniques as an alternative treatment.

2. Assessment of production activities that leads to the generation of wastewater in the selected industries

To determine wastewater characteristics, generation disposal processes and assessment of production activities that lead to wastewater generation within the pharmaceutical and petroleum industries were conducted through interviews, observation, and experimental analysis. From the information gathered, most pharmaceutical industries in a country located in West Africa produced more syrups than any other form of drugs because it is cheaper to produce. Production of syrups takes more than 50% of production water and an average of 48,000 L of water per day is being discharged as effluent during the production of syrups. Syrups can be analgesic, antacids, or pain relievers. Effluent discharge may contain codeine phosphate, paracetamol, chlorpheniramine maleate, ephedrine HCl, parabens, etc. which are regarded as active pharmaceutical compounds. PAHs and PCBs can be found within the effluent as a result of chemical metabolites. Most of the pharmaceutical industries around the States visited in that country discharge their effluents through the drains into the surface water.

The petroleum industries visited were located within the Delta area of the country. The industries have two sources of liquid waste which are produced water and water pollutant with crude spills. Averagely about 30 million barrels of produced water are said to be discharged per day into the environment. Most of the oil and gas industries in the Delta region utilize the common treatment technique like the Induced Gas Flotation (IGF) or the Induced Air Flotation called WEMCO with the Enviro-cell as the latest technology of the group. The orthodox separation technique used the standard of gravity with variances in density between oil and water.

The oil and grease in the produced water can be classified as free, dispersed, and dissolved oil [21, 22]. The conventional method is seen to be effective in the removal of dispersed oil and grease and cannot be used in the removal of the dissolved oil and grease which are the PAHS (**Figure 1**).

2.1 Production of local Adsorbent using local Technology

The production of local adsorbent from the natural agricultural material using local Technology is a crucial alternative to commercial absorbent since adsorbent is used in carrying out adsorption which is better than any other wastewater treatment methods due to its insensitivity to toxic substances and economically

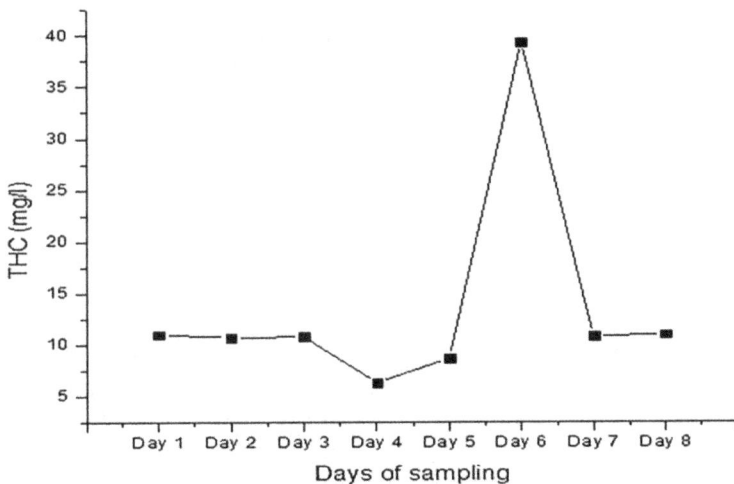

Figure 1.
Concentrations of oil and grease in produced water at different days of sampling.

based on types of materials employed as adsorbents [37]. But, the complete use of adsorption processes in purifying water is impeded by the insufficiencies of the commercial adsorbents like activated carbon and synthetic polymer resins, synthetic Nanomaterial. Hence, there is a need to develop a low-cost adsorbent for environmental research. So the adsorbent of agricultural products is becoming the popular alternative for commercial and synthetic adsorbents due to the hydrophobic-oleophilic potential that is needed for bioremediation processes [38]. The Activated Charcoal which is also known as activated carbon is a kind of carbon treated under heat to be extremely porous as processes very large surface area making room for adsorption or chemical reactions [39]. The activated carbons that are employed were formed from two different species of fresh bamboo culms that were cut at the height of 20 cm from the soil level and were chopped into 20 cm each as the peripheral materials were detached. The chopped bamboo culms were left to dry at the normal ambient temperature of 26-28°C and later cut down to 5 cm. The bamboo species were weighted and the aluminum foil was used to tightly cover in preparation for carbonization, the wrapping with the foil was done to complete deoxygenated processes. It was carbonized at 350°C for 2 hrs in an electric muffle furnace. Carbons were cooled and oven-dried at 105°C for 360 min. The carbonated samples were granulated and sieved to 1.18 m size and stored. Activation was done with Phosphoric acid (H_3PO_4) and Potassium chloride (KCl) as dehydrating agents. 26.25w/w of activator was used in the activation of the carbonated samples. Characterization was done chemically using Point of zero charge (pH pzc) and The Scanning Electron Microscopy (SEM) was used to view the surface structure of the samples at a magnification of 100, 300, 500, 2000, and 5000 times the original size to view the pore space development and reveal other information such as texture (external morphology) and structural orientation see **Figure 2** [10]. The point of zero charge (pH pzc) was used for further determination of pore and adsorption capacity. The point of zero charge is the point where the pH of the net total particle charge is zero. It is important in describing variable charge surface and it indicates the approximation equilibrium time at which the carbon is required to adsorb. Points of zero charge of the adsorbents were determined by measuring 20 ml of 0.01 m Nacl solutions into 9 separate beakers. The pH of each solution was adjusted to between 2 and 10 by adding 0.1 M of HCl or NaOH solution to each of the flasks. The flasks were thereafter placed in a water bath shaker at 25°C. The suspensions were agitated for 30 min and allowed to equilibrate for 48 hrs to ensure equilibrium point (pH pzc) after which the final pH(s) were measured see **Figure 3**. The differences between the initial and final were calculated as:

$$\Delta pH = pH_i - pH_f \qquad (1)$$

Where ΔpH = Change in pH, pH_i = Initial pH, and pH_f = Final pH.

The values of changes in pH were then plotted against the initial pH values.

From **Figure 3** above the pH pzc for carbon activated with salt is greater than pH value of the activated with acid, i.e., pH pzc < pHi COA KCl and pH pzc for CBV H3PO4 is lower than that pHi implying that pH > pH pzc This implies that the surfaces of these carbons are positively charged and this arises from the basic site that combines with protons from the medium. These results also confirmed [40, 41]

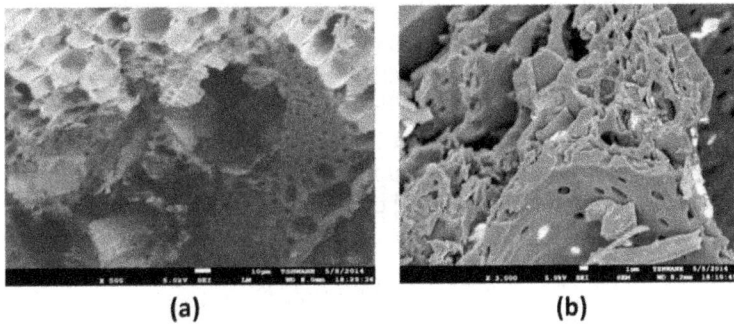

Figure 2.
SEM image of (a) CBV 350°C H$_3$PO$_4$ and (b) COA 350°C KCl at magnification of 2000.

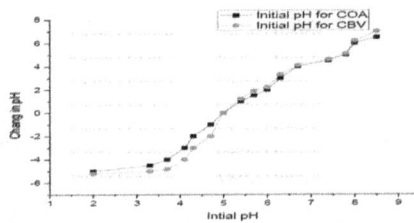

Figure 3.
Potentiometric titration curves (pH pzc) for CBV H$_3$PO$_4$ and COA KCl.

studies that a positive surface charged adsorbent would strongly attract to acidic compound in any polluted water, while a negative surface charge would strongly attract pollutant in a natural media.

2.2 Adsorption of PAH and PhACs from industrial wastewater

The behavior of Adsorption for PhACs in pharmaceutical effluents and polycyclic aromatic hydrocarbon (PAHs) in petroleum wastewater onto activated carbon produced from bamboo was studied in a batch process. Experiments were done at room temperature and the adsorption efficiency of activated carbon made from bamboo was determined using contact time. Half a liter of Pharmaceutical effluent was poured into conical flasks with a capacity of 600 ml. Selected bamboo activated carbon of two grammes each was weighed into conical flasks to make an adsorbent/solute solution. Solutions were mixed at a stirring speed of 160 rpm to ensure propped contact of the adsorbent and solute in the solution. 6 hrs contact time was used to observe each solution before reaching a dynamic equilibrium. Thereafter, solutions were filtered with filter paper 0.45 μm size. 300 ml filtrate was poured into sampling bottles with a tie cap sealed with aluminum foils and kept at a temperature of 4°C for further analysis of extraction, clean-up, and Vis-UV. For accuracy, all experimental analysis was repeated. Similarly, 200 ml of petroleum wastewater simulated, was poured into different conical flasks of 250 ml capacity. 1 g of each selected bamboo activated carbon was weighed into the conical flasks to form an adsorbent/solute solution. Solutions were mixed at a stirring speed of 160 rpm to ensure propped contact of the adsorbent and solute in solution while observing each solution in equilibrium for 5 hrs contact-time to attain dynamic equilibrium. After 5 hrs of clean up and solutions were filtered with filter paper 0.45 μm and the filtrate of 150 ml was poured into amber bottles with a Teflon cap and kept at a temperature of 4°C for further analysis of and extraction were done before analying with GC-MS. The 5 hrs contact time was informed based on the experiment performed at the terminal station of the oil and gas industry. To obtain accuracy, all experimental analysis was duplicated [10]. The amount of PhACs (qe) and PAHs (qe) adsorbed by bamboo activated carbons can be expressed mathematically as:

$$qe = ((Co - Ce)/M) \times v \qquad (2)$$

The percentage removal is evaluated using.

$$\% Removal = ((Co - Ce)/M) \times 100 \qquad (3)$$

Where V is the volume of PAHs and PhACs in solution (L), Co is initial concentrations of PAHs and PhACs ($mg\ L^{-1}$), Ce is equilibrium concentrations of PAHs and PhACs ($mg\ L^{-1}$), M is the mass of the adsorbent (g).

The Spectra in **Figure 4** shows, 16 priority PAHs in Simulated Petroleum wastewater. The efficiency of adsorbent based on contact time in the removal of PAHs from simulated petroleum wastewater by COA KCl and CBV H$_3$PO$_4$ are stated in **Figures 5** and **6**. The various contact times used were 30 min, 2 hrs, and 12 hrs, and there were otnotany changes differences in the adsorption rate with time. It was deduced that the percentage removal efficiency of PAHs with time by COA KCl was not consistent. At 30 min of adsorption rate; the percentage removal efficiency was 49.8% of total PAHs, while 39.1% of total PAHs were adsorbed at 2 hrs and about 72.3% of total PAHs were adsorbed at 12 hrs contact time.

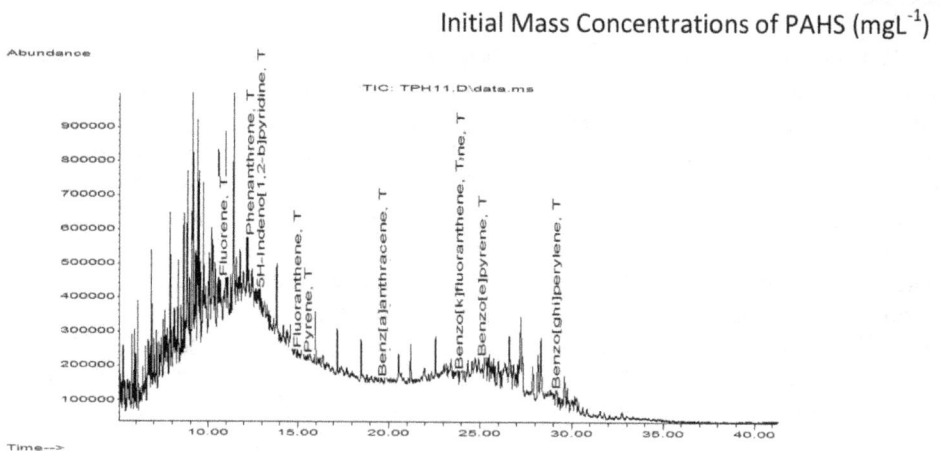

Central Research Laboratory, University of Lagos

Figure 4.
Spectra of 16 priority PAHs in Simulated Petroleum wastewater.

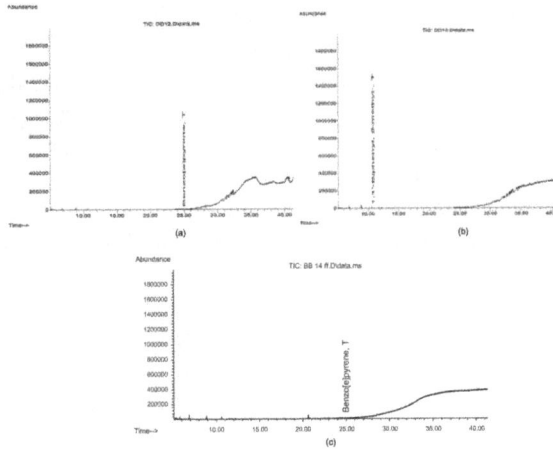

Figure 5.
Effect of contact time on adsorption rate of PAHs by COA KCl at (a) 30 mins, (b)2 hrs and (c) 12 hrs.

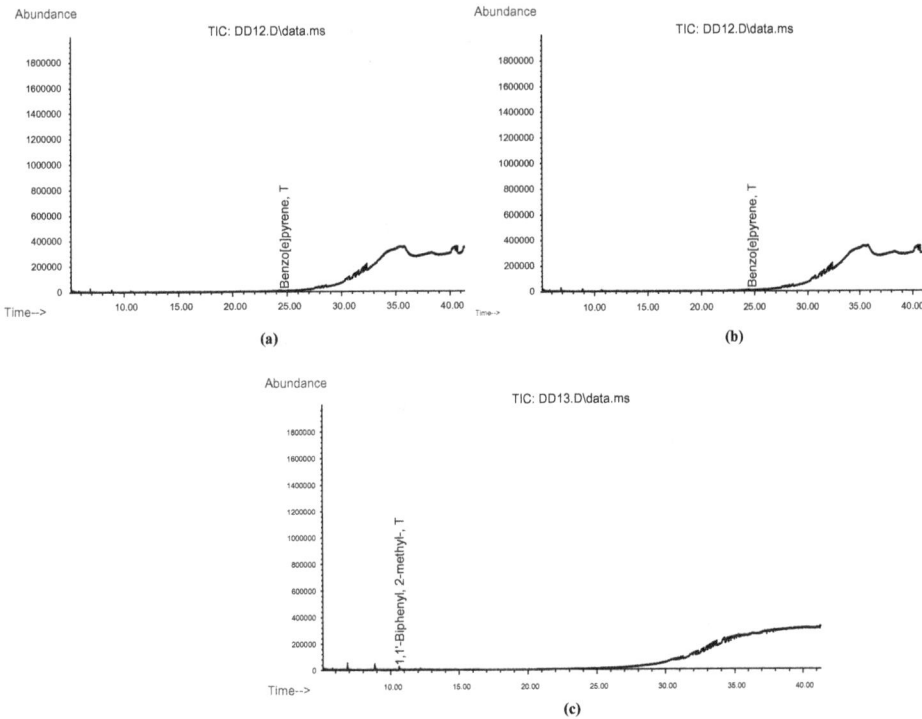

Figure 6.
Effect of contact time on adsorption rate of PAHs by CBVH3PO4 at (a) 30 mins, (b) 2 hrs and (c) 12 hrs.

Also, the adsorption rate of PAHs by CBV H_3PO_4 was seen not to follow the adsorption pattern wherein the adsorption rate increased with time. About 85.1% was adsorbed at 30 min; with an increase in contact time to 2 hrs its shows a reduction in adsorption rate to 25.1% but a further increase in time to 12 hrs increases adsorption efficiency to 87.7% which is the maximal contact time for CBV H_3PO_4 in adsorbing PAHs. It was deduced that adsorption patterns do not follow the norms of adsorption, hence adsorption of PAHs is in two stages. In the first stage, the PAHs were adsorbed easily onto the accessible hydrophobic site within the adsorbent or granular activated carbon matrix for the first 30 min. These may have resulted from the chemical interaction between the PAHs and the adsorbent. The reduction in adsorption rate implies that in the second stage, the adsorption rate may be restricted by the slow movement of PAHs to less available sites associated with the microspores within the adsorbents matrix which could take hours [10].

Figure 7a-c revealed the adsorption trend of PhACs, it was deduced that a slight reduction of adsorption rate was observed before an increase after which equilibrium was observed for COA KCl adsorbent. Similarly, the adsorption trend of CBV H_3PO_4 shows a sharp reduction to a level, and thereafter an increase was seen before an equilibrium point was reached. These observations are similar to [42–45] findings. It can be explained that absorption of PhACs with COA KCl and CBV H_3PO_4 occurred in two different stages. The first stage occurred during the first 30-360mins

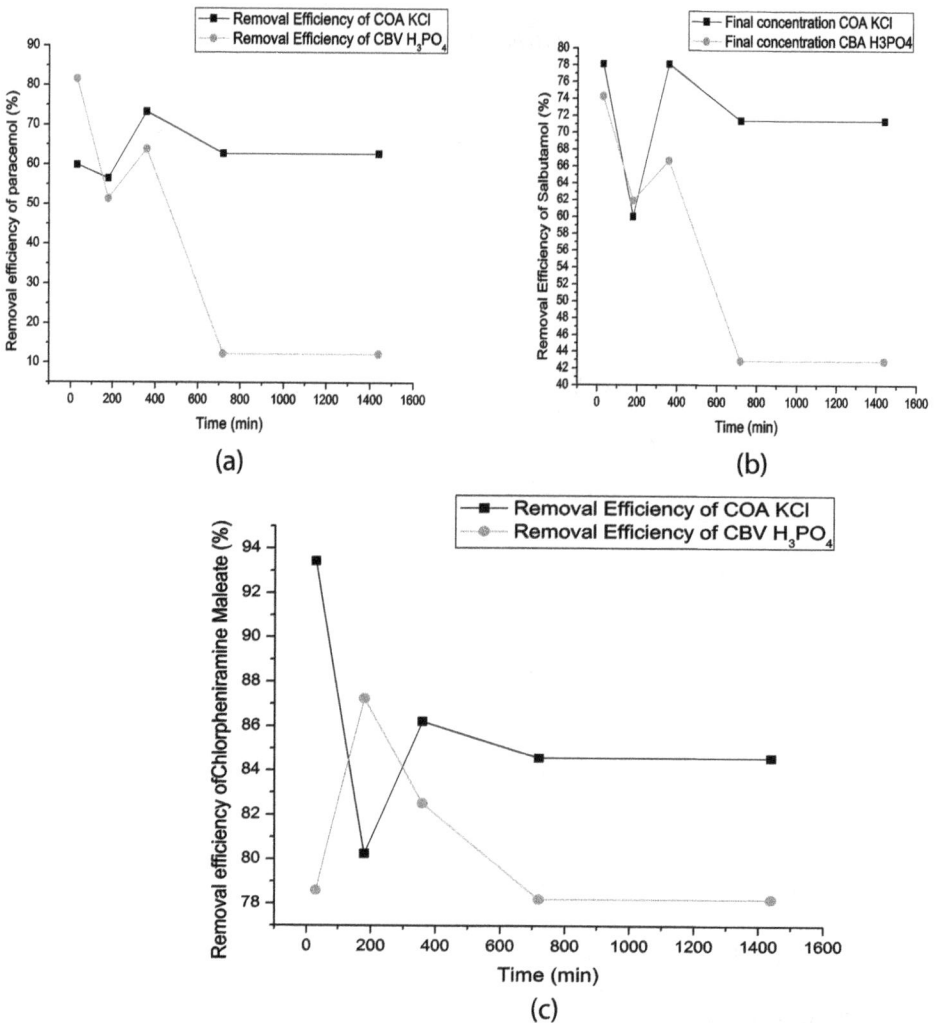

(a)

(b)

(c)

Figure 7.
(a-c). Effect of contact time on adsorption rate of PhACs by COA KCl and CBVH₃PO₄ at 30, 180, 360, 720, and 1440 mins.

contact time, with a high number of active binding sites on the adsorbent's surfaces. The adsorption rate is rapid in this stage and the points to adsorption are being controlled by diffusion processes of paracetamol, salbutamol, and chlorphenira-mine molecules from the bulk phase to the adsorbent surface. The second stage of adsorption is an attachment-controlled process due to a decrease in the number of the active sites available for Paracetamol, salbutamol, and chlorpheniramine. Adsorptions graphs showing the rate and removal efficiency of PhACs by COA KCl and CBV H₃PO₄ at varying contact times of 30, 180, 360, 720, and 1440 min are shown below.

3. Conclusion

The ever-growing human population cannot do without water; hence clean water becomes a critical issue. Therefore, the various conventional ways of treating water have been established by existing researchers, most of which are said to be inadequate in the removal of organic contaminants. This study showed that adsorbents made from Oxytenanthera Abyssinia and Bambusa vulgaris can efficiently adsorb selected PhACs and PAHs in industrial Contaminated Water. *Oxytenanthera abyssinica* (COA 350°C KCl) had a Removal efficiency of (73.3%, 78.1%, and 86.2%) for PhACs while *B. vulgaris* (CBV 350°C H₃PO₄) had (63.9%, 66.7%, and 82.2%) in remediating Pharmaceutical actives contaminants such as paracetamol, salbutamol, and chlor-pheniramine, respectively. For polycyclic aromatic hydrocarbons (PAHs) Removal efficiency of COA and CBV ranged from 42.5–81.2% and 8.9–65.5% respectively. The adsorption mechanism of trace organics followed the same pattern though with little differences. For all organic pollutants, adsorption rate is in two stages viz.: optimiza-tion and reduction followed by equilibrium. In comparison, COA showed the highest removal efficiency for PhACs and PAHs. The characterization of the adsorbent devel-oped from agricultural materials was also revealed by Scanning Electron Microscopy (SEM) and point of zero charge (pH pzc)

Acknowledgements

The author expresses her heart of gratitude to all organizations and industries that afford her access to their facilities and environment to carry out the study.

Conflict of interest

The author declares no conflict of interest

Author details

Opololaoluwa Oladimarun Ogunlowo
Faculty of Engineering, Department of Civil Engineering, Federal University Otuoke,
Ogbia, Bayelsa State, Nigeria

*Address all correspondence to: opololaoluwaijaola121@gmail.com;
ijaolaoo@fuotuoke.edu.ng

IntechOpen

References

[1] Rowell M. Removal of metal ions from contaminated water using agricultural residues. In: Second International Conference on Environmental compatible Forest production. Oporto Portugal: Fernando passoa university; 2006. pp. 241-247

[2] Ijaola OO, Ogedengbe K, Sangodoyin AY. On the efficacy of activated carbon derived from bamboo in the adsorption of water contaminants. International Journal of Engineering Inventions. 2013;**2**(4):29-34

[3] Olafadehan OA, Aribike DS. Treatment of industrial wastewater effluent. Journal of Nigerian Society of Chemical Engineers. 2000;**19**:50-53

[4] Ademiluyi FT, Amadi SA, Amakama NJ. Adsorption and treatment of organic contaminants using activated carbon from Waste Nigerian Bamboo. Journal Application Science Environment Management. 2009;**13**(3):39-47

[5] Ahmad AL, Loh MM, Aziz JA. Preparation and characterization of activated carbon from oil palm wood and its evaluation on methylene blue adsorption. Dyes and Pigments. 2007;**75**:263-272

[6] Igwe JC, Abia AA. A bioseparation process for removing heavy metals from wastewater using biosorbents. African Journal of Biotechnology. 2006;**5**(12):1167-1179

[7] Achmad A, Kassim J, Suan TK, Amat RC, See TL. Equilibrium, kinetic and thermodynamic studies on the adsorption of direct dye onto a novel green adsorbent developed from Uncaria gambir extract. Journal of Physical Science. 2012;**23**(1):1-13

[8] Fourest E, Canal C, Roux JC. Improvernment of heavy metals biosorption by mycelia dead biomasses (Rhizopus arrhizus, Mucor miehei, and Penicillium chrysogenum): Ph control and cationic activation. FEMS Microbiology Reviews. 1994;**14**(4):325-332

[9] Daughton GC. Non-regulated water contaminants: Emerging research. Journal of Environmental Research. Imp. Ass. 2004;**24**:711-734

[10] Ijaola OO, Sangodoyin AY. Remediation of emerging pollutants in industrial contaminated water using Oxytenanthera abyssinica and Bambusa vulgaris in a treatment media. IOP Conference Series: Materials Science and Engineering. 2021;**1036**(01):2012

[11] Pampanin DM, Sydne MO. Polycyclic aromatic hydrocarbons a constituent of petroleum: Presence and influence in the aquatic. Environment. 2013;**5**:83-118

[12] USEPA. The United States Environmental Protection Agency (USEPA). National Recommended Water Quality Criteria-Correction: EPA 822/Z-99-001. Washington DC: USEPA; 1999

[13] World Health Organization & International Programmed on Chemical Safety. Selected Non-Heterocyclic Polycyclic Aromatic Hydrocarbons. World Health Organization; 1998. Available from: https://apps.who.int/iris/handle/10665/41958

[14] Haritash AK, Kaushik CP. Biodegradation aspects of polycyclic aromatic hydrocarbons (PAHs): A review. Journal of Hazard Materials. 2009;**69**(1-3):1-15

[15] Comerton AM, Andrew RC, Bagley DM. A practical overview of

analytical methods for endocrine-disrupting compounds, pharmaceuticals, and personal care products in water and wastewater. Philosophical Transactions of the Royal Society A Mathematical Physical & Engineering Sciences. 2009;**367**(1904):3923-3939

[16] Stackelberg PE, Furlong ET, Meyer MT, Zaugg SD, Henderson AD, Reissman DB. Persistence of pharmaceutical compounds and other organic wastewater contaminants in a conventional drinking-water treatment plant. Journal of Science of the Total Environment. 2004;**329**:99-113

[17] D'ıaz-Cruz MS, Lo'pez de Alda MJ, Barcelo D. Emerging organic pollutant in wastewater and sludge. Trends Analytical Chemistry. 2003;**22**:340-351

[18] Ncube EJ. Selection and Prioritization of Organic Contaminants for Monitoring in the Drinking Water Value Chain [Ph.D. diss.]. Faculty of Health Sciences, University of Pretoria; 2009. pp. 10-40

[19] WHO (World Health Organization). Guidelines for Drinking Water Quality. 3rd ed. WHO; 2004. Available from: www.who.int/water_sanitation_health/dwq/en/ Accessed: December 25, 2021

[20] Noelia RG. Organic Contaminants in Environmental Atmospheres and Waters [Ph.D. diss.]. Department de Química Analítica i Química Orgànica, Universitat Rovira; 2011. pp. 6-20

[21] Jeffrey L. Towards a Novel Methodology for Environmental Remediation of Oil-polluted Aqueous Systems [Ph.D. Thesis]. Aberdeen, United Kingdom: University of Aberdeen; 2010. pp. 8-30

[22] John AV, Markus GP, Deborah E, Robert JR. A White Paper Describing Production of Crude oil, Natural gas, and Coal Bed Methane. U.S. Department of Energy, National Energy Technology Laboratory, prepared under contact W-31-109-Eng 38.21; 2004

[23] Goldsmith R, Hessian S. Ultrafiltration Concept for Separating Oil from Water. Washington: U.S. Coast Guard; 1997

[24] Choong HR, Paul CM, Jay GK. Removal of Oil and Grease in Oil Processing Wastewaters. Prepared for the Sanitation District of Los Angeles Country; 1986

[25] Foo PYL, Lee LY. Preparation of activated carbon from *Parkia Speciosa* pod by chemical activation. In: *Proceedings of the World Congress on Engineering and Computer Science*. Vol. 2. San Francisco USA: WCECS; 2010

[26] Amuda OS, Ibrahim AO. Industrial wastewater treatment using natural material as adsorbent. African Journal of Biotechnology. 2005;**5**(16):1483-1148

[27] Ajayi-Banji A, Sangodoyin A, Ijaola O. Coconut husk char biosorptivity in heavy metal diminution from contaminated surface water. Journal of Engineering Studies and Research. 2015;**21**(4):7-13

[28] Nwabanne JT, Igbokwe PK. Mechanism of copper (II) removal from aqueous solution using activated carbon prepared from different agricultural materials. International Journal of Multidisciplinary Sciences and Engineering. 2012;**3**(7):46-52

[29] Ademiluyi FT, Ujile AA. Kinetics of batch adsorption of iron II ions from aqueous solution using activated carbon from Nigerian bamboo. International Journal of Engineering and Technology. 2013;**3**(6):18-25

[30] Essien OE, John IA. Impact of crude-oil spillage pollution and chemical remediation on agricultural soil properties and crop growth. *J. Appl. Sci. Environ*. Manage. 2010;**14**(4):147-154

[31] Mbaneme FCN, Okoli CG. Occurrence and level of mononuclear aromatic hydrocarbons (MAHS) in deep groundwater aquifers of Okirika Mainland, Rivers State. International Journal of Environmental Science, Management and Engineering Research. 2012;**1**(6):219-236

[32] El-Nafaty UA, Muhammad IM, Abdulsalam S. Biosorption and kinetic studies on oil removal from produced water using banana peel. Civil and Environmental Research. 2013;**3**(7):2013-2125

[33] Adewuyi GO, Ogunneye AL. UV-spectrophotometry and RP-HPLC methods for the simultaneous estimation of acetaminophen: Validation, comparison, and application for marketed tablet analysis in South West, Nigeria. Journal of Chemical and Pharmaceutical Research. 2013;**5**(5):1-11

[34] Obiora-Okafor IA, Onukwuli OD. Utilization of sawdust (*Gossweilerodendron balsamiferous*) as an adsorbent for the removal of total dissolved solid particles from wastewater. International Journal of Multidisciplinary Sciences and Engineering. 2013;**4**(4):45-53

[35] Jones OAH, Voulvoulis N, Lester JN. the occurrence and removal of selected pharmaceutical compounds in a sewage treatment works utilising activated sludge treatment. Journal of Environmental Pollution. 2007;**145**:738-744

[36] Chang EE, Wan J, Kim H, Lian C, Dai Y, Chiang PC. Adsorption of selected pharmaceutical compounds onto activated carbon in dilute aqueous solutions exemplified by acetaminophen, diclofenac, and sulfamethoxazole. Scientific World Journal. 2015;**186**(501): 1-11

[37] Aslam R, Sharif F, Baqar M, et al. Source identification and risk assessment of polycyclic aromatic hydrocarbons (PAHs) in air and dust samples of Lahore City. Scientific Reports. 2022;**12**:2459. DOI: 10.1038/s41598-022-06437-8

[38] Czaplicka M. Photodegradation of chlorophenols in the aqueous solution. Journal Hazard Material B. 2006;**134**:45-59

[39] Xuejiang W, Xia L, Yin W, Xin W, Mian L, Daqiang Y, et al. Adsorption of copper (II) onto activated carbons from sewage sludge by microwave-induced phosphoric acid and zinc chloride activation. Desalination. 2011;**278**:231-232

[40] Park J. The pH Buffering Effect and Charging Behavior of Oxides in Aqueous Solution. Chicago: Heckman Binery Inc; 1995. pp. 1-345

[41] Park J, John R. A Simple, accurate determination of oxide PZC and the strong buffering effect of oxide surfaces at incipient wetness. Journal of Colloid and Interface Science. 1995;**5**(23):176-180

[42] Vergilli I, Barlas H. Removal of selected pharmaceutical compounds from water by an organic resin. Journal of Scientific and Industrial Research. 2009;**68**(2):17-425

[43] Meenakshisundaram M, Srinivasagan G, Rejinis J. Novel eco-friendly adsorbents for the removal of victoria blue dye. Journal of Chemical and Pharmaceutical Research. 2011;**3**(6):584-594

[44] Al-Khateeb LA, Almotiry S,
Salam MA. Adsorption of
pharmaceutical pollutants onto graphene
nanoplatelets. Chemical Engineering
Journal. 2014;**248**(2):191-199

[45] Zhangab Y, Peic C, Zhangd J,
Chengab C, Chenab XLM, Huangd B.
Detection of polycyclic aromatic
hydrocarbons (PAHs) using a high
performance-single particle aerosol mass
spectrometer (HP-SPAMS). Journal of
Environmental Science. 2022;**22**(3):
121-222. DOI: 10.1016/j.jes.2022.02.003

Chapter 6

Study of Change Surface Aerator to Submerged Nonporous Aerator in Biological Pond in an Industrial Wastewater Treatment in Daura Refinery

Omar M. Waheeb, Mohanad Mahmood Salman and Rand Qusay Kadhim

Abstract

Daura refinery, with a capacity of 140,000 barrel per stream day as a refining capacity, wastewater discharged from refining and treatment processing units, polluted water as foul water, drainages, oil spills, blowdown of boilers and cooling towers, and many other polluted water sources, aims to remove pollutants and reject clean water to the river; wastewater treatment system takes place in this treatment process. Wastewater treatment system suffers from many problems and specifically biological stage; at this stage, activated sludge with bacteria, should be supplied with oxygen, aeration system done by surface aerators with four surface fans; these fans suffer from high vibration, loss support, and in consequence, lack in oxygen supply to aerobic bacteria less than 4 ppm. The nonporous aerator is suggested as an oxygen source for the biological pool. The pilot plant builds the aim to study the ability to apply the new aeration system at the biological pool, pilot plant build with 1 cubic meter capacity tank and continuous overflow of wastewater of 10 liters.min^{-1}, air injected with the pressure of (0.5–0.75) bar(g), and airflow of (7.6–9.7) liter.min^{-1} respectively. Oxygen concentration was recorded as (3.4–6.0) ppm; in terms of consumption power, changing the aeration system reduces it to less than 20%.

Keywords: wastewater treatment, submerged aeration, fin bubbles aeration, biological pool, activated sludge

1. Introduction

Industrial wastewater treatment in refineries just receives water infected with a large number of pollutants; these pollutants are hydrocarbons (light oil products – heavy oil products), phenols [1–4], solvents, high turbidity water, saline water, foul water, drainages, and water may be discharged from equipment under maintenance or test [2]. Water with all of these pollutants is treated with a train of processes as follows:

IntechOpen

• **API separator:** Hydrocarbon cuts, free oil, and greases can be removed by skimming from the surface of the water [2, 5], oil skimmed from the surface is sent to the tank, settled, and dehydrated, the water is sent back to the API separator; it may contain a low percentage of oil suspension, separated oil pumped to the slop tanks in refinery [6].

• **Clarifiers and flocculation:** Emulsified oil can be removed by using clarifier and flocculation (air flotation), which are attached to a dissolved air floatation system (DAF) [5]; emulsified oil is separated by flocculation and sludge sent to the sludge dehydration and loading system [5, 6].

• **Biological reactor:** Dissolved oil can be removed by using a biological tank reactor (biological digestive pool) [5]; this stage uses the activated sludge [1, 7], which uses aerobic bacteria aiming to break down organic material into carbon dioxide (CO_2) and biomass with aid of oxygen (O_2).

Oxygen supply can be done by surface aeration or direct-contact air injection inside the water (porous or nonporous aeration system). Water overflows from the biological tank to another stage, which contains a secondary clarifier, overflowed water from biological reactor is treated in this clarifier (called secondary clarifier), water is separated as treated water and sludge divided into two portions, portion disposal as sludge and another portion of sludge circulated back to the bioreactor [7, 8]. The aeration tank (Biological Reactor) in an industrial wastewater treatment system needs urea and phosphoric acid, which are added to the biological reactor as a feed to bacteria, but in the municipal system, it is not required due to the high content of urea and phosphoric acid in the feed. Treated water is discharged to the river as clean treated water (**Figure 1**) [6].

• **Aeration system:**

This system is responsible for supplying oxygen into a biological reactor, which gives the bacteria the ability to digest and oxidize wastes (oil and all other undesirable materials) [9–11].

Oxygen concentration can be estimated in a theoretical way as in equation no (1) [6, 12]

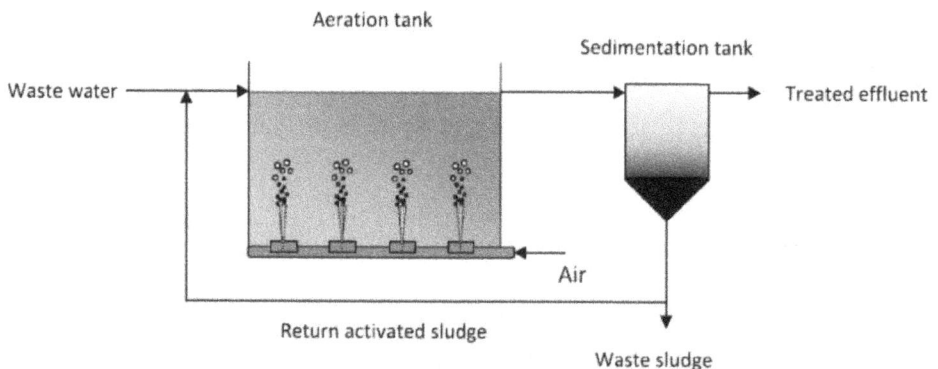

Figure 1.
Activated sludge system.

$$\ln\frac{C_S - C_O}{C_S - C_t} = K_{la}t \tag{1}$$

Where:

C_s: Oxygen concentration at saturation mg.l^{-1}.

C_o: Oxygen at time 0 mg.l^{-1}.

C_t: Oxygen at time minute.

K_{la}: Transfer coefficient.

t: Time in minutes.

There are many types of aeration systems to supply oxygen to the biological pool as follows:

1. **Surface aeration:** This type of aeration is a mechanical-type aerator, which supplies oxygen by introducing air into the water in the (biological reactor), by making turbulence at the surface of the liquid inside the pool with a depth of not more than 3.5 m, types of rotor (blades type, brush type) vertical or horizontal position [11, 13, 14]. A mechanical aerator supplies the biological reactor with a sufficient amount of air (oxygen) and mixes the content, oxygen supply will promote the biological activities and digest wastes, remove carbon dioxide and other undesirable gases released due to the biological activities [15].

2. **Subsurface aeration:** This type of aeration injects air inside the biological reactor pool directly; there are many types of methods as follows:

   • Fine bubble aerator (nonporous diffusers): This type of aeration supplies oxygen to the bioreactor with a small bubble size (fine bubble), with a high rate of oxygen transfer in terms of efficiency, low consumption power [11]; there are many types of these diffusers (such as membrane diffuser, coarse diffuser) [14, 16].

   • Porous diffusers: Classified into four classes: disc diffuser, dome diffuser, tube flexible sheath diffuser, and plate diffuser [11, 17]. These types of diffusers are always made of membrane, ceramic, and plastic [11, 14]; this type of diffusers supply oxygen to the biological reactor at high rate and efficiency [6, 18].

The oxygen transfer rate for each aeration system can be estimated in terms of horsepower as in Eq. (2) [6].

$$hp = \frac{Q * d * L}{24 * q} \tag{2}$$

Where:

hp.: Horsepower required.

Q: Liquid flow rate million gallons per day (mgd).

d: Density of liquid 8.314 lb.gal^{-1} for water.

L: BOD – biological oxygen demand (PPM).

q: Oxygen transfer rate in lbO$_2$.hp-h.$^{-1}$

Sludge treatment system: sludge treatment needs a train of processes with the aim of treating sludge as follows:

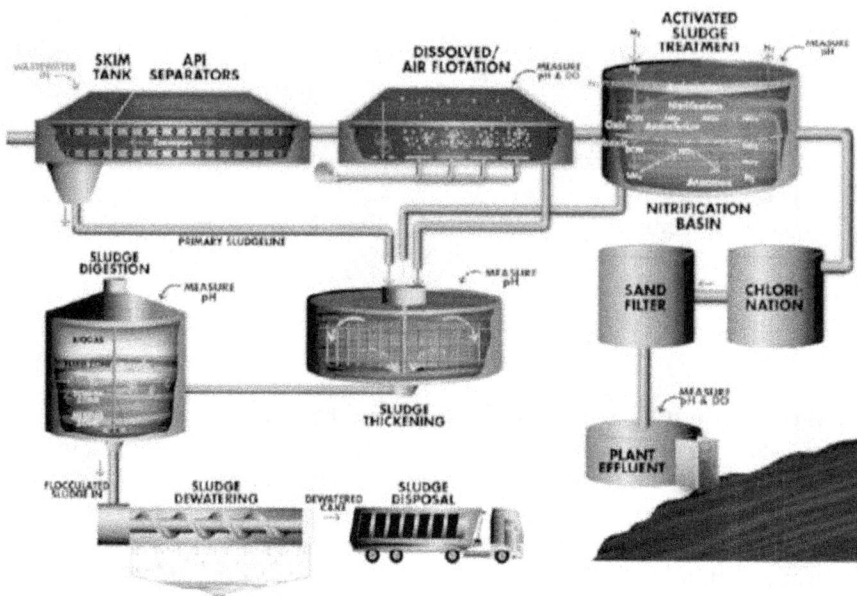

Figure 2.
Wastewater treatment system in general [5].

Sludge decanting separates water from sludge, sludge from decanting system is sent to the incineration system, ash and other nonhydrated sludge send to the rotary drum vacuum filters with the aim to separate the maximum amount of water from sludge, water is separated and sent to the API separator again (**Figure 2**) [5].

1.1 Wastewater treatment in Daura Refinery (DR)

Wastewater treatment system in DR, designed with a capacity of 850 $m^3.h^{-1}$ and operating capacity of 750 $m^3.h^{-1}$, and 1450 $m^3.h^{-1}$ in stormy weather for 2 hours only. Polluted water is received from many sources as follows: sewer water, drainages, foul water out of desalters, saline water from reverse osmosis units (RO), blowdown of boilers, cooling towers, condensate, equipment washing or the hydrostatic test, oil spills, and stormy weather [19].

The wastewater treatment system consists of the following operations:

1. API separator: all the wastewater in the refinery is collected in the header and entered the API separator to separate hydrocarbons from water by stormy water basin, and precipitated sludge at the bottom is removed by gravity separation.

Oil removed from the surface of water is sent to a slop tank in the DR, and collected sludge at the bottom of the API separator is sent to the sludge treatment unit (thickener).

2. Flocculation and flotation: at this stage, de-oiled water out of the API separator passes to the flocculation basin; at this stage, pH is controlled from 7 to 8 by adding sulfuric acid, or adding caustic soda, aluminum, or ferrous basin also contain mixer to homogenize the mixture.

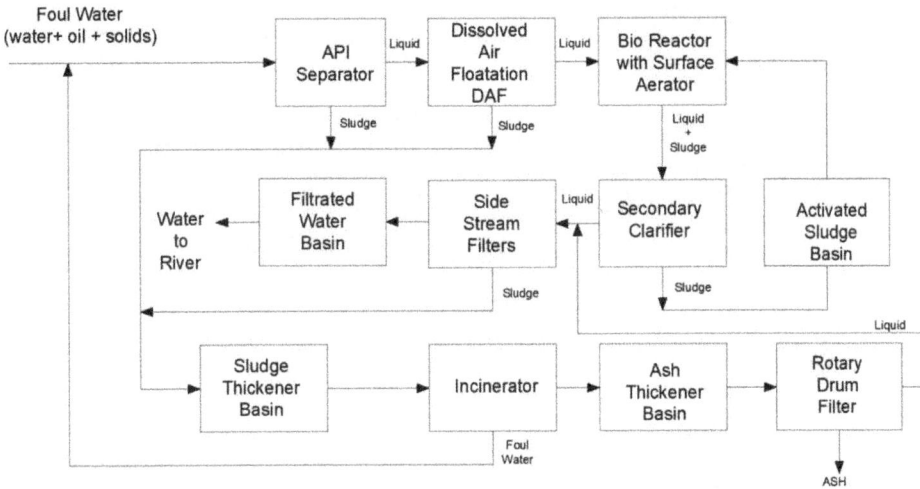

Figure 3.
Wastewater treatment system in Daura Refinery.

3. Bioreactor unit

4. Clarification and filtration

5. Final check basin (filtrated water basin)

6. Sludge treatment unit (sludge thickener, incinerator, ash thickener basin, rotary drum filter) (**Figure 3**).

1.2 Problem

Aeration system installed in biological pond in wastewater treatment in Daura refinery with dimensions of (16,000 X 32,000) mm two pools, type of aeration is surface aeration of mechanical fan aerator fixed at the concrete supports.

Four fans were fixed at the top of the pond, these fans were installed in the middle of each quarter of the pond, Fins of the fan were fiberglass type, fins of each fan were corrupted and replaced with stainless steel fins.

These fins are heavier than fiberglass, and the vibration generated is more than that generated by the original fan.

Stainless steel fan installed in 2004, due to the continuous operation of fans, cracks in concrete foundations of each fan appear; cracks in the foundation as a result of excess vibration, cracks in the foundation as in **Figure 4**.

Cracks in the bearer foundations make run of these fans' type of imagination, due to the risk of failure of the concrete foundations.

Suggested solution:

Maintenance measures to solve vibration problems or fix the bearer foundation, this type of solution does not pass away, and the problem is just raised to the surface. Replace the surface aeration system with another type, such as a submerged aeration system, this system will be a suitable type of aeration in terms of solving the problem and avoiding vibration and foundation failure.

Figure 4.
Surface with corrupted foundation due to the vibration.

1.3 Methodology

The pilot plant was just built to study the performance of air injection non-porous diffuser (aeration system) in an activated sludge tank with a continuous overflow system of wastewater out of a flocculation system with the aim to simulate a biological tank.

The apparatus is just built as in **Figure 5** from the following items:

1. Isocontainer with dimensions of 1X1X1 m with a capacity of 1 m³.

2. ¾" high-pressure flex rubber hoses ended with connection adapters, six hoses.

3. Foul water flow measurement 0–50 liter.min^{-1} (Nippon – Japan)

4. Pressure gauge 0–6 bar (Wika – Germany).

5. Airflow measurement 0–50 liter.min^{-1} (Yokogawa – Japan).

6. Air distributor with a diameter of 200 mm, 89 holes, 5 concentric circles with a hole diameter of 0.8 mm, ½" female threaded end connection.

7. ½" Polyvinyl chloride pipe SCH-80 (Al Amal Al Sharief)

8. ½" ball valve

9. Air source 0–4 bar (g).

10. Oxygen concentration indicator

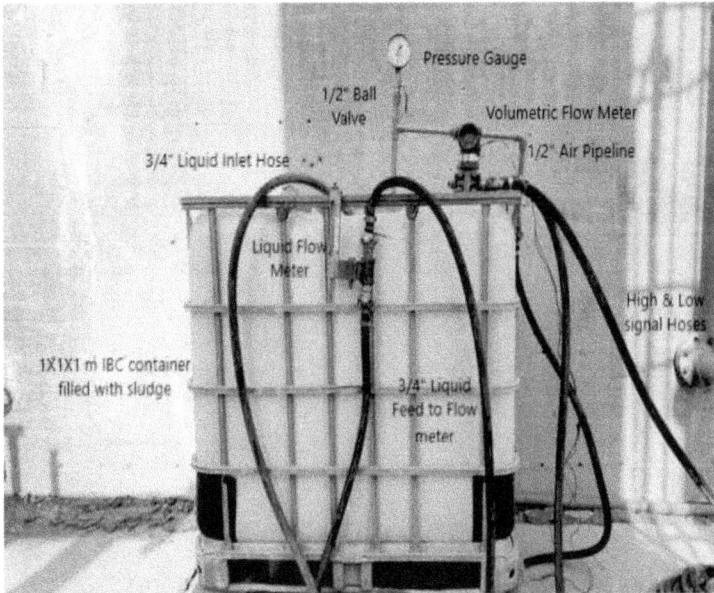

(a)

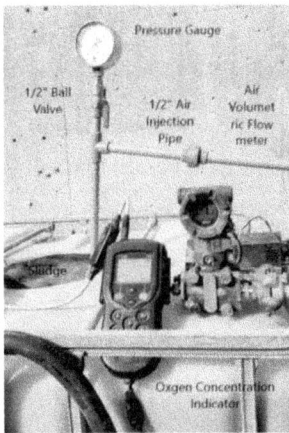

(b)

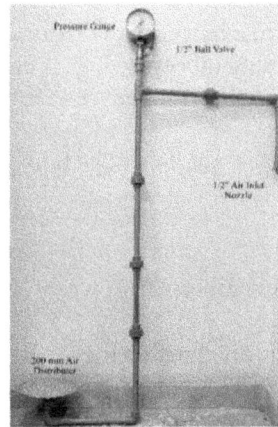

(c)

Figure 5.
Test apparatus arrangement: (a) sludge tank. (b) oxygen indicator device, (c) air diffuser.

The procedure of the experiment:

1. All the previous parts are connected as in **Figure 6**.

2. The air valve opened and the air passed to the air distributor inside the sludge tank, the pressure was just set at 0.5 bar(g) with a flow rate measured as 7.5 liter. min^{-1} at 30°C.

3. Oxygen concentration was recorded for each minute.

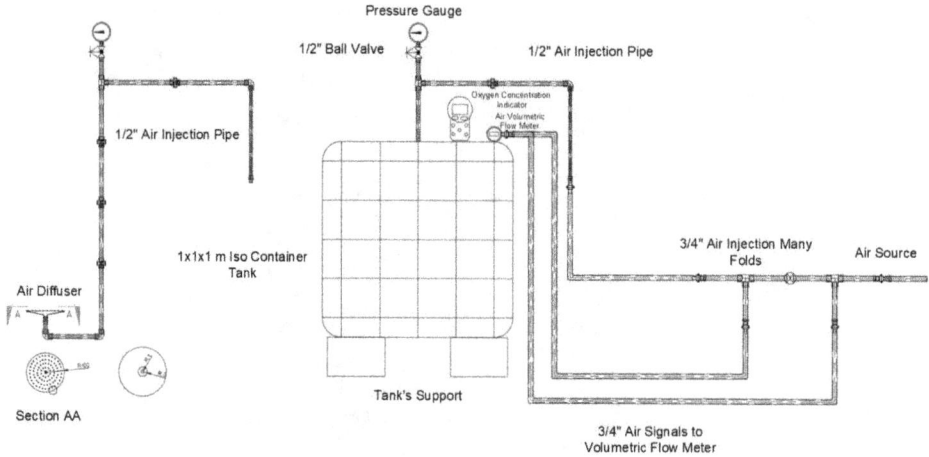

Figure 6.
Apparatus diagram.

4. The second experiment with an air pressure of 0.65 bar(g) and 30°C, i.e., air pressure increased by 0.1 bar(g), and airflow of 8.5 liter.min^{-1} and oxygen concentration was recorded for each minute.

5. The third experiment with an air pressure of 0.75 bar(g) and 30°C, i.e., the air pressure increased by 0.1 bar(g), and airflow of 9.7 liter.min^{-1} and oxygen concentration was recorded for each minute.

6. Foul water inlet to sludge tank just set to 10 liter.min^{-1}, constant at all three experiments.

2. Result and discussion

Air was injected inside the tank through the air diffuser, and the feed of wastewater to be treated, which was pumped into the tank, activated sludge placed in the experiment tank.

The oxygen transfer rate was recorded through the experiment and the oxygen concentration was measured with an oxygen detector.

Oxygen concentration was measured through the first experiment and recorded with 0.45 PPM at the beginning and 3.2 PPM after 33 minutes as in **Figure 7**.

The temperature of the system raised from 26.7°C at the beginning of the experiment to 27.7°C after 33 minutes as in **Figure 7**.

Oxygen concentration was measured through the second experiment, oxygen concentration was recorded at 3.23 PPM 34 minutes from the beginning and 4.46 PPM after 18 minutes from the beginning of the second experiment as in **Figure 8**.

The temperature of the system raised from 27.8°C at the beginning of the experiment to 28°C after 18 minutes as in **Figure 8**.

Oxygen concentration at the last experiment was recorded as 4.8 PPM after 41 minutes from the beginning of the experiment, concentration of oxygen was recorded at the end of the experiment at 6.01 PPM, Temperature increased from 28 to 28.2°C at this experiment.

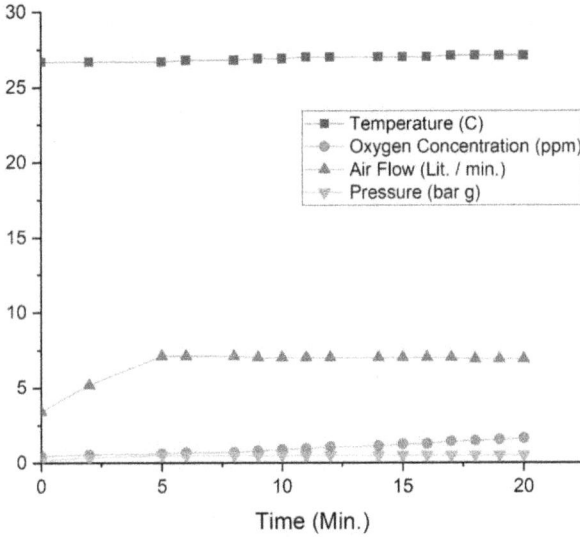

Figure 7.
Concentration of oxygen and temperature of liquid with time at 0.5 bar(g) and air flow of 7.5 L.min⁻¹.

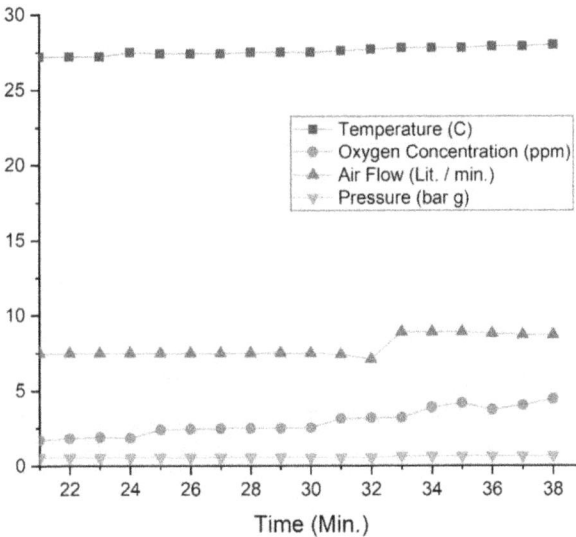

Figure 8.
Concentration of oxygen and temperature of liquid with time at (0.5–0.65) bar (g) and air flow of 8.7 L.min⁻¹.

Increasing the temperature can relate to the biological activities of bacteria due to the digesting of wastes in wastewater [20].

The rate of oxygen (air), supplied with an air diffuser in the experiment tank was high compared with the calculated air required by Eq. (2) [6] and as in **Table 1**.

The amount of air supplied was more than that estimated in the third experiment, because the short path of air from the injection point to the surface leads to air bubbles escaping out of the liquid tank, [21], the pressure of injected air increased with the aim to increase the flow rate of air and reach the required concentration of 6.01 PPM.

Pressure bar (g)	Air flow rate L.min^{-1} (required)	Air flow rate L.min^{-1} (Measured) (Maximum)	Oxygen concentration PPM	Temperature °C of tank
0.5	9.4	7	1.46	27.1
0.55	8.642	7.5	3.23	27.7
0.65	8.428	8.5	4.46	28
0.75	7.940	9.7	6.01	28.2

Table 1.
Final results and estimated amount of air required.

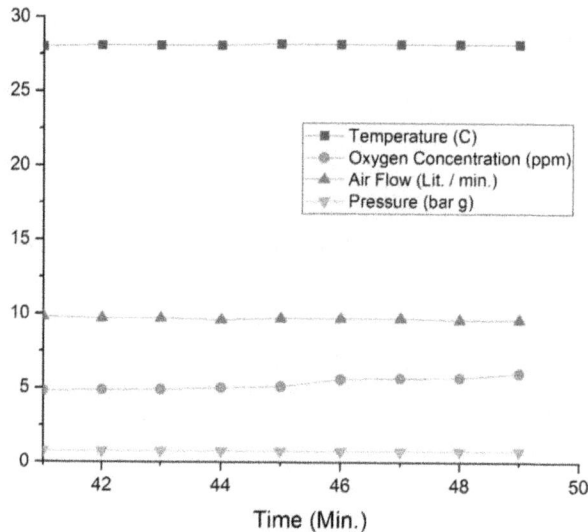

Figure 9.
Concentration of oxygen and temperature of liquid with time at 0.75 bar (g) and air flow of 9.7 L.min^{-1}.

All three experiments were continuous and pressure increased with the aim to increase the flow rate of air, this was when the oxygen concentration did not increase and stopped increasing and sometimes decreased when the temperature of the liquid increased [22], and this was observed very clearly through the experiments data (**Figures 7, 8** and **9**).

New Aeration System Proposed for Biological Reactor of Waste Water Treatment Complex in DR:

The biological reactor in wastewater treatment complex in DR will be changed from mechanical aeration to air direct injection by nonporous diffusers, air distributor applied at 1cubic meter capacity can be applied at the biological reactor.

The power duty and oxygen transfer rate (air flow rate) required can be estimated as in Eq. (2) [6]. power duty with 10% over design is 200 hp. at (BOD 250 PPM as a design condition), the amount of air required to realize 6 PPM oxygen concentration is not less than 587 kg.h^{-1} (454 Nm^3.h^{-1}) at a liquid flow rate of 500 m^3.h^{-1} as a charge to the biological reactor.

The total number of nonporous diffusers required to cover the required amount of air is 604 with a diffuser capacity of 0.9735 kg.h^{-1} (12.532 NL.min^{-1}), which was used in the experiments.

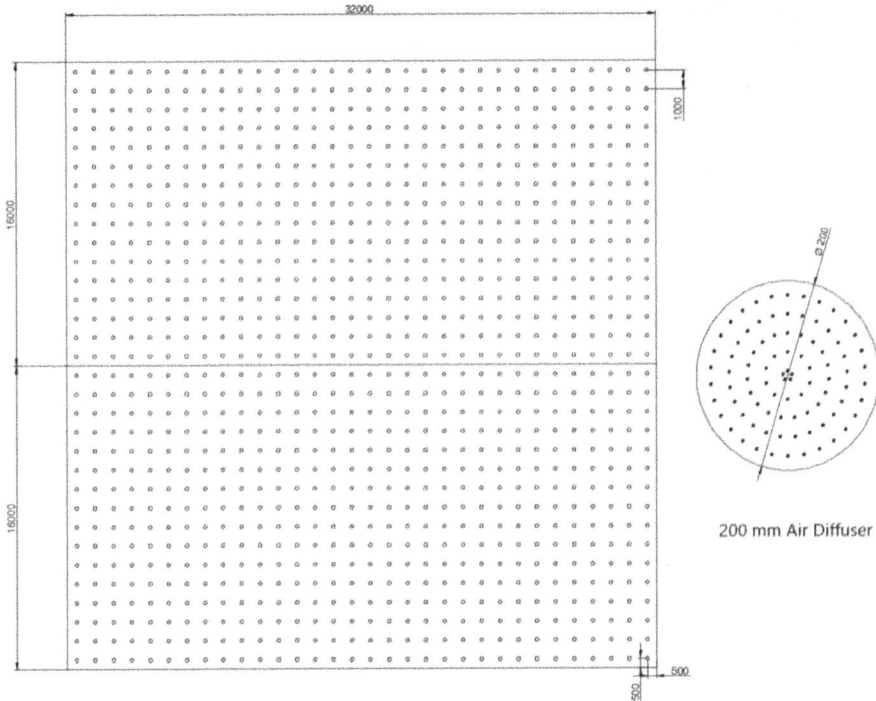

Figure 10.
Arrangement of air diffusers in biological pool.

The number of diffusers will be used more than that calculated according to the capacity of the diffuser, in terms of providing the required amount of oxygen required.

The number of diffusers will be 1024, distributed for two sides of the biological reactor of dimensions 16,000 X 32,000 mm, the distance between each diffuser and the other is 1000 mm, and the distance from the wall is 500 mm as in **Figure 10**.

Installing excess numbers of diffusers can provide well mixing and at the same time avoid the dead zones in the biological reactor; dead zones in the biological reactor activate anaerobic bacteria and in consequence reduce the efficiency of waste digesting [22].

The consumption power through the new proposed system of fine bubbles (diffuser nonporous type) will be reduced from 241.4 hp. in surface aeration to 200 hp, and this is due to the reduction in the amount of air required in the aeration process.

3. Conclusions

1. Oxygen concentration achieved is 6.01 PPM, which is close to the optimum required concentration when 0.9735 kg.h^{-1} of air is injected through a fine bubble diffuser (nonporous type) in a 1 m^3 liquid tank filled with sludge.

2. Biological reactor (the liquid tank) temperature increased with increasing oxygen concentration due to bacteria's biological activities. Even if the temperature rise affects the oxygen concentration in an aqueous solution, airflow must be increased.

3. The type of diffuser used in the experiments can be applied at the biological reactor with an excess aim to supply the oxygen demand, which realizes well mixing, avoids dead zones, and avoids the growth of anaerobic bacteria.

4. Replace the surface aeration system with a direct injection fine bubble diffuser (nonporous diffuser) will reduce consumption power from 241.4 hp. to 200 hp. due to the reduction in air supply, which can be considered more economic and power saving.

Acknowledgements

I would like to thank Mr. Waleed H. Mhaseb Head, division of wastewater treatment for his support and cooperation, and also I would like to thank Miss Aza A. Fiadh, senior instrument engineer for her cooperation.

Author details

Omar M. Waheeb*, Mohanad Mahmood Salman and Rand Qusay Kadhim
Chemical Engineering, Midland Refineries Co, Iraq

*Address all correspondence to: omsuch@gmail.com

IntechOpen

References

[1] Al-Attabi AWN. Treatment of Petroleum Refinery Wastewater in an Innovative Sequencing Batch Reactor. Ph.D. Thesis. Liverpool: Liverpool John Moores University; Jan 2018

[2] Al-Ani FH. Treatment of oily wastewater produced from old processing plant of north oil company. Tikrit Journal of Engineering Sciences. Mar 2012;**19**:23-34

[3] Albarazanjy MG. Treatment of wastewater from oil refinery by adsorption on fluidized bed of stem date. 2017

[4] Pombo F, Magrini A, Szklo A. 22 Technology Roadmap for Wastewater Reuse in Petroleum Refineries in Brazil. Available from: www.intechopen.com

[5] Corporation, Yokogawa Electric. Oprex analyzer, Refinery Waste Water Oil and Grease Removal. Yokogawa Electric Corporation; 2020. Available from: https://www.yokogawa.com/an/ [Accessed December 27, 2021]

[6] Water, Ecolab Company NALCO. The NALCO Water Handbook. 4th ed. In: Flynn DJ, editor. Section 4, Chapter 29: Energy Use in Effluent systems/Aeration. United States of America: McGraw-Hill; 2017

[7] Jassby D, Xiao Y, Schuler AJ. Biomass density and filament length synergistically affect activated sludge settling: Systematic quantification and modeling. Water Research. 2014;**48**(1):457-465. DOI: 10.1016/j.watres.2013.10.003

[8] Ziauddin S, David A, Graham W, Dolfing J. Wastewater treatment: Biological. 2013. DOI: 10.1081/ E-EEM-120046063

[9] Ahansazan B, Afrashteh H, Ahansazan N, Ahansazan Z. Activated sludge process overview. International Journal of Environmental Science and Development. 2014;**5**(1):81-85. DOI: 10.7763/IJESD.2014.V5.455

[10] Alkhalidi A, Amano RS. Bubble deflector to enhance fine bubble aeration for wastewater treatment in space usage. 2012. DOI: 10.2514/6.2012-999

[11] Agency, Environmental Protection. Waste Water Technology Fact Sheet Fine Bubble Aeration. Washington D.C) EPA- 832- F -99-065: Office of Water; 1999. pp. 1-6

[12] Hongprasith N, Dolkittikul N, Apiboonsuwan K, Pungrasmi W, Painmanakul P. Study of different flexible aeration tube diffusers: Characterization and oxygen transfer performance. Environmental Engineering Research. 2016;**21**(3): 233-240. DOI: 10.4491/eer.2015.082

[13] Drewnowski J, Remiszewska-Skwarek A, Duda S, Łagód G. Aeration process in bioreactors as the main energy consumer in a wastewater treatment plant. Review of solutions and methods of process optimization. Processes. 2019;**7**(5). DOI: 10.3390/pr7050311

[14] Miletta BA, Amano RS, Alkhalidi AAT, Li J. Study of air bubble formation for wastewater treatment. In: Proceedings of the ASME Design Engineering Technical Conference. Vol. 2011, 2. pp. 275-280. DOI: 10.1115/ DETC2011-47065

[15] Gillot S, Capela-Marsal S, Roustan M, Héduit A. Predicting oxygen transfer of fine bubble diffused aeration systems: Model issued from dimensional analysis.

Water Research. 2005;**39**(7):1379-1387. DOI: 10.1016/j.watres.2005.01.008

[16] Kaliman A, Rosso D, Leu SY, Stenstrom MK. Fine-pore aeration diffusers: Accelerated membrane ageing studies. Water Research. 2008;**42**(1-2):467-475. DOI: 10.1016/j.watres.2007.07.039

[17] Papadakis E, Klejn K, Stobbe P. Oxygen Mass Transfer in Liquids. CerCell ApS, Malmmosevej 19C, Holte, Denmark; 2017. Available from: https://www.semanticscholar.org/paper/Oxygen-mass-transfer-in-liquids-Papadakis-Klejn/fba27c5945052405288916c5eae7d82a9d3070c5. Corps ID: 215770948

[18] Oprina GC, Florentina B, Băran G, Oprina G, Bunea F. The bubbles emission stability generated by porous diffusers New types of actuators specific for space applications View project Innovative aeration system of the water used by hydraulic turbines, for preservation of the aquatic life View projectThe bubbles emission stability generated by porous diffusers. In: Scientific Bulletin of the Politehnica University of Timisoara Transactions on Mechanics Special issue Workshop on Vortex Dominated Flows-Achievements and Open Problems. 2005. Available from: https://www.researchgate.net/publication/237662183

[19] Altch Group, Ingico Lang International S.A. Waste Water Treatment System. Geneve, Switzerland: Operating Manual; 1981

[20] Lin Q et al. Microorganism-regulated mechanisms of temperature effects on the performance of anaerobic digestion. Microbial Cell Factories. 2016;**15**(1)

[21] Cahyadi A, Zulkarnain R. Application of aeration injection to increase dissolved oxygen of surface water in the floating net cage. IOP Conference Series: Earth and Environmental Science. 2021;**934**(1)

[22] Sabry T, Alsaleem S, Sabry TI, Hamdy W, Alsaleem SS. Application of different methods of natural aeration of wastewater and their influence on the treatment efficiency of the biological filtration. In: Performance Management of Small to Medium Sized Water Utilities View project Treatment of Stormwater for Artificial Groundwater Recharge-Application of a Low-cost Ceramic Filter View project Application of Different Methods of Natural Aeration of Wastewater and their Influence on the Treatment Efficiency of the Biological Filtration. 2010. Available from: http://www.americanscience.orghttp://www.americanscience.org

Section 3

Biological and Chemical Processes in Wastewater Treatment

Electro-Peroxone and Photoelectro-Peroxone Hybrid Approaches: An Emerging Paradigm for Wastewater Treatment

Tatheer Fatima, Tanzeela Fazal and Nusrat Shaheen

Abstract

Electrochemical advanced oxidation practices (EAOPs), remarkably, electro-peroxone (EP), photoelectro-peroxone (PEP), and complementary hybrid EP approaches, are emerging technologies on accountability of complete disintegration and elimination of wide spectrum of model pollutants predominantly biodegradable, recalcitrant, and persistent organic pollutants by engendering powerful oxidants in wastewater. A concise mechanism of EP and PEP approaches along with their contribution to free radical formation are scrutinized. Furthermore, this chapter provides a brief review of EP, PEP, and complementary hybrid EP-based EAOPs that have pragmatically treated laboratory-scale low- and high-concentrated distillery biodigester effluent, refractory pharmaceutical, textile, herbicides, micropollutant, organic pollutant, acidic solution, landfill leachates, municipal secondary effluents, hospital, and industries-based wastewater. Afterward, discussion has further extended to quantitatively evaluate energy expenditures in terms of either specific or electrical energy consumptions for EP and PEP practices through their corresponding equations.

Keywords: electro-peroxone, photoelectro-peroxone, wastewater, complementary hybrid EP approaches, energy consumption

1. Introduction

In current scenario, diverse industrial setups have been expanded very rapidly. Consequently, numerous industrial effluents particularly textile, oil, and gas, pharmaceutical, paint, fertilizer, petrochemical, metal, and mining industries have made major contribution to wastewater. These industrial effluents contain toxic dyes, nitrates, 2,4-dichlorophenoxyacetic acid herbicide (2,4-D herbicide), toxic heavy metals, pharmaceutical waste, organic waste, total ammonia nitrogen (TAN), micropollutants, and so on [1]. Wastewater comprising these noxious chemicals is

IntechOpen

lethal to humans as well as aquatic life. In this frame of reference, several techniques have been exploited for wastewater treatment in the literature explicitly, biological techniques, chemical procedures, and physical methods. Since biological techniques were constrained due to toxic contaminants, long processing time, as well as insufficient degradation of pollutants, that is, perfluorinated compounds is devoid of biological disintegration owing to 533 kJ mol^{-1} energy content required to fracture C—F bond [2, 3], and physical adsorption is a nondestructive method, which could not oxidize pollutants entirely, solely accountable for shifting pollutants from one phase to another as well as pricey method for a powerful adsorbent, which cannot regenerate [4], and chemical methods, which increase cost as well as generate toxic sludge [5].

Over the last few decades, diverse advanced oxidation processes, namely peroxone, ozonation, and electro-oxidation, have been carried out for wastewater treatment through hydroxyl-free radical (HȮ) production [6, 7]. Peroxone technique is a blend of hydrogen peroxides (H_2O_2) and ozone (O_3), and on this account several free radicals are produced, which oxidize waste organic compound present in water, but requirement of hydrogen peroxide enhances its cost as well as its storage and transport problem [8]. Furthermore, peroxone has shortcomings of low oxygen ozone conversion rate and been suppressed under neutral and acidic environments [9]. Likewise, ozonation is highly resourceful practice for treatment of large bio-recalcitrant-based wastewater; such type of waste usually required high quantity of energy to decay and has the ability to resist microbes; being an oxidizing agent, a large number of intermediates are generated by ozone, which initiated chain reaction and hence degraded waste. On the contrary, ozone reacts with naturally occurring bromide ions in water to form carcinogenic bromates as side products [10] and has less oxidation potential of 2.07 as well as inadequacy of degrading ozone refractory compounds [11]. Although electro-oxidation process has been provoked in treatment of refractory compounds [12] and micropollutants-based wastewater, nevertheless it has a drawback of more energy consumption (3–5 V) during electrolysis [13]. Additionally, electrochemical-based electrocoagulation techniques are where current is passed across wastewater solution containing electrodes, and metallic ions released from dissolution of anode result in coagulation *via* counter ions in corresponding solution and suspended waste particle made cluster at bottom, it has drawback of electrode encapsulation *via* oxide layer, and hence, it was not a continuous technique [14].

To overcome these dilemmas of traditional advanced oxidation techniques, researchers have been devising various electrochemical advanced oxidation practices notably, electro-Fenton, photoelectro-Fenton, electro-peroxone, and photoelectro-peroxone for wastewater treatment. Nonetheless, homogeneous electro-Fenton and photoelectro-Fenton techniques catalyzed degradation of persistent organic pollutants only under acidic media, and its alternative heterogeneous techniques could conduct full mineralization of same pollutants under neutral pH [15]. In this circumstance, hybrid electro-peroxone (EP) and photoelectro-peroxone (PEP) have been accredited for wastewater treatment under alkaline, neutral, acidic pH, posed good disintegration, and mineralization rates [16–18]. As a matter of fact, EAOPs are hybrid approaches, which have been constructed by integrating two or more practices for enhanced ȮH formation to accelerate abatement of pollutants in wastewater [19]. As a matter of fact, ȮH species is the second strongest oxidant with 2.8 V oxidation potential usually prompting nonselective attacks on C—H bond to oxidize and mineralize pollutants very swiftly as demonstrated through Eq. (6) [20]. Additionally, ȮH could randomly demolish refractory pollutants when existing satisfactorily in water and exploited admirable degradation rate of 10^8 to 10^{10} M^{-1}s^{-1} [21].

Similarly, electro-peroxone is basically hybrid of two elementary approaches, which includes ozonation and electrolysis. In this context, all these techniques were taken into an account to mitigate their drawbacks and develop a novel method named electro-peroxone by putting all together [22]. Solely, oxygen was injected into ozone generator, which interleaved its inlet sparged effluent within cathode at electrolytic cell, where oxygen reduction *via* two electrons at cathode was main culprit of *in situ* hydrogen peroxide generation founded on Eq. (1). Electrochemically formed H_2O_2 subsequently catalyzed transformation of ozone into $\dot{O}H$ by means of peroxone reaction as discussed *via* Eq. (2). Henceforth, electrochemical formation of H_2O_2 and peroxone reactions are the two key reactions of hybrid electro-peroxone approach [23]. Other reactions could have taken place *via* EP process as elaborated with Eqs. (3)–(5) [24, 25]. Major gratification of EP technique is to produce low sludge, comparatively cost-effective, manageable, and continuous production of H_2O_2, alleviate energy intake owing to good rate flow within the system, which promotes mass transfer and convection [26].

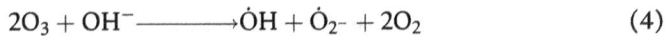

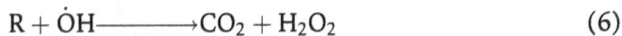

$$O_2 + 2H^+ + 2e^- \longrightarrow H_2O_2 \tag{1}$$

$$H_2O_2 + O_3 \longrightarrow \dot{O}H + \dot{O}_2^- + H^+ + O_2 \tag{2}$$

$$2H_2O_2 + 2O_3 \longrightarrow \dot{O}H + H_2O + H\dot{O}_2\text{-} + 3O_2 \tag{3}$$

$$2O_3 + OH^- \longrightarrow \dot{O}H + \dot{O}_2\text{-} + 2O_2 \tag{4}$$

$$H_2O + O_3 + e^- \longrightarrow \dot{O}H + O_2\text{-} + OH^- \tag{5}$$

$$R + \dot{O}H \longrightarrow CO_2 + H_2O_2 \tag{6}$$

Even though EP is an expedient approach, its rate of degradation of pollutants usually diminishes with acidity of solution; these acids further make complex with ions, thereby preventing their oxidation. Furthermore, much quantity of O_3 is consumed during EP process [27]. Therefore, existing techniques were modified by incorporating UV light as energy source into electro-peroxone to devise hybrid PEP approach. Photo-electro-peroxone is fundamentally hybrid of three elementary approaches, which include ozonation, electrolysis, and photolysis; these methodologies were coupled to endorse full abatement of pollutants by $\dot{O}H$ formation, which could be proceeded either through Eq. (7) or through (8) *via* PEP approach [28]. This process is expedience with elegant performance even at acidic media where photo-synthesized electron within conduction band of a semiconductor bismuth oxychloride (BiOCl) interacts with ozone to yield ozone-free radicals ($\dot{O}_3^-$) based on Eqs. (10) and (11); afterward, $\dot{O}_3^-$ subsequently will take H^+ and then finally convert into $\dot{O}H$ as discussed in Eqs. (12) and (13) [2, 29]. Moreover, activation of ozone and H_2O_2 is being abetted by PEP. Likewise, PEP technique has demonstrated 98% efficiency for decontamination of total organic carbon (TOC) from wastewater with specific energy consumption of 0.66 kWh $(gTOC_{removed})^{-1}$, while same amount of pollutants could be refined *via* UV/O_3 and electro-peroxone with specific energy consumption of 3.56 and 1.07 kWh $(gTOC_{removed})^{-1}$ sequentially, *via* low reaction rate. That is way photoelectro-peroxone and electro-peroxone are privileged over conventional hybrid advanced oxidation techniques such as UV-integrated electrolysis (UV/electrolysis) and ozone (UV/O_3) for wastewater treatment [30].

$$H_2O + O_3 + h\nu \longrightarrow 2\dot{O}H + O_2 \tag{7}$$

$$H_2O_2 + h\nu \longrightarrow 2\dot{O}H \tag{8}$$

$$O_3 + h\nu \longrightarrow O + O_2 \tag{9}$$

$$BiOCl + h\nu \longrightarrow BiOCl - h^+ + BiOCl - e^- \tag{10}$$

$$O_3 + BiOCl - e^- \longrightarrow \dot{O}_3^- \tag{11}$$

$$H^+ + \dot{O}_3^- \longrightarrow H\dot{O}_3 \tag{12}$$

$$H\dot{O}_3 \longrightarrow H\dot{O} + O_2 \tag{13}$$

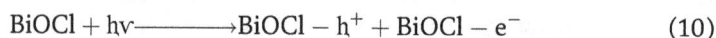

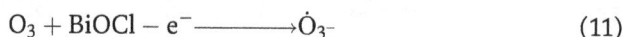

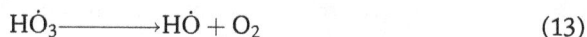

This chapter study aimed to theoretically probe environmentally friendly, cost-effective, comparatively less energy consuming, no secondary toxic side product instigating, and highly versatile novel techniques for wastewater treatment. In this context, recently EAOPs-based hybrid EP and PEP approaches have been discussed for wastewater treatment. Photo-electroperoxone and EP have vividly treated distillery biodigester effluent [31], refractory pharmaceutical [32], hospital [33], ballast water [34], herbicides [18], micropollutants [35], organic pollutant [30], acidic [2], landfill leachates [36], industrial [37], and municipal secondary effluents [26]-based wastewater. Degradation rate of pollutants could be written in terms of rate law to demonstrate chemical kinetic of pollutants during wastewater treatment by electroperoxone approach. When uniform current is provided to reactor, $H\dot{O}$ formation rate also turns out to be constant and $k[H\dot{O}]$ in Eq. (14) becomes equal to k_{app} based on Eq. (15) pseudo-first-order rate constant. Here, $\frac{-d[P]}{dt}$ is rate of disintegration of pollutants in solution; while [P], [$H\dot{O}$], k, and k_{app} denote concentration of pollutants and hydroxyl-free radicals in wastewater, absolute rate constant, and apparent rate constant, respectively [38, 39].

$$Rate = \frac{-d[P]}{dt} = k[P][H\dot{O}] \tag{14}$$

$$Rate = \frac{-d[P]}{dt} = k_{app}[P] \tag{15}$$

2. Setup devised for electro-peroxone (EP) and photoelectro-peroxone (PEP)

Experimental setup has been devised for degradation of pollutants in wastewater through EP and PEP approaches, which is illustrated in **Figure 1**. Wastewater treatment was processed in air-proof semi-batch reactor [40]. One-liter wastewater was incorporated inside the reactor, and pairs of electrodes, that is, cathode and anode 1 cm apart, were interleaved in the middle of the reactor. Quartz jacket-enclosed UV lamp was perpendicularly immersed in reactor for UV photolysis during PEP process. Bubble diffuser and magnet stirrer bar were placed to diffuse mixture of ozone and O_2 gases in aqueous solution and to mix content inside the reactor. Electrolytic operations such as EP and PEP were performed under galvanostatic conditions *via* direct current power supply in the presence of supporting electrolyte [36]. Additionally, constant temperature was maintained *via* water flow around reactor chamber. Ozone generator was operated to attain ozone and oxygen mixture from inlet oxygen supplied through oxygen cylinder. It was connected to ozone meter to estimate ozone concentration within reactor's inlet and outlet channels, which subsequently attached with gas flow

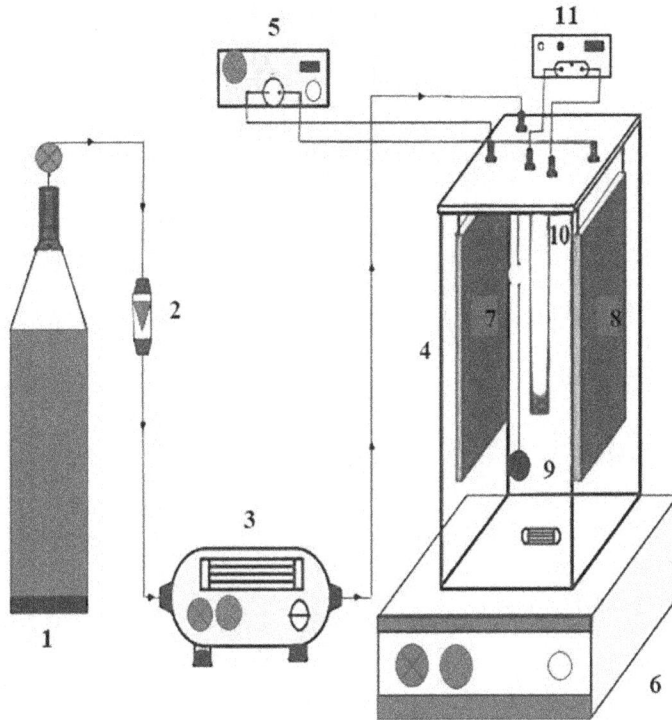

Figure 1.
Schematic illustration of reactor devised for electro-peroxone and photoelectro-peroxone for wastewater treatment image reproduced from ref. [18]. 1-oxygen cylinder, 2-rotameter, 3-ozone generator, 4-reactor, 5-power supply, 6-stirrer, 7-anode, 8-cathode, 9-fine bubble diffuser, 10-UV-lamp, 11-UV source.

meter [40]. Desired quantity of ozone was incorporated inside reactor by modifying flow rate of inlet ozone gas.

3. Electro-peroxone approaches for diverse wastewater treatment

In the literature, a wide spectrum of wastewater applications such as textile, pharmaceutical, biodigester effluents, refractory compounds, and real wastewater treatments have been successfully conducted by researchers as demonstrated by **Table 1**.

Textile industries are producing huge volume of wastewater nearly 30–50 cm^3 water volume ton^{-1}dyes. Subsequently, dyestuff effluents (10–20 mg L^{-1}) have been discharged into sewages and rivers [49, 50]. Treatment of textile wastewater is inappropriate through biological treatment meanwhile it culprits toxic secondary by products at the end, alternatively several oxidants have been reported which were being restricted on accountability of structural intricacy of dyestuffs. Consequently, textile wastewater treatment is prompted by EAOPs. [51–53]. Anionic dye Acid Orange 7 ($C_6H_{11}N_2NaO_4S$) containing wastewater (500 mg L^{-1}) has been fully decontaminated with 90% and 99% exclusion of TOC and chemical oxygen demand (COD), respectively, within 90 minutes through EP approach carried out in cylindrical reactor [54]. Likewise, Acid Orange 7 was pulverized in a cylindrical baffled reactor to boost exchange among reactants and well-organized electrode arrangement by EP approach. Acid Orange 7 (500 mg L^{-1}) mineralization cleared out 92% TOC and 99% COD were

Category of wastewater	Anode	Cathode	Standard reaction conditions	References
Secondary effluent from coal industry	Dimension-ally stable anode (DSA)	Natural air diffusion electrode	pH: 4, current: 200 mA, electrolyte: 0.3 M Na_2SO_4, treatment time: 3 h, inlet ozone dose: 6 mg min^{-1}, flow rate: 100 mL min^{-1}	[37]
BDE of rice grains	Al	Polytetrafluoroethylene (PTFE)	pH: 6, current 0.032 mA m^{-2}, electrolyte: 0.15 M Na_2SO_4, treatment time: 50 min, inlet ozone dose: 135 mg L^{-1}, flow rate: 70 L min^{-1}	[31]
Hospital wastewater	Pt sheet	Activated carbon fiber	pH: 9, current: 400 mA, electrolyte: 0.05 M Na_2SO_4, inlet ozone dose: 5 g h^{-1}, flow rate: 1 L min^{-1}	[33]
Ballast water	Perforated DSA	PTFE and carbon black-modified graphite felt (GF)	pH: 7, current: 50 mA, aeration rate: 50 mL min^{-1}, temperature: 25°C, *E. coli*: 10^6–10^7 CFU mL^{-1}, flow rate: 7 mL min^{-1}	[34]
Refractory OMPs	Mixed oxides of ruthenium and iridium (RuO_2/IrO_2) coated Ti plate	Stainless steel plate	Current: 60 mA, treatment time: 15 min, inlet ozone dose: 7 mg L^{-1}, concentration of model pollutant: 150 µg L^{-1}, flow rate: 0.15 L min^{-1}	[35]
LEV	Pt mesh	Carbon fiber composite	pH: 6.8, current: 140 mA, electrolyte: 0.05 mol L^{-1} Na_2SO_4, treatment time: 15 min, inlet ozone dose: 47 mg L^{-1}, temperature: 25°C, flow rate: 0.08 mg L^{-1}	[41]
TC and disinfections	Perforated dimension stable anode	Carbon black and PTFE-modified GF	pH: 7, current: 50 mA, electrolyte: 0.05 M Na_2SO_4, aeration rate: 50 mL min^{-1}, E coli: 1000 CFU mL^{-1}, concentration of model pollutant: 700 µg L^{-1}, flow rate: 35 mL min^{-1}	[26]
Multiple FQs	Pt plate	Fe-modified carbonized MOF	pH: 4.2, current: 210 mA, treatment time: 10 min, inlet ozone dose: 40.2 mg L^{-1}, temperature: 25°C, concentration of model pollutant: 20 mg L^{-1}, flow rate: 50 mL min^{-1}	[42]
Carbamazepine	Carbon rod	CeO_x/GF	pH:5, current: 0.05 mA, electrolyte: 0.05 Na_2SO_4 mol L^{-1}, treatment time: 60 min, ozone output: 50 mg h^{-1}, temperature: 25°C, concentration of model pollutant: 10 mg L^{-1}, flow rate: 0.5 L min^{-1}	[43]
Antibiotics and biocides	Pt	Carbon-PTFE	Phosphate buffer: 50 mM, current: 35 mA, electrolyte: Na_2SO_4 50 mM, inlet ozone dose: 4.5 mg L^{-1}, temperature: 15°C, concentration of model pollutant: 10 µg L^{-1}, flow rate: 0.35 L min^{-1} for each	[44]

Category of wastewater	Anode	Cathode	Standard reaction conditions	References
Acid orange 7	Graphite (4 cm^2)	Graphite (4 cm^2)	pH:7.7, current: 0.5 A, anode to cathode ratio: 6: 6, electrolyte: 0.1 M Na$_2$SO$_4$, treatment time: 10 min, temperature: 25°C, concentration of model pollutant: 500 mg L^{-1}, ozone flow rate: 8.5 L min^{-1}	[45]
AR14	Pt sheet	Carbon-PTFE (XC-72 carbon powder)	pH:10, current: 0.7 A, electrolyte: 0.1 M Na$_2$SO$_4$, treatment time: 30 min, temperature: 25°C, concentration of model pollutant: 400 mg L^{-1}, flow rate: 0.25 L min^{-1}	[46]
AV19	Ti|IrSnSb-oxide plate	3D GDE	pH: 3, current density: 20 mA cm^{-2}, electrolyte: 0.05 M Na$_2$SO$_4$, inlet ozone dose: 14.5 mg L^{-1}, temperature: 25°C, concentration of model pollutant: 40 mg TOC L^{-1}, electrolyte flow rate: 2 L min^{-1}, pressure exerted at GDE: 3 psi	[47]
CV	Pt rod	Stainless steel wool	pH: 9, current: 0.1 A, electrolyte: 100 mg L^{-1} NaCl, inlet ozone dose: 2 mg L^{-1}, temperature: 22°C, concentration of model pollutant: 50 mg L^{-1}, peroxide: 15 mmol L^{-1}	[48]

Table 1.
Diverse EP approaches are exemplified for diverse wastewater treatment under standard reaction conditions.

declined within 90 minutes at pH 7.7 with large electrode surface area ratio (6:6) by which degradation was considerably enhanced [45]. Similarly, Acid red 14 (AR 14) wastewater (400 mg L^{-1}) has been disintegrated in Ep-based Box-Behnken experimental setup. Full disintegration of AR 14 was accomplished, and 69% COD exclusion was achieved within 60 minutes at 10 pH [46]. Another attempt has been made on decomposition of crystal violet (CV) with K$_{app}$ of 2.69×10^{-2} and 2.87×10^{2} min^{-1}, for 100 and 200 mg L^{-1} CV in wastewater, respectively. About 98% CV was eliminated at pH range of 7–9 within 5 minutes through combination of electrolysis/peroxone/H$_2$O$_2$, and corresponding treated wastewater was manifested no toxicity for microbes. Electroperoxone approaches were provoked decolorization at alkaline media [48]. A novel approach has been made to smash Acid Violet 19 (AV19) in a lab-scale filter-press-based plant employing 3-D gas diffusion electrode as a cathode to attain better oxygen reduction reaction *via* EP. This led to 60% mineralization and 100% decolorization, at 3 psi pressure that was employed to gas diffusion electrode with acidic pH medium (3). It was demonstrated that AV19 disintegration was consequence of *in situ* generated peroxide coupled with ozonation as well as an anodic oxidation [47].

Recurring detection of personal care and pharmaceutical compounds in water has increased health heedfulness and environmental considerations [55, 56]. Wastewater treatment plant could not completely eliminate antibiotics through traditional activated sludge and sedimentation techniques; as a result, these have been monitored in secondary wastewater effluents in certain quantity [56–58]. In contrast to traditional ozonation practices, ozone recalcitrant micropollutants notably chloramphenicol, ibuprofen, and clofibric acid have been effectively pulverized and accelerated degradation kinetics through triggering HȮ production by means of EP approach [59, 60]. Numerous advanced oxidation techniques have been launched to smash biorecalcitrant paracetamol (PCT). EP approach has exhibited good efficiency in full disintegration of PCT with rate constant of 0.1662 min^{-1} [32]. Levetiracetam (LEV) removal was a bit challenging owing to polar structures, which was not susceptible to ozone degradation [61]. Extremely water-soluble antiepileptic drug LEV has been manifested pseudo-first-order degradation kinetics *via* EP approach by means of promoted synergy between O$_3$ and H$_2$O$_2$ for actively electrochemical generation of HȮ and led to withdrawn of 53.4% LEV at 15 minutes from wastewater [41]. Numerous antibiotics notably ciprofloxacin, norfloxacin, ofloxacin, and trimethoprim have been degraded with EP and tracked pseudo-first-order kinetics. Outcomes revealed that EP technique effectively eradicates antibiotics and ozone inert biocides within short time and lessen energy consumption than that of ozonation process [44]. Similarly, tetracycline (TC) and microbes were smashed into organic acids and after mineralization totally excluded from wastewater by EP approach [44]. Highly stable polyacrylonitrile-based carbon fiber cathode was designed for mineralization of phenol, where oxidation promoted transformation of pyridinic-N of polyacrylamide into pyridonic-N through EP, which endorsed cathode for oxygen reduction reaction. Major gratification of this procedure was 30-fold recyclability of fabricated cathode as well as cathodic potentials declined energy expenditure from 91.5% (simple cathode) to 48.2% (current cathode) for H$_2$O$_2$ formation [62]. Electro-peroxone approach was conducted by fabricating iron-modified carbonized metal organic framework (MOF)-based cathode for treatment of biostatic drugs such as fluoroquinolones (FQs); these were selectively 99% decomposed by O$_3$ and the rest of its ozone-reluctant transformed intermediates and side products were removed by HȮ. Moreover, HȮ was also liable to overall TOC removal, and its formation was promoted by surface functionalities of MOF-based cathode through synergic effect of adsorption and activation [42]. Aforementioned hybrid EP approaches have been successfully applied for the treatment of synthetic wastewater.

Real wastewater in contrast to synthetic wastewater is more complex, having abundant organic micropollutants with varieties of molecular structures accompanying physicochemical properties [63, 64]. Biochemical oxygen demands (BODs)/COD ratio has been commenced for probation of biodegradability prospects and wastewater encompassing 0.4 or its onward ratio has manifested good biodegradability as well as decline in bio-toxicity [65]. Electro-peroxone approach was carried out for processing of reverse osmosis concentrate obtained from industrial coal wastewater. As a result, 92% decolorization, 89% UV$_{254}$, and 71.2% TOC have been eradicated within 6 hours [37]. After treatment, 91.3% color reduction and 99.9% COD elimination were detected in distillery biodigester effluents (BDE) through EP approach, and low cost of 1 m^3 BDE/2$ and less sludge formation are major gratification of this approach [31]. Chloramphenicol and clofibric acid such as ozone obstinate micropollutants in surface water have been oxidized by EP system and subsequently resulting in hypochlorous and hypobromous acids, which are the main culprit of engendering chlorinated and brominated derived byproducts were efficiently quenched by electrochemically produced H$_2$O$_2$ in surface water [35]. Several antibiotics conspicuously ciprofloxacin, norfloxacin, ofloxacin, and trimethoprim have been eliminated from secondary wastewater effluents through EP [44]. Electro-peroxone practice was also employed to process municipal secondary effluents with negligible disinfection side products as well as 65% COD and 44% BOD were declined [26]. Diverse 89 pharmaceutical compounds were examined in terms of organic micropollutants (OMPs) exclusively existing in a real wastewater to evaluate their ozone reactivities and physiochemical properties by means of quantitative structure activity relationship (QSAR) for the sake of kinetic assessment. Pharmaceutical compounds having partial charge moieties, and branched, electrophiles, and lowest unoccupied molecular orbital energies were categorized as ozone-resistant compounds with ozone rate constant (k$_{O3}$) < 10^2 M^{-1} s^{-1} and were degraded by sole EP. Conversely, ozone reactive pharmaceuticals accompanying with nucleophilic species, highly occupied molecular orbital energies, and conformation contingent charge descriptors were deemed to be ozone reactive with rate constant greater than 10^2 M^{-1} s^{-1} would be rapidly eradicated by EP and ozonation [66]. Additionally, ultrasound coupled EP system and virgin EP has been applied for textile industry effluent at 5.8 pH. 93% and 99% decolorization have been accomplished through virgin EP and the integrated ultrasound EP process within 60 minutes after treatment [67]. Similarly, real pesticide wastewater has been treated *via* 3D/EP system with elimination of 97.5% TOC and 95.1% COD up to 500 and 300 minutes sequentially; consequently, long reaction time contributes to more cost although 3-D/EP is more cost effective to that of 2-D EP. Moreover, 30 minutes later BODs/COD ratio has been incremented from 0.049 to 0.571 [68].

In context of acidic wastewater treatment, few attempts have been made to mend EP process. Tannic acid has been oxidatively smashed by EP approach in two phases firstly tannic acid pulverized *via* O$_3$ and HȮ, which were accountable for carboxylic acid like intermediates. Consequently, more intermediate formation lowered pH of wastewater, and thereby HȮ was also declined. Somehow it has been overcome *via* increasing inlet ozone as well as adjusting current and unlike ozonation more than 10% efficacy has been achieved *via* EP [69]. Electrode-separated compartmental-based EP approach was carried out to eliminate para-aminobenzoic acid (PABA) from wastewater solution. As a result, 63.6%–89.5% PABA has been abolished by contribution of cathodic and anodic side reactions, respectively, at 10 minutes as well as incremented pseudo-first-order reaction kinetics and HȮ formation [70].

3.1 Miscellaneous electrode texture-based electro-peroxone approaches for wastewater treatment

Carbon nanotubes (CNTs) have exhibited brilliant adsorption to pollutants, and this tendency along with adsorption kinetic was further enhanced in terms of electro-sorption by employing them as electrodes [71]. Furthermore, CNTs have been demonstrated good electrochemical oxidation of pollutants, good chemical stability, electrical conductivity, and noteworthy mechanical strength during electrolysis [72] and photolysis [73]. Advanced oxidation approach has been integrated with adsorption to construct hybrid system for actively pulverization of pollutant in wastewater [74, 75]. Pharmaceutical compounds, particularly diclofenac sodium (DS), were completely fragmented by carbon nanotubes-polytetrafluoroethylene (CNTs-PTFE) electrode over five consecutives cycles exploiting pseudo-second-order kinetics. Where negatively charge diclofenac sodium was exhibited electro-sorption to CNTs-PTFE anode afterward, adsorption phenomena switched this anode into cathode and corresponding adsorbed pollutants were subsequently disintegrated by EP approach within 10 minutes and 99% TOC were eliminated after 1 hour [76]. Likewise, copper ferrite-modified carbon nanotubes ($CuFe_2O_4$/CNTs) were used as catalysts having brilliant recyclability to decomposed fluconazole (FLC) wastewater through EP. Catalyst has adsorbed FLC on its sphere and enhanced FLC mass transfer to electrode surface and thereby eliminated 89% FLC and integrated adsorption-EP technique has contributed 10% efficiency to virgin EP approach [77].

Similarly, carbon nitride-multiwall carbon nanotubes-based nanocomposite (n-C_3N_3/MWCNT) catalyst has actively smashed sodium oxalate in wastewater by endorsing adsorption of pollutant and accelerating electron transfer, which trigger O_3 and O_2 electro reduction [78]; consequently, H_2O_2 and $\dot{O}_3^-$ were generated, which has further enhanced $H\dot{O}$ formation [79]. On account of large surface area, activated carbons are good in elimination of micropollutants (MPs); on the contrary, MPs saturated activated carbons having high affinity for adsorbates pose a major challenge in regeneration of electrode, which was overwhelmed by oxidation of MPs through ozonation process [80] but some sorts of MPs were inert toward ozonation reaction [64]. In this frame, EP coupled with ozonation to exclude diverse MPs, namely trimethoprim, ciprofloxacin, perfluorooctanoic acid, carbamazepine, diclofenac, and benzotriazole from wastewater and efficiently pulverized MPs from ozone. Afterward, ozone-resistant MPs were disintegrated *via* EP with simultaneous regeneration of powdered activated carbons (PAC). In contrast to virgin PAC, all MPs have been exploited more than 100% efficacy for PAC regeneration except diclofenac and perfluorooctanoic acid (PFOA) [81].

Electro-peroxone approaches have been well organized at alkaline and neutral pH; on the contrary, its progress was constrained at acidic pH, which bounds rate constant of H_2O_2 as of 9.6×10^6 to $0.01\,M^{-1}\,s^{-1}$ for 11 to 3 pH, respectively. Hence, reaction between ozone and deprotonated peroxide has no more yielded reactive oxygen species [22]. Manganese carbon nitride-carbon nanotubes (C_3N_4-Mn/CNT) composite catalyst overcomes drawbacks of disintegrating pollutants in strongly acidic solution *via* EP reaction. Moreover, C_3N_4-Mn/CNT heterogeneous catalyst has been accelerated peroxone reaction between H_2O_2 and O_3, and decomposed oxalic acid within 30 minutes at pH 3 for up to 5 cycles [82]. Additionally, C_3N_4-Mn/CNT-integrated EP system has been evaluated for disintegration efficiency of oxalic acids at a wide range of pH, Outcomes reveal that 57.1- and 2.6-fold increments have been achieved in integrated system, at 3 and 9 pH as compared with virgin EP [82].

Traditional EP approaches were mostly carried out by commencing 2-D electrode system, which have been demonstrated low mass transfer; therefore, to boost electrode performances for additional optimization of conducted treatment were suggested for forthcoming generation [78, 83]. Reticulated enamel carbon, graphite felt, polytetrafluoroethylene, and carbon felt-based cathodic materials were manifested O_2 reduction for H_2O_2 formation [84]. Unlike conventional 2D-electrode in EP approach, 3D-electrode system could considerably promote the electrochemical efficiency of reactor owing to large surface area, which boosted H_2O_2 formation [85]. TiO_2-loaded granular-activated carbon (TiO_2-GAC) as a 3D electrode in EP system was applied for decomposition of diuron, which is a phenyl urea herbicide wastewater, hybrid 3-D/EP system was demonstrated two times more pseudo-first-order disintegration rate (effectiveness) than those of sole EP system. Diurons were adsorbed by TiO_2-GAC and later polarized to synthesize microelectrodes, which yields ȮH. Moreover, TiO_2-GAC has considerably enhanced H_2O_2 formation in a corresponding solution [68]. Being a 3-D activated carbon system, carbon felt (CF) has been shown elegant electrolytic proficiency, good mechanical stability, and cost effective [86]. N-doped-reduced graphene oxides (N-rGOs) supported carbon as well-designed cathode was demonstrated to improve oxygen reduction feedback for H_2O_2 generation, better conductivity, boosted electrocatalysis, and electron transfer rate [87, 88]. Diuron was completely smashed at 9 pH within 15 minutes through EP approach using versatile N-rGO/CF-based cathode electrode. Furthermore, N-rGO/CF exploited good efficiency in H_2O_2 formation and lessen energy expenditures for 10 cycles continuously to that of sole CF cathode. This system has led to processed real pesticide wastewater having COD of 3680 mg L^{-1} after processing till 360 minutes, and COD was declined to 47.7 mg L^{-1}. Moreover, BOD/COD ratio of 0.4 and 0.04 has been obtained for processed and unprocessed real pesticide wastewater, respectively [88]. Another attempt was made in which a filter-press flow cell integrated with three-dimensional air diffusion electrode-based lab scale plant was devised to disintegrate levofloxacin and 63% mineralization accomplished at 3 pH [89]. Likewise, GF was modified with cerium oxides (CeOx) to well-designed cathode, and H_2O_2 exhibited chemisorption with CeOx; consequently, it will prompt reaction with O_3 as compared with bulk H_2O_2. Consequently, CeOx/GF-EP system has been manifested 69.4% TOC exclusion in disintegration of carbamazepine within 60 minutes at pH range of 5–9 with upright fivefold recyclability. In contrast to traditional EP, this strategy is perquisite for degradation of refractory organic pollutants under acidic media by proficiently activating ozone, upgraded surface hydrophilicity, and lessen energy expenditure for electro-generation of H_2O_2 [43].

A novel hybrid approach comprising three electrodes in EP system for oxalate-containing wastewater has been developed. After elaboration of reaction mechanism, it was suggested that all reactions in combination subsidizes HȮ formation in EP. In contrast to two electrode systems, three electrodes system could be comparatively privilege in providing precise control and purifying salt-rich wastewater [79].

3.2 Complementary hybrid electro-peroxone approaches for wastewater treatment

To proficiently mineralize and eliminate a wide range of biodegradable contaminants along with refractory pollutants from wastewater by low electrical energy requirement in cost effective and easy ways, some conventional approaches conspicuously biological treatment, ultrasound, electrocoagulation, and low-pressure filtration were coupled with electro-peroxone to devise novel complementary hybrid electro-peroxone system [67, 90, 91]. In this circumstance, synergy overcome constrained individual approaches that have low efficiency independently and accelerated attenuation of wastewater in terms of complementary hybrid electro-peroxone system.

Bio-electroperoxone (Bio-EP) approach has been devised for a two-way treatment of pharmaceutical wastewater, where microbes biodegrade some compounds at electrically bound biofilm reactor (EBBR), and the rest of all compounds that did not undergo biological oxidation was pulverized *via* EP approach. Integrated Bio-EP has been eliminated 89% TOC, 84% suspended solids, 99.99% deactivated microbes, and 92.20% decolorized wastewater [39]. Another attempt was made to pulverize recalcitrant contaminants particularly methylene blue. In this system, self-sustained energy achieved from microbial fuel cell-based cathode by Bio-EP process was supplied and 83% methylene blue has been eliminated within 30 minutes by exploiting pseudo-first-order kinetics with 2.05 h^{-1} rate constant during pulverization [90].

Hydroxyl-free radicals could be synthesized by fracturing bubbles cavitation in aqueous medium through ultrasound (US) based on Eq. (16) [92]. Moreover, US also splits up ozone and peroxide based on Eqs. (17) and (18) [93, 94]. Integrated US/EP approach was applied to fragmentize acid orange 7 at pH 7, which has manifested 88% mineralization, 99% decolorization, and 85% COD elimination with pseudo-first-order kinetics [67].

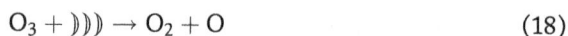

$$H_2O +))) \rightarrow H\dot{O} + \dot{H} \tag{16}$$

$$H_2O_2 +)))) \rightarrow H\dot{O} + H\dot{O} \tag{17}$$

$$O_3 +)))) \rightarrow O_2 + O \tag{18}$$

Shale gas fracturing flowback water (SGFFW) was processed with electroperoxone-integrated electrocoagulation (EC/EP) or ECP approach and led to 82.5% COD exclusion up to 90 minutes, with 29.9% average current effectiveness. In ECP technique, coagulant hydroxyl-aluminum at anode eliminates colloids and suspended items [95] as well as catalyzed H$\dot{O}$ formation by reaction with O_3 to breakdown pollutants *via* EP at cathode [96]. Likewise, peroxi-coagulation was integrated with EP; thereby, high efficiency with less reaction time was obtained than that of virgin EP with full decolorization, and 92.2% UV$_{254}$ and 72.2% TOC were eliminated [37].

Low-pressure filtration was coupled with EP approach to design hybrid electro-peroxone filtration (EPF) system, which was continuously eliminated 64.87% ibuprofen (IBU) at less filtration pressure (0.8 kPa) within 8 seconds in its very low concentration of 1 mg L^{-1}. IBU elimination efficacy was comparatively three times than that of individual efficiencies achieved from electrochemical filtration and ozone filtration. In contrast to sole EP, EPF was promoted mass transfer of H$\dot{O}$ and O_3 owing to membrane permeation drift [91].

Downflow bubble column electrochemical reactor (DBCER) has led to incremented mass transfer and contact area owing to energetically liquid inflow, and small bubble formation takes place to cause commotion as well as did not let out electrochemically synthesized *in situ* oxygen *via* Eq. (19) rather it was dispersed within cell [97, 98]. As a matter of fact, under large current density and low pH, H$\dot{O}$ production would favor based on Eq. (20), which subsequently subsidizes oxidation at electrode surface *via* Eq. (21) [99]. DBCER with boron-doped diamond (BDD) electrode system has also been prerequisites for *in situ* formation of O_2, O_3, and H_2O_2 to accomplish EP process with 75% TOC exclusion at pH 3 phenol smashed by ozone and 65% TOC with drawn at pH 7 phenol pulverized by H_2O_2 within 6 hours [98].

$$2H_2O \rightarrow O_2 + 4H^+ + 4e^- \tag{19}$$

$$BDD + H_2O \rightarrow BDD(H\dot{O}) + H^+ + e^- \tag{20}$$

$$BDD(H\dot{O}) + R \rightarrow BDD + CO_2 + H_2O + H^+ + e^- \tag{21}$$

4. Photoelectro-peroxone practices for wastewater treatment and their comparison with EP

Simple ozonation, photolysis, and electrolysis mechanisms were integrated to fashion novel hybrid PEP approach for wastewater treatment to overcome shortcoming of low mineralization rate during EP at acidic pH and dwindle corresponding electrical energy consumption. Henceforth, PEP and EP approaches have been contributed to synergistic effect, which has been quantitatively determined through enhancement factor calculated by Eqs. (22) and (23) [55].

$$Enhancement = \frac{(k_{PEP})}{(k_O) + (k_{UV}) + (k_E)} \tag{22}$$

$$Enhancement = \frac{(k_{EP})}{(k_O) + (k_E)} \tag{23}$$

Where k_{EP}, k_{PEP}, k_O, k_{UV}, and k_E denote rate constants during pollutant disintegration for EP, PEP, ozonation, photolysis, and electrolysis, respectively. Enhancement factors along with degradation rate constants have been comparatively incremented during PEP approaches than those of EP for same wastewater treatment (e.g., **Table 2**).

Organic pollutants containing plentiful wastewater have been magnificently treated by PEP. In this framework, derivatives of benzene particularly nitrobenzene, chlorobenzene, and benzaldehyde containing wastewater were processed through electro-peroxone and photoelectro-peroxone approaches. Although both approaches have been drawn out 98% TOC, PEP exhibited good degradation kinetics, and consumed less energy than that of EP and other advanced oxidation processes, which have been exploited slow degradation kinetics and used up high energy [30]. Similarly, 1,4-dioxane, a major contributor to refractory organic pollutant, is exclusively found in industrial wastewater and landfill leachates and was disintegrated with 33 times proficient pseudo-first-order rate constant *via* PEP as compared with UV photolysis, ozone, and electrolysis. Photoelectro-peroxone approach has discharged 98% TOC after 1,4-dioxane mineralization in wastewater solution. Solely 37% TOC was drawn out *via* EP owing to reliant on pH, which gradually lower consequently interfere with $\dot{O}H^-$ due to resulting intermediates of 1,4-dioxane decomposition notably, carboxylic acid [100]. Furthermore, 4-nitrophenol comprising wastewater has been processed by PEP technique along inserting BDD electrode as an anode that accelerates numerous free radical creations on its sphere and other conventional electrochemical-based advanced oxidation processes. As a result, all mineralized TOC was excluded from wastewater solution in 45 minutes *via* PEP [40]. Likewise, TOC elimination was taken into an account for

Category of wastewater	Conducted approach	Anode	Cathode	Apparent degradation rate constant K_{app}	Enhancement factor	Ref.
Nitrobenzene	PEP	12 cm^2 platinum sheet	40 cm^2 carbon-PTFE	93.0[a]	5.8	[30]
—	EP	—	—	86.7[a]	5.5	[30]
Chlorobenzene	PEP	—	—	212[a]	8.4	[30]
—	EP	—	—	164[a]	6.7	[30]
Benzaldehyde	PEP	—	—	112[a]	4.7	[30]
—	EP	—	—	98.0[a]	4.1	[30]
1,4-Dioxane	PEP	Platinum/titanium (Pt/Ti)	Carbon-PTFE	2.237[b]	33.4	[100]
—	EP	—	—	1.749[b]	28.2	[100]
Nitrophenol	PEP	50 cm^2 carbon belt	50 cm^2 BDD	0.145[c]	13.2	[40]
—	EP	—	—	0.074[c]	6.7	[40]
2,4-D herbicide	PEP	Titanium	Graphene	0.0009[d]	ND	[18]
MY	ZVI/PEP	Platinum	Graphite felt	0.1926[e]	1.54	[101]
Reactive Yellow F3R	PEP	Ti/TiO$_3$	Graphite	1.176[e]	1.38	[102]
—	EP	—	—	0.834[e]	1.28	[102]
PFOA	PEP	rGO/BiOClfilm-based photoanode	Graphite	17.5[f]	4.73	[2]
CBZ	EP	Carbon rod	CeOx/GF	2×10^{-2}[g]	3.12	[43]
Organic contaminant	Bio-EP	Nematic liquid crystal display electrode	Pt coated titanium	0.0177[g]	ND	[39]
Methylene blue	Bio-EP	Activated carbon granules	XC-72 carbon black	0.237[g]	ND	[90]

*rGO/BiOCl = reduced graphene oxide/bismuth oxy-chloride, **ND** = not determined, **b** = $K_{app} \times 10^3\ s^{-1}$, **a** = $K_{app} \times 10^2\ min^{-1}$, **c** = $K_{app}^{NP}\ min^{-1}$, **d** = $(min)^{-1}$ reaction rate constant (K_{obs}), **e** = h^{-1}, **f** = $K_{app} \times 10^4\ min^{-1}$, **g** = min^{-1}.*

Table 2.
Comparison between PEP and EP techniques has been demonstrated on basis of enhancement factors along with degradation rate constants for wastewater treatment.

mineralization degree of 4-chlorophenol, benzotriazole, metanil yellow (MY), TC, and carmoisine with 85, 84.2, 65.6, 62.4, and 60.2% TOC removal, respectively. It could be considered that pollutants or compounds with high carbon content were deemed to have low TOC removal owing to compacted structure [101]. Additionally, PFOA was hardly smashed by advanced oxidation techniques and HȮ is also somewhat inactive for PFOA [103, 104]. Therefore, PFOA has been 56.1% decomposed by PEP within 3 hours manifesting pseudo-first order kinetics [2].

Similarly, PEP approaches have been eliminated herbicides at both alkaline and neutral pH. In this context, 2,4-D herbicide was entirely degraded within 25 minutes and its degradation kinetics has exploited first-order reaction rate by PEP approach having rate constant of about 2.5-folds higher than the rate constant of EP. Furthermore, 58.9% TOC has been wiped out during 2,4 D mineralization at pH of 7 from wastewater solution. On contrary to stainless steel and graphite felt, cathodic-activated carbon has promoted reaction rate by engendering H_2O_2 [28]. Likewise, another attempt has been made to boost 2,4-D herbicide disintegration through UV-assisted PEP. Complete fragmentation of 2,4-D herbicide in solution (58 mg L^{-1}) was obtained at 5.6 pH in 112 minutes along with its 91% elimination; moreover, 76% TOC has been withdrawn during 2,4-D herbicide mineralization after 2 hours. Low pH and 85% COD along with trapping assessment revealed that both species $\dot{O}H$ and $\dot{O}_2^-$ have contributed to wastewater treatment [18]; hence, approximately at slightly acidic pH reaction could be proceeded between H^+ ions and $\dot{O}_3^-$ to produce reaction active species ($H\dot{O}^-$) *via* Eq. (24) [105].

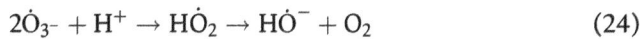

$$2\dot{O}_{3^-} + H^+ \rightarrow H\dot{O}_2 \rightarrow H\dot{O}^- + O_2 \tag{24}$$

$$O_2 + 2H^+ + 2e^- \rightarrow H_2O_2 \tag{25}$$

Furthermore, PEP-based some attempts have been made in textile wastewater treatment. In this circumstance, MY dye containing wastewater has been processed by incorporating zero-valent iron (ZVI) as a nano-catalyst in the solution, which was further followed by PEP process and accelerated wastewater treatment. This hybrid PEP/ZVI approach has successfully decolorized wastewater solution (50 mg L^{-1}) at acidic pH 3 within 25 minutes, as acidic media promotes H_2O_2 electrolytically based on Eq. (25) [101]. Moreover, reactive yellow F3R (RY F3R) wastewater was pulverized by PEP manifesting first-order kinetics and 97.66% decolorization and 84.64% TOC has been excluded with 14 and 1.4 times more degradation rate constant as compared with photolysis and EP sequentially.

Moreover, real textile wastewater also has been treated by PEP effectively by withdrawing TOC [102] and decolorization rate could be promoted by incorporating transition metals that in turn produce Fenton reagent. Fe^{+2} triggers ozone activation and hydroxyl-free radical formation as discussed in Eqs. (26) and (27) [28, 106].

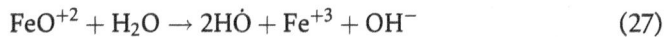

$$Fe^{+2} + O_3 \rightarrow FeO^{+2} + O_2 \tag{26}$$

$$FeO^{+2} + H_2O \rightarrow 2H\dot{O} + Fe^{+3} + OH^- \tag{27}$$

$$COD_{exclusion\%} = \frac{C_f - C_0}{C_0} \times 100 \tag{28}$$

In addition, COD parameter was applied to analyze pollutant concentration in landfill leachate and lower the pollutant concentration, and lesser oxidant would be acquired; hence, lower COD exclusion would be attained. Percentage of COD exclusion could be calculated by Eq. (28) where C_0 and C_f denote quantity of COD that has been consumed by leachate before and after its treatment [107]. In this frame, 83% COD exclusion has been achieved at 5.6 pH through PEP [36].

5. Energy expenditures for electro-peroxone, complementary hybrid EP, and photoelectro-peroxone approaches

Some amount of energy has required to perform electrochemical oxidation of wastewater. In this framework, specific energy is mandatory to disintegrate pollutants in innumerable wastewater treatment. Therefore, Eqs. (29) and (30) have been proposed to estimate energy supplied during EP, PEP, and complementary hybrid EP approaches [94].

$$SEC_{EP} = \frac{U \times I \times t + r \times CO_3}{(TOC_0 - TOC_t) \times V} \tag{29}$$

$$SEC_{PEP} = \frac{P_{UV} \times t + U \times I \times t + r \times CO_3}{(TOC_0 - TOC_t) \times V} \tag{30}$$

$$SER_{PEP} = \frac{UIT \times U_{photolysis} \times rCO_3}{(COD_0 - COD_t) \times V} \tag{31}$$

$$SEC_{EP} = \frac{U \times I \times t + r \times CO_3}{([PCT]_0 - [PCT]_t) \times V} \tag{32}$$

$$\text{Energy consumption} = \frac{U \times I \times t}{V} \tag{33}$$

$$EC = \frac{U \times I \times t}{V \times \Delta(TOC_{exp})} \tag{34}$$

$$EC = \frac{(V \times I + \text{ozone generator energy}) \times t \times 1000}{C_{dye\ removal} \times \text{cell volume}} \tag{35}$$

$$SEC_{EP} = \frac{U \times I \times t + C \times Q \times t \times R}{(C_0 - C_t)V} \tag{36}$$

$$EEC = \frac{U \times I \times t + Q_{gas} \times a \times CO_3}{V} \times 10^{-3} \tag{37}$$

$$EEC = \frac{1000 \times U \times I \times t + Q_{gas} \times a \times CO_3}{(TOC_0 - TOC_t)V} \tag{38}$$

Where SEC_{EP} and SEC_{PEP} are the specific energy consumptions for EP and PEP sequentially measured in kWh $(gTOC_{removed})^{-1}$, and SER_{PEP} is the specific energy consumption or electrical energy requirement measured in kWh $(gCOD_{removed})^{-1}$. U is an average cell voltage (V), I denotes current (A), t represents reaction time (h), r symbolizes energy requirements for ozone formation (kWh $(kgO_3)^{-1}$), C_{O3} designates ozone quantity consumed during EP and PEP approaches, TOC_0 and TOC_t indicate total organic carbon in the solution at 0 time and any time t (mg L^{-1}), $[PCT]_0$ and $[PCT]_t$ symbolize concentrations of unprocessed and processed paracetamol (PCT), respectively, V shows solution volume (L), P_{UV} denotes power of UV lamp (W) [108], $\Delta(TOC_{exp})$ represents change in the concentration of TOC [89], $(C_0 - C_t)$ is the concentration of LEV in untreated and treated wastewater sequentially, R denotes energy expenditure for ozone formation, C symbolizes concentration of inlet ozone, Q indicates flow rate of gaseous ozone [41], a represents energy attained for ozone formation, Q_{gas} designates feed gas volume, and C_{O3} reveals feed gas comprising ozone concentration [26].

Total organic carbon was effectively discarded from wastewater during mineralization of benzene derivatives through SEC_{EP} and SEC_{PEP}, of 1.07 and 0.66 kWh $(gTOC_{removed})^{-1}$ respectively [30]. Similarly, SEC_{EP} and SEC_{PEP} of 0.22 and 0.30 kWh $(gTOC_{removed})^{-1}$ have been achieved by removing TOC from 1,4-dioxane containing wastewater sequentially [100]. Nitrophenol decomposition has been expended SEC_{EP} and SEC_{PEP}, of 7.5 and 4.1 kWh $(gTOC_{removed})^{-1}$, respectively, for entire elimination of TOC. In addition to PEP, BDD electrode dramatically enhanced reaction kinetic; therefore, deducted energy requirement [40] SER_{PEP} of 1.5 kWh $(gCOD_{removed})^{-1}$ has been consumed in landfill leachate treatment using Eq. (31) [36]. Entire PCT breakdown *via* EP has expended SEC_{EP} of 0.1164 kWh $(gPCT_{removed})^{-1}$ based on Eq. (32) [32]. 1.676 and 22.86 kWh m^{-3} energy have been expended during hybrid bio-EP and solely EP, respectively, calculated through Eq. (33) [39]. Electrolytic energy consumption (EC) of 0.27 kWh $(gTOC_{removed})^{-1}$ was obtained *via* Eq. (34) during levofloxacin mineralization employing 3-D perforated electrode by EP [89]. 37.7% and 41.1% COD have been excluded via EP and 3-D TiO_2/GAC system within 90 minutes by exhausting electrical energy of 0.1 and 0.08 kWh $(gCOD_{removed})^{-1}$, respectively [68]. Additionally, 39.2% and 43.6% COD have been eliminated from real pesticide wastewater with energy expenditures of 0.088 and 0.079 kWh $(gCOD_{removed})^{-1}$ *via* CF-EP and N-rGOs/CF-EP system sequentially [88]. Eq. (35) was taken into an account; afterward, 53 kWh kg^{-1} (kg denote weight of removed dye) energy has been estimated during Acid Orange 7 disintegration with 99% COD exclusion during EP in cylindrical reactor [54]. To diminish SEC, another attempt was made in which Acid Orange 7 was entirely pulverized (99% $COD_{removal}$) through EP with EC of 8 kWh kg^{-1} founded on Eq. (35) [45]. AV19 dye has been 60% mineralized with energy expenditure of 0.085 kWh $(gTOC_{removed})^{-1}$ by laboratory-scale EP plant equipped with 3-D electrode based on Eq. (34) [47]. Similarly, LEV drug was smashed through EP approach with SEC_{EP} of 0.326 kWh $(gLEV_{removed})^{-1}$ on the basis of Eq. (36) [41]. Electrical energy consumption (EEC) of 0.47 kWh m^{-3} has been calculated through Eq. (37) in decontamination of municipal wastewater and TC disintegration *via* EP approach [26]. In the same way, IBU elimination through EPF system used up 0.16 kWh m^{-3} energy [91]. Sequential EEC of 1136.8 and 828.4 kWh $(kg_{TOC})^{-1}$ have been achieved for virgin EP and hybrid peroxi-coagulation/EP system founded on Eq. (38) [37]. Likewise, 99.9% COD exclusion has been achieved during treatment of BDE of rice grains through EP approach *via* EEC of 3.8 kWh m^{-3} [31].

Overall, PEP dwindled almost 45% specific energy consumption than that of EP approaches for a same category of wastewater under unchanged reaction conditions; nevertheless, some exceptions may be commenced conspicuously in degradation of 1,4-dioxone. Complementary hybrid EP approaches have foremost expedience of comparatively reducing energy expenditures to that of a virgin EP as well as enhanced abatement of pollutants in wastewater treatment. In this milieu, bio-electroperoxone system offered much indulgence by in taking very low energy. In contrast to conventional 2-D electrodes, latest 3-D electrodes-based EP approaches have been manifested less energy consumption.

6. Conclusions

High-operating cost-advanced oxidation processes on accountability of derisory performance to wastewater treatment have been exploited sundry shortcomings, which urge necessity for EAOPs-based alternative techniques. In this framework, to exaggerate traditional 2-D EP system has been transformed into 3-D EP by modification in electrode texture as a result, more peroxide formation was catalyzed by large electrode surface area as well as considerably SEC were also dwindled. Notwithstanding PEP approaches were established to overcome dilemma of existing EP techniques under harsh conditions, where UV accelerated further prevailing hydroxyl-free radicals and synergistic effect of individual mechanisms involved in PEP have been substantially boosted enhancement factor along with degradation kinetics of pollutants in wastewater thereby diminishing energy expenditures in the form of SEC_{PEP}. Additionally, to improve some conventional methods more conspicuously filtration, electrocoagulation and biological treatments were coupled with EP to devise novel complementary hybrid EP-based EAOPs, which have demonstrated pragmatic mineralization effectiveness and declined required electrical energy consumption.

Over the last decade, EAOPs in wake of nonselective oxidation and prohibition of secondary products have been acquainted for wastewater treatment. Henceforth, EAOPs more conspicuously novel complementary hybrid EP and PEP approaches could be more economical option for wide spectrum of synthetic and real wastewater treatment along with reducing energy expenditures, which could be fruitful from laboratory to large scale.

Conflict of interest

The authors declare no conflict of interest.

Author details

Tatheer Fatima, Tanzeela Fazal* and Nusrat Shaheen
Department of Chemistry, Abbottabad University of Science and Technology (AUST), Abbottabad, Pakistan

*Address all correspondence to: tanzeelafazal@yahoo.com

IntechOpen

References

[1] Prabakar D et al. Pretreatment technologies for industrial effluents: Critical review on bioenergy production and environmental concerns. Journal of Environmental Management. 2018;**218**: 165-180. DOI: 10.1016/j.jenvm an.2018.03.136

[2] Li Z et al. Highly efficient degradation of perfluorooctanoic acid: An integrated photo-electrocatalytic ozonation and mechanism study. Chemical Engineering Journal. 2020;**391**:123533. DOI: 10.1016/ j.cej.2019.123533

[3] Kundu D, Hazra C, Chaudhari A. Biodegradation of 2, 6-dinitrotoluene and plant growth promoting traits by *Rhodococcus pyridinivorans* NT2: Identification and toxicological analysis of metabolites and proteomic insights. Biocatalysis and Agricultural Biotechnology. 2016;**8**:55-65. DOI: 10.1016/j.bcab.2016.08.004

[4] Wu Y et al. Effects of composition faults in ternary metal chalcogenides $(Zn_xIn_2S_{3+x}, x= 1–5)$ layered crystals for visible-light-driven catalytic hydrogen generation and carbon dioxide reduction. Applied Catalysis B: Environmental. 2019;**256**:117810. DOI: 10.1016/j.apcatb.2019.117810

[5] Gunatilake S. Methods of removing heavy metals from industrial wastewater. Methods. 2015;**1**(1):14. ISSN: 2912-1309

[6] Deng Y, Zhao R. Advanced oxidation processes (AOPs) in wastewater treatment. Current Pollution Reports. 2015;**1**(3):167-176. DOI: 10.1007/ s40726-015-0015-z

[7] Umamaheswari J et al. A feasibility study on optimization of combined advanced oxidation processes for municipal solid waste leachate treatment. Process Safety and Environmental Protection. 2020;**143**:212-221. DOI: 10.1016/j.psep.2020.06.040

[8] Chow C-H, Leung KS-Y. Removing acesulfame with the peroxone process: Transformation products, pathways and toxicity. Chemosphere. 2019;**221**: 647-655. DOI: 10.1016/j. chemosphere.2019.01.082

[9] Ding Y et al. Oxygen vacancy of CeO_2 improved efficiency of H_2O_2/O_3 for the degradation of acetic acid in acidic solutions. Separation and Purification Technology. 2018;**207**:92-98. DOI: 10.1016/j.seppur.2018.06.027

[10] Yang J et al. Enhancement of bromate formation by pH depression during ozonation of bromide-containing water in the presence of hydroxylamine. Water Research. 2017;**109**:135-143. DOI: 10.1016/j.watres.2016.11.037

[11] Yao W et al. Comparison of methylisoborneol and geosmin abatement in surface water by conventional ozonation and an electro-peroxone process. Water Research. 2017; **108**:373-382. DOI: 10.1016/j.watres.2016. 11.014

[12] Särkkä H, Bhatnagar A, Sillanpää M. Recent developments of electro-oxidation in water treatment—a review. Journal of Electroanalytical Chemistry. 2015;**754**:46-56. DOI: 10.1016/j. jelechem.2015.06.016

[13] Wang C et al. Insights of ibuprofen electro-oxidation on metal-oxide-coated Ti anodes: Kinetics, energy consumption and reaction mechanisms. Chemosphere. 2016;**163**:584-591. DOI: 10.1016/ j.chemosphere. 2016.08.057

[14] Syam Babu D et al. Industrial wastewater treatment by electrocoagulation process. Separation Science and Technology. 2020;**55**(17): 3195-3227. DOI: 10.1080/ 01496395.2019.1671866

[15] Ye Z et al. Mechanism and stability of an Fe-based 2D MOF during the photoelectro-Fenton treatment of organic micropollutants under UVA and visible light irradiation. Water Research. 2020;**184**:115986. DOI: 10.1016/j. watres.2020.115986

[16] Wang Y et al. The electro-peroxone process for the abatement of emerging contaminants: Mechanisms, recent advances, and prospects. Chemosphere. 2018;**208**:640-654. DOI: 10.1016/j. chemosphere.2018.05.095

[17] Ghalebizade M, Ayati B. Acid Orange 7 treatment and fate by electro-peroxone process using novel electrode arrangement. Chemosphere. 2019;**235**: 1007-1014. DOI: 10.1016/j. chemosphere.2019.06.211

[18] Kermani M, Mehralipour J, Kakavandi B. Photo-assisted electroperoxone of 2,4-dichlorophenoxy acetic acid herbicide: Kinetic, synergistic and optimization by response surface methodology. Journal of Water Process Engineering. 2019;**32**:100971. DOI: 10.1016/j.jwpe. 2019.100971

[19] Oturan MA, Brillas E. Electrochemical advanced oxidation processes (EAOPs) for environmental applications. Portugaliae Electrochimica Acta. 2007;**25**(1):1. DOI: 10.4152/ pea.200701001

[20] Asgari G et al. Sonophotocatalytic treatment of AB113 dye and real textile wastewater using ZnO/persulfate: Modeling by response surface methodology and artificial neural

network. Environmental Research. 2020; **184**:109367. DOI: 10.1016/j.envres. 2020.109367

[21] Buxton GV et al. Critical review of rate constants for reactions of hydrated electrons, hydrogen atoms and hydroxyl radicals ($\cdot$OH/$\cdot$O$-$) in aqueous solution. Journal of Physical and Chemical Reference Data. 1988;**17**(2):513-886. DOI: 10.1063/1.555805

[22] Li X et al. Electro-peroxone treatment of the antidepressant venlafaxine: Operational parameters and mechanism. Journal of Hazardous Materials. 2015;**300**:298-306. DOI: 10.1016/j.jhazmat.2015.07.004

[23] Lin Z et al. Perchlorate formation during the electro-peroxone treatment of chloride-containing water: Effects of operational parameters and control strategies. Water Research. 2016;**88**: 691-702. DOI: 10.1016/j.watres. 2015.11.005

[24] Bakheet B et al. Electro-peroxone treatment of Orange II dye wastewater. Water Research. 2013;**47**(16):6234-6243. DOI: 10.1016/j.watres.2013.07.042

[25] Xia G et al. The competition between cathodic oxygen and ozone reduction and its role in dictating the reaction mechanisms of an electro-peroxone process. Water Research. 2017;**118**:26-38. DOI: 10.1016/j.watres.2017.04.005

[26] Zhang Y et al. Simultaneous removal of tetracycline and disinfection by a flow-through electro-peroxone process for reclamation from municipal secondary effluent. Journal of Hazardous Materials. 2019;**368**:771-777. DOI: 10.1016/j.jhazmat.2019.02.005

[27] Wang H et al. Mechanisms of enhanced total organic carbon elimination from oxalic acid solutions by

electro-peroxone process. Water Research. 2015;**80**:20-29. DOI: 10.1016/j.watres.2015.05.024

[28] Jaafarzadeh N, Barzegar G, Ghanbari F. Photo assisted electro-peroxone to degrade 2,4-D herbicide: The effects of supporting electrolytes and determining mechanism. Process Safety and Environmental Protection. 2017;**111**:520-528. DOI: 10.1016/j.psep.2017.08.012

[29] Agustina TE, Ang HM, Vareek VK. A review of synergistic effect of photocatalysis and ozonation on wastewater treatment. Journal of Photochemistry and Photobiology C: Photochemistry Reviews. 2005;**6**(4): 264-273. DOI: 10.1016/j.jphotochemrev.2005.12.003

[30] Frangos P et al. A novel photoelectro-peroxone process for the degradation and mineralization of substituted benzenes in water. Chemical Engineering Journal. 2016;**286**:239-248. DOI: 10.1016/j.cej.2015.10.096

[31] Dubey S et al. Electro-peroxone treatment of rice grain based distillery biodigester effluent: COD and color removal. Water Resources and Industry. 2021;**25**:100142. DOI: 10.1016/j.wri.2021.100142

[32] Öztürk H, Barışçı S, Turkay O. Paracetamol degradation and kinetics by advanced oxidation processes (AOPs): Electro-peroxone, ozonation, goethite catalyzed electro-fenton and electro-oxidation. Environmental Engineering Research. 2021;**26**(2):1-13. DOI: 10.4491/eer.2018.332

[33] Zheng H-S et al. Electro-peroxone pretreatment for enhanced simulated hospital wastewater treatment and antibiotic resistance genes reduction. Environment International. 2018;**115**: 70-78. DOI: 10.1016/j.envint.2018.02.043

[34] Zhang Y et al. Disinfection of simulated ballast water by a flow-through electro-peroxone process. Chemical Engineering Journal. 2018;**348**: 485-493. DOI: 10.1016/j.cej.2018.04.123

[35] Yao W et al. The beneficial effect of cathodic hydrogen peroxide generation on mitigating chlorinated by-product formation during water treatment by an electro-peroxone process. Water Research. 2019;**157**:209-217. DOI: 10.1016/j.watres.2019.03.049

[36] Kermani M et al. Optimization of UV-electroproxone procedure for treatment of landfill leachate: The study of energy consumption. Journal of Environmental Health Science and Engineering. 2021;**19**(1):81-93. DOI: 10.1007/s40201-020-00583-9

[37] Jiao Y et al. Treatment of reverse osmosis concentrate from industrial coal wastewater using an electro-peroxone process with a natural air diffusion electrode. Separation and Purification Technology. 2021;**279**:119667. DOI: 10.1016/j.seppur.2021.119667

[38] Guivarch E et al. Degradation of azo dyes in water by electro-Fenton process. Environmental Chemistry Letters. 2003; **1**(1):38-44. DOI: 10.1007/s10311-002-0017-0

[39] Srinivasan R, Nambi IM. Liquid crystal display electrode-assisted bio-electroperoxone treatment train for the abatement of organic contaminants in a pharmaceutical wastewater. Environmental Science and Pollution Research. 2020;**27**(24):29737-29748. DOI: 10.1007/s11356-019-06898-x

[40] Bensalah N, Bedoui A. Enhancing the performance of electro-peroxone by

incorporation of UV irradiation and BDD anodes. Environmental Technology. 2017;**38**(23):2979-2987. DOI: 10.1080/ 09593330.2017.1284271

[41] Wu D et al. Elimination of aqueous levetiracetam by a cyclic flow-through electro-peroxone process. Separation and Purification Technology. 2021;**260**: 118202. DOI: 10.1016/j. seppur.2020.118202

[42] Yao J et al. Interfacial catalytic and mass transfer mechanisms of an electro-peroxone process for selective removal of multiple fluoroquinolones. Applied Catalysis B: Environmental. 2021;**298**: 120608. DOI: 10.1016/j. apcatb.2021.120608

[43] Wang X et al. Electro-catalytic activity of CeOx modified graphite felt for carbamazepine degradation via E-peroxone process. Frontiers of Environmental Science & Engineering. 2021;**15**(6):122. DOI: 10.1007/ s11783-021-1410-x

[44] Wang H et al. Oxidation of emerging biocides and antibiotics in wastewater by ozonation and the electro-peroxone process. Chemosphere. 2019;**235**:575-585. DOI: 10.1016/j.chemosphere.2019. 06.205

[45] Ghalebizade M, Ayati B. Investigating electrode arrangement and anode role on dye removal efficiency of electro-peroxone as an environmental friendly technology. Separation and Purification Technology. 2020;**251**: 117350. DOI: 10.1016/j. seppur.2020.117350

[46] Shokri A, Karimi S. Treatment of aqueous solution containing acid red 14 using an electro peroxone process and a box-Behnken experimental design. Archives of Hygiene Sciences. 2020;**9**(1): 48-57. DOI: 10.29252/ArchHygSci.9.1.48

[47] Cornejo OM et al. Degradation of Acid Violet 19 textile dye by electro-peroxone in a laboratory flow plant. Chemosphere. 2021;**271**:129804. DOI: 10.1016/j. chemosphere.2021.129804

[48] Abdi M et al. Degradation of crystal violet (CV) from aqueous solutions using ozone, peroxone, electroperoxone, and electrolysis processes: A comparison study. Applied Water Science. 2020; **10**(7):1-10. DOI: 10.1007/s13201-020-01252-w

[49] Paździor K, Bilińska L, Ledakowicz S. A review of the existing and emerging technologies in the combination of AOPs and biological processes in industrial textile wastewater treatment. Chemical Engineering Journal. 2019;**376**:120597. DOI: 10.1016/ j.cej.2018.12.057

[50] Cetinkaya SG et al. Comparison of classic Fenton with ultrasound Fenton processes on industrial textile wastewater. Sustainable Environment Research. 2018;**28**(4):165-170. DOI: 10.1016/j.serj.2018.02.001

[51] El-Desoky HS et al. Oxidation of Levafix CA reactive azo-dyes in industrial wastewater of textile dyeing by electro-generated Fenton's reagent. Journal of Hazardous Materials. 2010;**175** (1-3):858-865. DOI: 10.1016/j. jhazmat.2009.10.089

[52] Gebrati L et al. Inhibiting effect of textile wastewater on the activity of sludge from the biological treatment process of the activated sludge plant. Saudi Journal of biological sciences. 2019;**26**(7):1753-1757. DOI: 10.1016/j. sjbs.2018.06.003

[53] Garcia-Segura S et al. Comparative degradation of the diazo dye Direct Yellow 4 by electro-Fenton, photoelectro-Fenton and photo-assisted

electro-Fenton. Journal of Electroanalytical Chemistry. 2012;**681**: 36-43. DOI: 10.1016/j.jelechem.2012. 06.002

[54] Ghalebizade M, Ayati B, Ganjidoust H. Capability study of electro-peroxone process in a cylindrical reactor in degrading Acid Orange 7. Linnaeus Eco-Tech. 2018:87-87. ISBN: 978-91-88898-28-9

[55] Yang Y et al. Occurrences and removal of pharmaceuticals and personal care products (PPCPs) in drinking water and water/sewage treatment plants: A review. Science of the Total Environment. 2017;**596**: 303-320. DOI: 10.1016/j. scitotenv.2017.04.102

[56] Michael I et al. Urban wastewater treatment plants as hotspots for the release of antibiotics in the environment: A review. Water Research. 2013;**47**(3): 957-995. DOI: 10.1016/j. watres.2012.11.027

[57] Östman M, Fick J, Tysklind M. Detailed mass flows and removal efficiencies for biocides and antibiotics in Swedish sewage treatment plants. Science of the Total Environment. 2018; **640**:327-336. DOI: 10.1016/ j. scitotenv.2018.05.304

[58] Liu W-R et al. Biocides in wastewater treatment plants: Mass balance analysis and pollution load estimation. Journal of Hazardous Materials. 2017;**329**:310-320. DOI: 10.1016/j.jhazmat.2017.01.057

[59] Wang H et al. Comparison of pharmaceutical abatement in various water matrices by conventional ozonation, peroxone (O_3/H_2O_2), and an electro-peroxone process. Water Research. 2018;**130**:127-138. DOI: 10.1016/ j.watres.2017.11.054

[60] Yao W et al. Pilot-scale evaluation of micropollutant abatements by conventional ozonation, UV/O_3, and an electro-peroxone process. Water Research. 2018;**138**:106-117. DOI: 10.1016/j.watres.2018.03.044

[61] Kovalova L et al. Elimination of micropollutants during post-treatment of hospital wastewater with powdered activated carbon, ozone, and UV. Environmental Science & Technology. 2013;**47**(14):7899-7908. DOI: 10.1021/ es400708w

[62] Xia G et al. Evaluation of the stability of polyacrylonitrile-based carbon fiber electrode for hydrogen peroxide production and phenol mineralization during electro-peroxone process. Chemical Engineering Journal. 2020; **396**:125291. DOI: 10.1016/j.cej.2020. 125291

[63] Lee Y et al. Prediction of micropollutant elimination during ozonation of a hospital wastewater effluent. Water Research. 2014;**64**: 134-148. DOI: 10.1016/j.watres.2014. 06.027

[64] Lee Y et al. Prediction of micropollutant elimination during ozonation of municipal wastewater effluents: Use of kinetic and water specific information. Environmental Science & Technology. 2013;**47**(11): 5872-5881. DOI: 10.1021/es400781r

[65] Preethi V et al. Ozonation of tannery effluent for removal of cod and color. Journal of Hazardous Materials. 2009; **166**(1):150-154. DOI: 10.1016/ j. jhazmat.2008.11.035

[66] Mustafa M et al. Identification of resistant pharmaceuticals in ozonation using QSAR modeling and their fate in electro-peroxone process. Frontiers of Environmental Science & Engineering.

2021;**15**(5):1-14. DOI: 10.1007/s
11783-021-1394-6

[67] Ghanbari F et al. Enhanced electro-
peroxone using ultrasound irradiation
for the degradation of organic
compounds: A comparative study.
Journal of Environmental Chemical
Engineering. 2020;**8**(5):104167.
DOI: 10.1016/j.jece.2020.104167

[68] Asgari G et al. Diuron degradation
using three-dimensional electro-peroxone
(3D/E-peroxone) process in the presence
of TiO$_2$/GAC: Application for real
wastewater and optimization using RSM-
CCD and ANN-GA approaches.
Chemosphere. 2021;**266**:129179.
DOI: 10.1016/j. chemosphere 2020.129179

[69] Dinc O. Tannic acid oxidation by
electroperoxone. Journal of the Faculty
of Engineering and Architecture of Gazi
University. 2020;**35**(1):51-60.
DOI: 10.17341/gazimmfd.425326

[70] Wu D et al. Removal of aqueous
para-aminobenzoic acid using a
compartmental electro-peroxone
process. Water. 2021;**13**(21):2961.
DOI: 10.3390/w13212961

[71] Wang S et al. Electrochemically
enhanced adsorption of PFOA and PFOS
on multiwalled carbon nanotubes in
continuous flow mode. Chinese Science
Bulletin. 2014;**59**(23):2890-2897.
DOI: 10.1007/s11434-014-0322-6

[72] Vecitis CD, Gao G, Liu H.
Electrochemical carbon nanotube filter
for adsorption, desorption, and
oxidation of aqueous dyes and anions.
The Journal of Physical Chemistry C.
2011;**115**(9):3621-3629. DOI: 10.1021/
jp111844j

[73] Sampaio MJ et al. Carbon nanotube–
TiO$_2$ thin films for photocatalytic
applications. Catalysis Today. 2011;

161(1):91-96. DOI: 10.1016/j.cattod.
2010.11.081

[74] Shan D et al. Preparation of ultrafine
magnetic biochar and activated carbon
for pharmaceutical adsorption and
subsequent degradation by ball milling.
Journal of Hazardous Materials. 2016;
305:156-163. DOI: 10.1016/j.jhazmat.
2015.11.047

[75] Quesada-Peñate I et al. Degradation
of paracetamol by catalytic wet air
oxidation and sequential adsorption–
catalytic wet air oxidation on activated
carbons. Journal of Hazardous Materials.
2012;**221**:131-138. DOI: 10.1016/j.
jhazmat.2012.04.021

[76] Huang Q et al. Enhanced adsorption
of diclofenac sodium on the carbon
nanotubes-polytetrafluorethylene
electrode and subsequent degradation by
electro-peroxone treatment. Journal of
Colloid and Interface Science. 2017;**488**:
142-148. DOI: 10.1016/j.jcis.2016.11.001

[77] Wu D et al. Adsorption and catalytic
electro-peroxone degradation of
fluconazole by magnetic copper ferrite/
carbon nanotubes. Chemical Engineering
Journal. 2019;**370**:409-419. DOI: 10.1016/
j.cej.2019.03.192

[78] Guo Z et al. High activity of g-C$_3$N$_4$/
multiwall carbon nanotube in catalytic
ozonation promotes electro-peroxone
process. Chemosphere. 2018;**201**:
206-213. DOI: 10.1016/j.chemosphere.
2018.02.176

[79] Guo Z et al. Towards a better
understanding of the synergistic effect in
the electro-peroxone process using a
three electrode system. Chemical
Engineering Journal. 2018;**337**:733-740.
DOI: 10.1016/j.cej.2017.11.178

[80] Östman M et al. Effect of full-scale
ozonation and pilot-scale granular

activated carbon on the removal of biocides, antimycotics and antibiotics in a sewage treatment plant. Science of the Total Environment. 2019;**649**:1117-1123. DOI: 10.1016/j.scitotenv.2018.08.382

[81] Mustafa M et al. Regeneration of saturated activated carbon by electro-peroxone and ozonation: Fate of micropollutants and their transformation products. Science of the Total Environment. 2021;**776**:145723. DOI: 10.1016/j.scitotenv.2021.145723

[82] Guo Z et al. C_3N_4–Mn/CNT composite as a heterogeneous catalyst in the electro-peroxone process for promoting the reaction between O_3 and H_2O_2 in acid solution. Catalysis Science & Technology. 2018;**8**(23):6241-6251. DOI: 10.1039/C8CY01517A

[83] Kishimoto N et al. Effect of separation of ozonation and electrolysis on effective use of ozone in ozone-electrolysis process. Ozone: Science & Engineering. 2011;**33**(6):463-469. DOI: 10.1080/01919512.2011.615282

[84] Turkay O, Ersoy ZG, Barışçı S. The application of an electro-peroxone process in water and wastewater treatment. Journal of the Electrochemical Society. 2017;**164**(6): E94. DOI: 10.1149/2.0321706jes

[85] Banuelos JA et al. Study of an air diffusion activated carbon packed electrode for an electro-Fenton wastewater treatment. Electrochimica Acta. 2014;**140**:412-418. DOI: 10.1016/j.electacta.2014.05.078

[86] Mi X et al. Enhanced catalytic degradation by using RGO-Ce/WO_3 nanosheets modified CF as electro-Fenton cathode: Influence factors, reaction mechanism and pathways. Journal of Hazardous Materials. 2019;

367:365-374. DOI: 10.1016/j.jhazmat. 2018.12.074

[87] Le TXH et al. High removal efficiency of dye pollutants by electron-Fenton process using a graphene based cathode. Carbon. 2015;**94**:1003-1011. DOI: 10.1016/j.carbon.2015.07.086

[88] Asgari G et al. Carbon felt modified with N-doped rGO for an efficient electro-peroxone process in diuron degradation and biodegradability improvement of wastewater from a pesticide manufacture: Optimization of process parameters, electrical energy consumption and degradation pathway. Separation and Purification Technology. 2021;**274**:118962. DOI: 10.1016/ j.seppur.2021.118962

[89] Cornejo OM, Nava JL. Mineralization of the antibiotic levofloxacin by the electro-peroxone process using a filter-press flow cell with a 3D air-diffusion electrode. Separation and Purification Technology. 2021;**254**: 117661. DOI: 10.1016/j. seppur.2020.117661

[90] Chen S et al. Enhanced recalcitrant pollutant degradation using hydroxyl radicals generated using ozone and bioelectricity-driven cathodic hydrogen peroxide production: Bio-E-peroxone process. Science of the Total Environment. 2021;**776**:144819. DOI: 10.1016/j.scitotenv.2020.144819

[91] Yang Q et al. Ibuprofen removal from drinking water by electro-peroxone in carbon cloth filter. Chemical Engineering Journal. 2021;**415**:127618. DOI: 10.1016/j.cej.2020.127618

[92] Mahamuni NN, Adewuyi YG. Advanced oxidation processes (AOPs) involving ultrasound for waste water treatment: A review with emphasis on cost estimation. Ultrasonics

Sonochemistry. 2010;**17**(6):990-1003. DOI: 10.1016/j.ultsonch.2009.09.005

[93] Kıdak R, Doğan Ş. Medium-high frequency ultrasound and ozone based advanced oxidation for amoxicillin removal in water. Ultrasonics Sonochemistry. 2018;**40**:131-139. DOI: 10.1016/j.ultsonch.2017.01.033

[94] Zhang H et al. Degradation of CI Acid Orange 7 by ultrasound enhanced ozonation in a rectangular air-lift reactor. Chemical Engineering Journal. 2008;**138**(1-3):231-238. DOI: 10.1016/j.cej.2007.06.031

[95] AlJaberi FY, Ahmed SA, Makki HF. Electrocoagulation treatment of high saline oily wastewater: Evaluation and optimization. Heliyon. 2020;**6**(6): e03988. DOI: 10.1016/j.heliyon.2020.e03988

[96] Wang Y-K et al. The synergistic effect of electrocoagulation coupled with E-peroxone process for shale gas fracturing flowback water treatment. Chemosphere. 2021;**262**:127968. DOI: 10.1016/j.chemosphere.2020.127968

[97] Comninellis CCG. Electrochemistry for the Environment. New York, London: Springer; 2010

[98] Santana-Martínez G et al. Downflow bubble column electrochemical reactor (DBCER): In-situ production of H_2O_2 and O_3 to conduct electroperoxone process. Journal of Environmental Chemical Engineering. 2021;**9**(4): 105148. DOI: 10.1016/j.jece.2021.105148

[99] Thiam A et al. Electrochemical advanced oxidation of carbofuran in aqueous sulfate and/or chloride media using a flow cell with a RuO_2-based anode and an air-diffusion cathode at pre-pilot scale. Chemical Engineering

Journal. 2018;**335**:133-144. DOI: 10.1016/j.cej.2017.10.137

[100] Shen W et al. Kinetics and operational parameters for 1, 4-dioxane degradation by the photoelectro-peroxone process. Chemical Engineering Journal. 2017;**310**:249-258. DOI: 10.1016/j.cej.2016.10.111

[101] Ahmadi M, Ghanbari F. Degradation of organic pollutants by photoelectro-peroxone/ZVI process: Synergistic, kinetic and feasibility studies. Journal of Environmental Management. 2018;**228**:32-39. DOI: 10.1016/ j.jenvman.2018.08.102

[102] Joy AC et al. Photoelectro-peroxone process for the degradation of reactive azo dye in aqueous solution. Separation Science and Technology. 2020;**55**(14): 2550-2559. DOI: 10.1080/ 01496395.2019.1634732

[103] Trojanowicz M et al. Advanced oxidation/reduction processes treatment for aqueous perfluorooctanoate (PFOA) and perfluorooctanesulfonate (PFOS)–a review of recent advances. Chemical Engineering Journal. 2018;**336**:170-199. DOI: 10.1016/j.cej.2017.10.153

[104] Wang S et al. Photocatalytic degradation of perfluorooctanoic acid and perfluorooctane sulfonate in water: A critical review. Chemical Engineering Journal. 2017;**328**:927-942. DOI: 10.1016/ j.cej.2017.07.076

[105] Nawrocki J, Kasprzyk-Hordern B. The efficiency and mechanisms of catalytic ozonation. Applied Catalysis B: Environmental. 2010;**99**(1-2):27-42. DOI: 10.1016/j.apcatb.2010.06.033

[106] Zeng Z et al. Ozonation of acidic phenol wastewater with O_3/Fe(II) in a rotating packed bed reactor: Optimization by response surface

methodology. Chemical Engineering and Processing: Process Intensification. 2012; **60**:1-8. DOI: 10.1016/j.cep.2012.06.006

[107] Hassan M, Zhao Y, Xie B. Employing TiO_2 photocatalysis to deal with landfill leachate: Current status and development. Chemical Engineering Journal. 2016;**285**:264-275. DOI: 10.1016/j.cej.2015.09.093

[108] Katsoyiannis IA, Canonica S, von Gunten U. Efficiency and energy requirements for the transformation of organic micropollutants by ozone, O_3/H_2O_2 and UV/H_2O_2. Water Research. 2011;**45**(13):3811-3822. DOI: 10.1016/j.watres.2011.04.038

Impact of the Spreading of Sludge from Wastewater Treatment Plants on the Transfer and Bio-Availability of Trace Metal Elements in the Soil-Plant System

Najla Lassoued and Bilal Essaid

Abstract

The spreading of sludge from sewage treatment plants increased the production of durum wheat and rapeseed. Their richness in nitrogen, phosphorus, and potassium gives them a beneficial effect on crops. However, the application of the sludge can induce increases in the concentration of metals in plant tissues. This increase can generate disturbances at the level of the cell and organelles, such as mitochondria and chloroplasts, which can be altered. Repeated applications of the sludge on the same site tend to increase the accumulation of heavy metals in the soil, so that an cause toxicities for soil microorganisms, animals, and humans, via the food chain. However, it is important to specify that these nuisances mainly concerned industrial sludge, but the use of this sludge is strictly prohibited. In addition, the high doses used in our field experiments are significantly higher than those authorized in agricultural practice. Finally, the risk assessment by calculating both the level of consumer exposure and the number of years for soil saturation shows that the use of urban sludge is safe, especially in the short and medium-term. Nevertheless, the quality of the sludge to be spread must be constantly monitored.

Keywords: sludge, trace metal elements, wheat, rapeseed, soil–plant system

1. Introduction

The constant increase in the production of sludge from wastewater treatment plants presents a major environmental problem. Compared to traditional means such as landfill or incineration, agricultural sludge spreading appears to be the most cost-effective option for sludge disposal [1]. The use of sludge in agriculture appears among the most sustainable environmental solutions in their disposal. In fact, sludge potential fertilizer and the high cost of mineral fertilizers promote sludge use in agriculture. Nevertheless, their metallic trace elements content (ETM) presents a real disadvantage in their use. Actually, metallic elements retained by the sludge during wastewater treatment can cause high metallic charges accumulation in soil [2]. Metals can be found in the form of sulphites, oxides, hydroxides, silicates, phosphates, carbonates and insoluble salts. They can also be adsorbed or associated with the organic matter of the sludge. The amount of metals in the sludge depends on the origin of the wastewater and the treatments it has undergone [3, 4]. It is, therefore, necessary to try to understand the mechanisms and factors involved in the transfer of these elements into the soil and their effects on the plant following the addition of sludge. The behavior of heavy metals in soils and their absorption by plants depend on the quality of the sludge, the nature of the metal, the physico-chemical properties of the soils and the plant species. Plants differ in their ability to absorb and accumulate metals [2, 3]. From the perspective of an agricultural recovery of sludge, we have tried to contribute to the study of the impact of sludge on the transfer of metallic trace elements in the sludge-soil–plant system. Therefore, a field experiment was carried out in Oued Souhil (Tunisia). In this context, we propose to study the effect of two types of urban and industrial sludge on the distribution and compartmentalization of metallic trace elements in the different organs of two species (durum wheat and rapeseed) chosen according to their absorption capacity.

2. Materials and methods

The experimental protocol was installed in the field to the Agricultural Experiment Station of Oued Souhil - Nabeul, situated about 60 kilometers from Tunis and belonging to the National Institute for Research in Rural Engineering Water and Forest.

The urban mud used in this study is taken from the wastewater treatment plant in Korba with a treatment system at low load activated sludge followed by maturation. Sludge from this station underwent a stabilization in aerobic followed by drying on beds. The dry sludge is removed from the drying bed.

The industrial mud is provided from wastewater treatment plant Bou Argoub which hosts two big companies, the Tunisian beverage manufacturing company (SFBT) specialized in the food industry, and Assad company specialized in the electrical industry. Sludge from this station underwent a stabilization in aerobic followed by drying on beds. This sludge is loaded with heavy metals especially lead and chromium.

The plant materials that were used in this experiment are the rapeseed (*Brassica napus*), which is an annual plant with yellow flowers of the family Brassicaceae and durum wheat (*Triticum turgidum*) that can be defined as a species of wheat characterized by its hard and glassy kernel. Rapeseed was chosen for its ability to accumulate metals and also because it is one of the three main sources of edible vegetable oil with sunflower and olive.

Experimentation was carried out on two juxtaposed plots reserved for each crop (wheat or rapeseed). For each type of sludge, four doses (5, 25, 50 and 100 t ha^{-1}) were used. Results were compared to a control soil without any treatment.

Sludges were manually dug into the soil. Before utilization, the sludge was analyzed.

The soil was analyzed before the application of sludge and after the harvest. Sampling was conducted between the lines using an auger at four depths (0–10, 10–20, 20–40 and 40–60 cm).

In the laboratory, soil samples were dried in open air and sieved to 2 mm or 0.2 mm depending on the type of analysis required. The main measured parameters were particle size, total calcium, conductivity, carbon, organic matter, total nitrogen and heavy metals concentration. For the particle size, we used the method of the International pipette Robinson, which is essentially based on the destruction of organic matter in the soil using H_2O_2 and the dispersion of clays by sodium hexam-etaphosphate. Clays and silts are measured in the suspension of land following the decay time that depends on particle diameter (NF X 31–107). The settling velocity was measured by the formula of Stokes. The Mud and soil samples were analyzed by XRF (X-Ray Fluorescence) and ICP-AES (Inductive Coupled Plasma Atomic Emission Spectrometry Activa–Horiba Jobin Yvon Spectrometer) in the Geosciences and environment Department of National School of Mines in Saint Etienne. The Soil pH was measured by using a 1:2 soil to water ratio. Plant samples were washed with tap water and rinsed three times with distilled water, then separated into leaves, stems and roots, dried at 40°C to constant weight, crushed and sieved at 2 mm. Moreover, the digestion of plant samples was performed using nitric concentrated acid, according to [5–8]. The plant extracts were analyzed by ICP-AES.

The sowings were performed with 50 seeds m^{2-1} for rapeseed and 350 seedsm$^{2-1}$ for wheat. The rapeseed harvest was performed after the formation of slices. We weighed the aerial part and the root. The same work was done to wheat. The samples were subsequently dried and crushed ore to determine the mix of metals in different parts of the plant. The different parts of the plant were dried at 80°C to constant weight and then crushed to a fine powder using a porcelain mortar to prevent metal contamination. Digestion is done at high temperature (70°C) with aqua regia. For histological analysis, preparing the samples carefully for transmission microscopy was essential for obtaining reliable results. Therefore, samples were set at 4°C with a solution of 20.5% glutaraldehyde, pH was maintained at 7.4 with a solution of sodium cacodylate (0.1 M). The samples were then washed with sodium cacodylate buffer (0.1 M) and post-fixed in a solution of 1% osmium tetroxide in veronal buffered (0.1 M) [9]. After several washes in distilled water, the samples were dehydrated with a graded ethanol series of increasing concentrations going from 30 to 100%. The final inclusions were made from a mixture of resin [10]. Only the sections with interference colors are gray or silver, that is to say (thickness of 600 to 900A° (1A° = 0.1 nm)) were collected and deposited on a copper grid with 3 mm diameter. The ultrathin sections were mixed using an alcoholic solution of uranyl acetate and 7 by 1% lead citrate [11]. On top of that, observations were made using a Hitatchi H800 electron microscope.

The data were subjected to analysis of variance. The comparison of means at 5% level of significance was performed by the Newman–Keuls test using the Statistica 7 software.

The amount of heavy metal in sludge is not a good indicator for metal availability for *T. turgidum* plant uptake; accumulation factors were calculated based on metal availability and its uptake by a particular plant. A calculation of biological concentration factor (BCF) was as in Eq. 1, biological accumulation factor (BAF) as in Eq. 2, and transfer factor (TF) as equation

BCF = Metal content (mg kg^{-1}) in root/metal content (mg kg^{-1}) in sludge (1)

BAF = Mean metal content (mg kg^{-1}) in shoot (root+straw+spike)/metal content (mg kg^{-1}) in sludge (2)

TF = Mean metal Content (mg kg^{-1}) in shoot (root+straw+spike) /metal content (mg kg^{-1}) in root (3)

3. Results and discussion

XRD analyzes have shown that industrial sludge has high levels of chromium and lead. These elements mainly exist as Daubreelite Cr_2FeS_4, Brezininaite Cr_3S_4, Wattersite Hg_5CrO_6, Crocoite PbC_2O_4, Pheonicochroite $Pb_2O(CrO_4)$ and lead oxalate PbC_2O_4. As for urban sludge, we note the absence of chromium and the presence of lead in the form of Macphersonite $Pb_4(CO_3)_2(SO_4)$ and Lanarkite $Pb_2O(SO_4)$.

The results of the XRD spectrum (**Figure 1**) were confirmed by those obtained by the chemical and mineralogical analysis (SEM) shown in **Table 1**. Thus, industrial sludge from BouArgoub has very high levels of chromium, lead and cadmium. These contents are higher than the limit values of the Tunisian standard NT-106, which is not the case of urban sludge of Korba. Both types of sludge are rich in organic matter and nutrients, especially nitrogen and phosphorus.

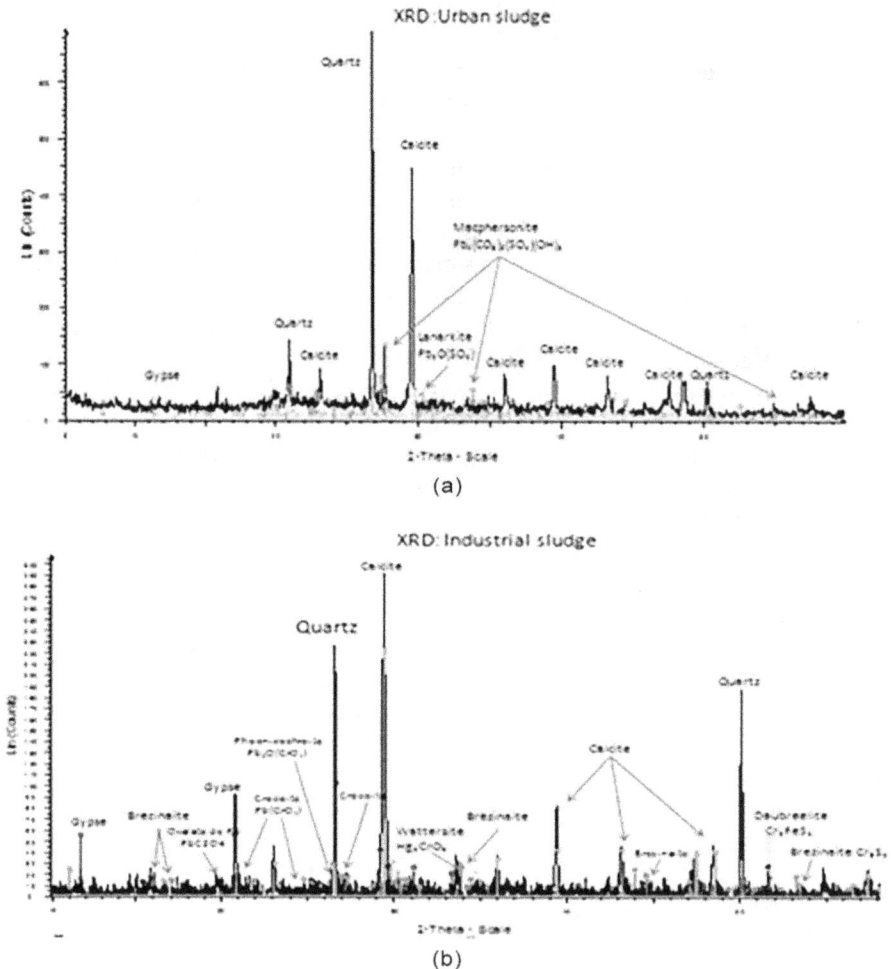

(a)

(b)

Figure 1.
XRD spectrum of sludge from urban (a) and industrial (b) wastewater treatment plants.

Contents	Industrial sludge	Urban sludge	Tunisian standard
pH	6.3	6.7	<6
MO %	57.9	66.6	50–70%
N %	4.3	5.2	3–9%
C %	31.9	39.0	ND
C/N	7.5	7.4	5–12
Fe_2O_3%	1.02	1.88	ND
MnO %	0.03	0.02	<1%
MgO %	0.8	1.22	ND
CaO %	8.51	13.4	ND
P_2O_5%	2.18	3.24	4–5%
Cd mg kg^{-1}	11	3	20
Co mg kg^{-1}	18	28	ND
Cu mg kg^{-1}	68	158	1000
Fe mg kg^{-1}	8300	10,700	ND
Mn mg kg^{-1}	81	152	ND
Ni mg kg^{-1}	49	78	200
Pb mg kg^{-1}	577	63	800
Zn mg kg^{-1}	360	440	2000
Cr mg kg^{-1}	8030	155	500

ND: not defined/Source ADEME.

Table 1.
Chemical contents (mg kg^{-1}) of industrial sewage sludge, urban sewage sludge and Tunisian standard values [12].

According to [13–18] sludge is a good source of nutrients for plant growth, it can improve the physical properties of the soil. Vlamis et al. [19, 20] reported that sludge can replace mineral fertilization. Phosphorus in sludge is as effective as phosphorus in fertilizers in increasing the extractable phosphorus in the soil to the level required for crop growth. Likewise, calcium carbonate in sludge increases soil pH more effectively than agricultural lime.

Our results made it possible to highlight the action of the sludge on the behavior of the plant in qualitative and quantitative terms. During the first year of experimentation, the productions obtained were generally higher than those obtained on the control plots. A dose-effect was very clear. The dose of 100 t ha^{-1} records the highest production whatever the crop. Similar results are obtained by other authors. A clear improving action of sludge on English Ray Grass production [21–23]. Zaier et al. [24] showed that the sludge significantly stimulates the biomass production of B. napus. Similar results have been demonstrated by [25] who showed that the biomass production of durum wheat was significantly improved by 18% with the addition of 40 t ha^{-1} of sewage sludge. Pasqualone et al. [26] found positive effects of increasing sludge doses on durum wheat productivity, 12 kg ha^{-1} of sludge was demonstrated to can effectively replace mineral fertilization. Boudjabi et al. [27] found a significant increase in the number of tillers, ears and kernels per ear of barley in amended soils. This has been linked to an improvement in the physical and chemical properties of soils. This beneficial effect has been observed on several crops such as wheat, sorghum, maize, chili peppers, barley and potatoes [21, 28–31].

For the two types of sludge, the results of the second application show a positive effect on rapeseed production at the doses of 5 t ha^{-1} and 25 t ha^{-1}. However, in this second application, industrial sludge causes an increase in the concentration of metals in the soil, especially chromium, cadmium and lead. The cumulative application of industrial sludge generates an excessive accumulation of heavy metals [32, 33] which could be harmful to soil fertility, affecting the ecosystem and human health [34]. Marchiol et al. [35] observed that *B. napus* grown on multi-contaminated soil was tolerant to heavy metals. They concluded that this species could eventually be used successfully in polluted soils without its growth being affected and that heavy metal extraction can be maintained at a satisfactory level. In our case, the tolerance index of this plant exceeds 1 for all treatments but during the second year for the dose of 100 t ha^{-1}, industrial sludge gave lower indices than urban sludge. Nonetheless, the cumulative application of urban sludge increases the production of rapeseed regardless of the dose.

During the experimentation, the plants presented a normal appearance but some rapeseed leaves cultivated in plots having received 200 t ha^{-1} of industrial sludge show spots of necrosis and a purplish color. In the literature, these symptoms are described as due to phosphate deficiency. Soil rich in iron or zinc can reduce the absorption capacity of phosphate ions, which could occur as a result of adding sludge to the soil [36]. In this context, [37] have also shown that the presence of high levels of lead can cause the formation of precipitate of lead phosphates, which cannot be assimilated by plants and consequently a phosphorus deficiency can occur. For rapeseed, we have seen a delay in germination in certain plots following a cumulative effect of industrial sludge. Also, germination appeared to be inhibited by the presence of urban sludge. Laboratory experiments confirmed this effect, but showed that germination was not permanently inhibited, but simply delayed. The latency period is dose-dependent, so it increases depending to the amount of sludge added. A similar effect was induced by heavy metals (Cu, Ni, Zn) in aqueous solution [38]. Also, germination is positively correlated with the degree of stability of the sludge and the organic matter contents [39]. The observed delay of germination was restored after one month and the development cycle resumed normally.

A study by [40] showed that sludge causes the late maturity of wheat. This was not the case in our experimentation where the wheat development cycle was not disrupted. An increase in crop yield resulted in an increase in the number of seed in our study. We also mentioned that urban sludge increases the oil yield of rapeseed after the 1st spreading. Similarly, [41] has shown that urban sludge considerably increases the productivity of sunflower oil. The use of urban sludge as a fertilizer has been considered for years given its richness in organic matter and nutrients [41, 42]. However, sludge contains phytotoxic metals which can cause certain problems at high levels [43]. The response of plants to these metals varies considerably from one species to another [44–46] and the results obtained are disparate. Other factors including edaphic can intervene. In our work, we have shown that whatever the site of culture (control, contaminated, heavily contaminated), wheat has a lower concentration of heavy metals than rapeseed. However, these differences are minimal on the control site and are accentuated with the addition of sludge, especially in the presence of industrial sludge. This finding is not the same for the seed since, on the control site, the wheat seems to concentrate more heavy metals than rapeseed, but this is only due to the high zinc contents of wheat seeds. If we disregard the zinc, we find the same result. In fact, Brassicaceae are generally considered to be metal accumulators that can tolerate high concentrations [47] unlike cereals [48]. However, the accumulating power of rapeseed remains low to consider phytoremediation [49]. It is important to note that both rapeseed and durum wheat were able to survive on a site treated with sludge loaded with heavy metals. This capacity may be due both to the existence of tolerance strategies in the plant making the assimilation of metal limited and also to the strong metal bonds in the sludge and the soil rich in matter organic. Most heavy metals can also be stored and detoxified in root tissues with minimal translocation to leaves whose cells are sensitive to phytotoxic effects [50–54].

Actually, the extent of metal contamination depends both on the concentration of the metal in the environment and on the intrinsic factors of the metal. The concentrations of the main tracemetalic elements are represented in **Figure 2**. In the control environments and those treated by urban sludge, the Zn contents are higher than the other metals while cadmium was the least abundant. The concentrations founded were as following: Zn > > Cr > Cu > Ni > Pb > Co > Cd. For the environments treated by industrial sludge, the amounts of trace elements were completely different, reflecting an increase in chromium and lead amounts. The concentrations founded in these environments were as following: Cr > > Zn > > Pb > Cu > Ni > Co > Cd.

The presence of metals in the soil can influence the uptake of essential nutrients for plant growth [55, 56]. The essential divalent cations such as Ca^{2+}, Mn^{2+}, Zn^{2+}, Mg^{2+} compete with toxic metals such as Cd. Therefore, the increase in the contents of these elements can reduce the absorption of metals such as Cd [57–59]. However, in our experimentation, for the rapeseed, we noted a synergy between a non-essential metal Cd and a trace element Zn. Also, the increase of Cd absorption decreases the absorption of iron and manganese by the root (**Figure 3**). These observations are explained by competitions between these different cations for surface complexation sites in the root and for the unspecific carriers of major cations or trace elements. Other works have shown that treatment with cadmium can cause deficiencies in iron, copper and manganese [57, 60].

For wheat, the absorption of iron, manganese and zinc increases according to the doses (**Figure 4**) of the sludge and probably plays a non-negligible beneficial role on the Phyto availability of non-essential elements such as cadmium, an

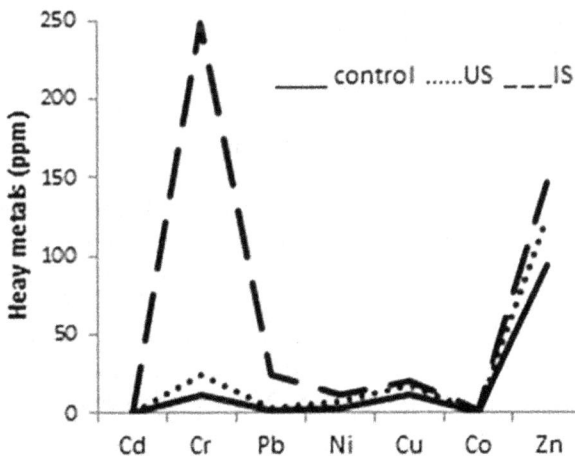

Figure 2.
Effect of metal on some of metal trace element concentrations in the plant.

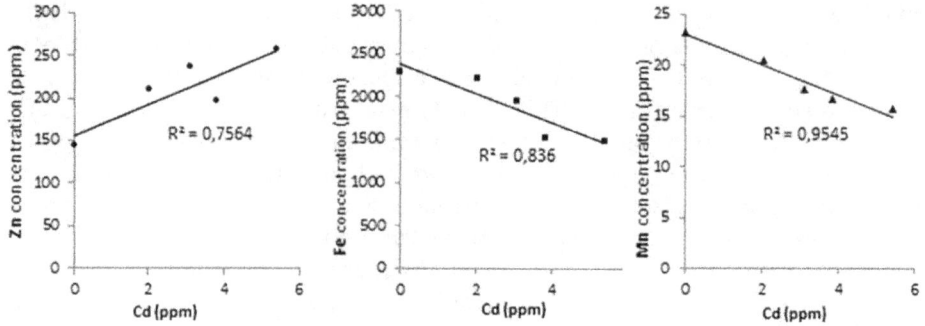

Figure 3.
Influence of cadmium on the absorption of zinc, iron and manganese in rapeseed roots.

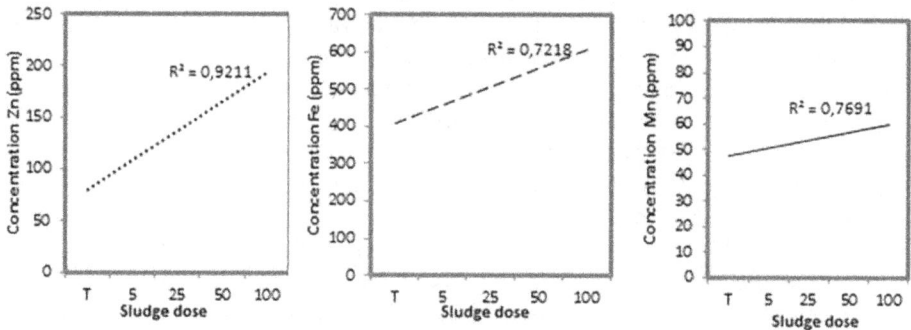

Figure 4.
Effect of industrial sludge on the absorption of zinc, iron and manganese in the roots of durum wheat.

antagonistic effect could have taken place, which could explain the negligible Cd contents found in wheat [61].

Sludge application causes a high accumulation of metallic trace elements in the soil, these metallic elements are then driven to the roots and finally to the aerial part. The addition of industrial sludge causes a significant contamination of the soil and the roots by heavy metals while the degree of contamination at the level of the aerial part is lower as shown in **Figure 5**. For urban sludge, the accumulation of heavy metals is high but much less than that of industrial sludge. The comparison between contaminated and uncontaminated environments shows that the more the environment is polluted with ETM (Trace Metal Elements), the higher the contents of these elements in the plant.

The difference in behavior between aerial organs and roots with respect to heavy metals is reflected in the ultrastructure. From a cytological point of view, the number of cells in apoptosis is higher in the roots. This suggests the existence of much more effective bio protection modalities in the root. The high doses of the sludge cause ultrastructural changes in the roots, but the nucleus retains its integrity and the chromatin is evenly distributed. As for the cytosol, it has a contracted appearance with grouping of ribosomes into polysomes. Mitochondria change their shape and have swollen ridges (**Figure 6**).

Fragmentation of the vacuole into many small vacuoles is also observed (**Figure 6**). The plasma membranes appear damaged. The best-known strategy is to interfere with the entry of heavy metals into root cells by trapping them in the apoplasm where

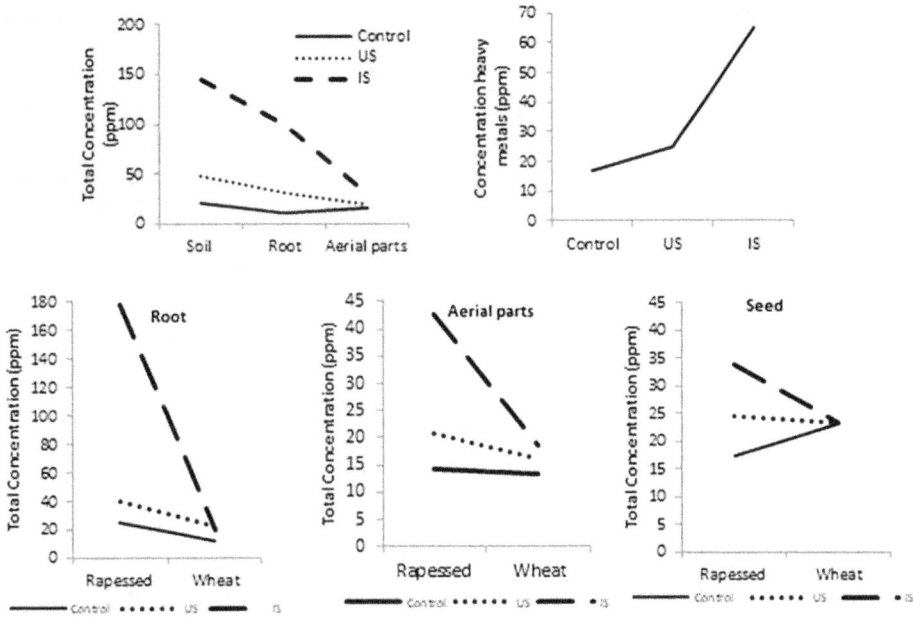

Figure 5.
Accumulation of metallic trace elements in the soil spread by sludges and in the different parts of the plant.

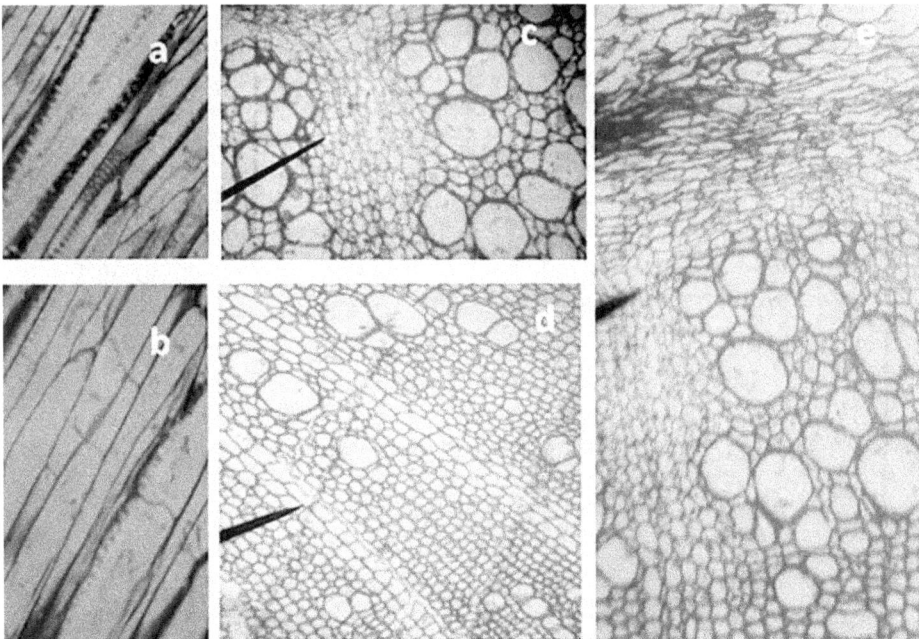

Figure 6.
(a) central cylinder of control roots in longitudinal section (Gr * 40). (b) central cylinder of a control root in cross section (c) Cross section of a root treated with 100 t ha^{-1} BU (Gr * 40). (d, e) Central cylinder of a root of treatment 100BI (Gr * 40). Cp: Parenchymal cell; X: xylem; RL: Woody rays; CC: central cylinder; P: periderm.

they associate with organic acids [62] or anionic groups in cell walls [63]. Once inside the plant, most heavy metals are held in deep cells, where they are detoxified by complexing with amino acids, organic acids or peptides and or they are sequestered in vacuoles [51]. This greatly limits translocation to aerial organs, thus protecting the leaf tissues, and in particular the metabolites of photosynthetic cells against possible damage. Another defense mechanism generally adopted by plants exposed to heavy metals is the improvement of cellular antioxidant systems which would limit oxidative stress [52, 53].

The export of metals to the aerial part is accompanied by physiological disturbances. A decrease in chlorophyll could be induced by excess zinc and cadmium [64]. Our results confirm this fact as shown in **Figure 7**. Likewise, a decrease in the level of carotenoids can be linked to an excess of cadmium [65] or to an excess of copper [66].

Metals alter electron transport and inhibit the activity of Calvin cycle enzymes [67]. High doses of industrial sludge also weaken photosynthetic activity [68, 69] and cause a gradual decrease in photochemical quenching (qp) [70, 71] accompanied by a significant increase in non-photochemical quenching (NPQ).

Chromium directly inhibits one of the key enzymes in chlorophyll biosynthesis NADPH [72]. Likewise, several metals such as Cd, Pb, Cu, Zn, and Ni can replace Mg in chlorophylls, resulting in inactive molecules [73–75]. In the leaf, excess metals can also induce changes in membrane stiffness, permeability and stability [76]. At the cellular level, rearrangements were observed (**Figures 8** and **9**). The damage produced at mitochondria and chloroplasts is more severe than that of the nucleus.

In the leaves of plants cultivated in the presence of high doses of industrial sludge loaded with metals, the number of chloroplasts has decreased. Similar results have been reported by [77]. Other structural damage is also frequently observed such as swelling of chloroplasts, rupture of the envelope, deformation of thylakoids. The thylakoid membranes lose their parallel arrangement, the grana become disorganized and the thylakoid surface is reduced. These ultra-structural changes are accompanied by an increase in the degree of membrane lipid peroxidation, appreciated by the production of malondialdehyde. We have also observed an enlargement of the mitochondria, the disintegration of the membranes, the disappearance of the ridges and a clear vacuolation. The intensity of toxicity actually varies from cell to cell.

The attenuation of toxicity could be due, for example, to the retention of metals on the cell wall [78] or their sequestration in the vacuole [79] or their storage in inactivated on specific proteins, amino amines or peptides.

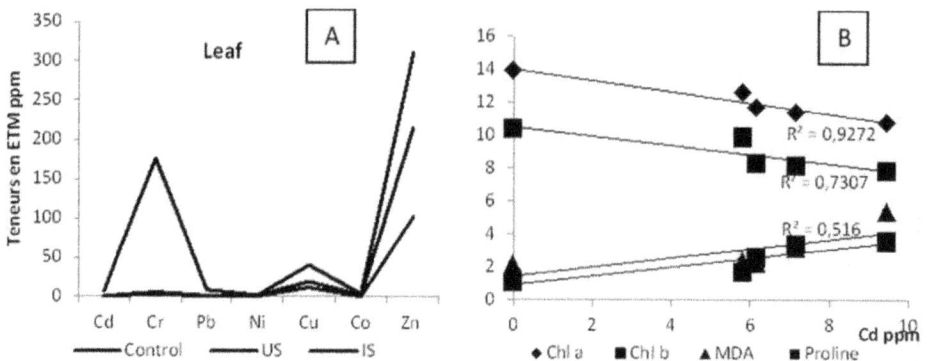

Figure 7.
(a) Variation of heavy metals in the leaf as a function of treatment. (b): Effect of cadmium on the biosynthesis of chlorophyll, proline, and MDA in the leaves of rapeseed.

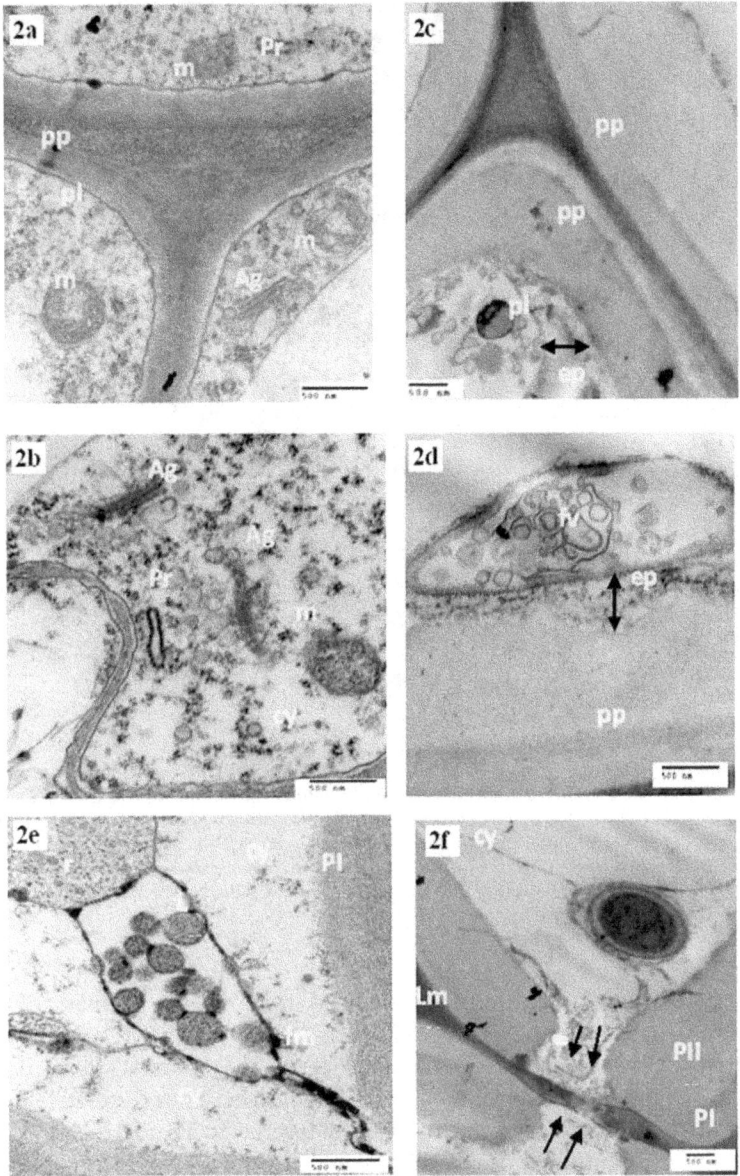

Figure 8.
Observation under a transmission electron microscope of cross sections of root cortical cells of a control plant (2a and 2b) and of a plant treated with 100 t ha⁻¹ industrial sludge (2c to 2f). 2 a: Plasmalemma (pl) of a cortical cell of a control root 2 b: Cell organelles of a control root with homogeneous distribution of ribosomes (r). 2 c: Retraction of the cytoplasm with detachment of the plasmalemma and formation of a periplasmic space (ep). 2 d-e: Vesicular formations (vf) and membrane formations (fm). 2 f: Degradation of the primary (PI) and secondary (PII) wall. Pp.: Pectocellulosic wall, m: Mitochondria Ag: Golgie apparatus, cy: Cytoplasm, pr: Polyribosome.

The study of the risks of ETM (Trace Metal Elements) associated with the spreading of sludge requires not only knowledge of the total metal content, but also of the metal content in the various compartments that make up the soil. Nevertheless, assessing the total stock of an element is a good approach to study the degree and extent of soil contamination by a metallic element. Our results have shown that in

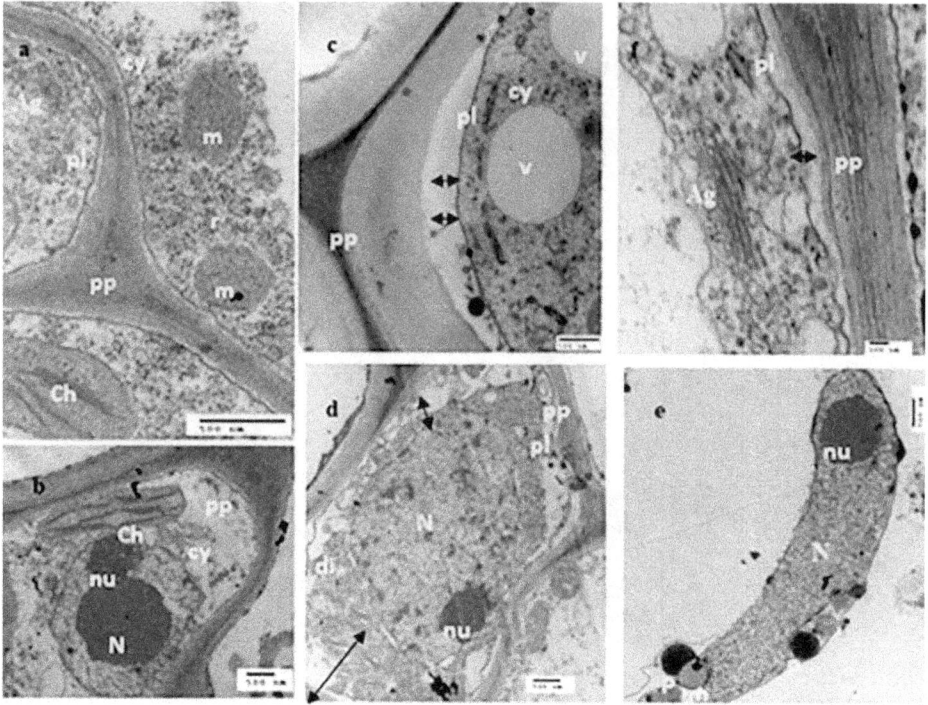

Figure 9.
Observation under a transmission electron microscope of the cortical zone of a rapeseed stem from the control treatment (a and b) (a: General appearance of the cell and b: Appearance of the nucleus (N) and after a contribution of 100 t ha⁻¹ BI (c, d and e) (c: Detachment of the plasma membrane (pl) and formation of periplasmic space; d: Irregularly shaped nucleus divided into small nucleoli (nu); e: Vesicle surrounded by a single membrane and containing certain cellular organelles (nucleus, mitochondria and peroxisome (p) and in the presence of 100 t ha⁻¹ BU (f) with pairing of Golgian saccules forming a dictyosome (Ag) and releasing Golgian vesicles giving rise to lysosomes. (di): Cytoplasmic digitation.

general, heavy metals are preferentially localized in the surface horizon and this for different media (**Figure 10**). However, on the control medium, the difference in accumulation between the 3 horizons is minimal and the contents are more or less comparable. The more the environment is polluted, the greater the difference in accumulation between the horizons. In fact, the level of accumulation of heavy metals depending on the pollution of the environment is much greater at the surface horizon than for the underlying horizons.

The second addition of sludge increases the ETM (Trace Metal Elements) content in the soil of the two crops. The surface layer (0-10 cm) appears the richest in cadmium, chromium and lead. After the second harvest, the cultivated wheat soil has higher ETM (Trace Metal Elements) contents than those of the rapeseed plots (**Figure 11**). This is due to the low extracting power of wheat compared to rapeseed. In a study on Brassica napus, [80] reported that in the presence of sewage sludge the extraction of heavy metals by this species increases significantly due to its hyperaccumulation power, but its use in phytoremediation is a very long process.

To assess the risk associated with heavy metals in the case of spreading waste products (sludge, compost, wastewater, etc.), one of the most widely used methods is to prevent the accumulation of trace elements in the soil [81, 82]. Only the soil and sludge contents, as well as the quantities of sludge added are taken into account.

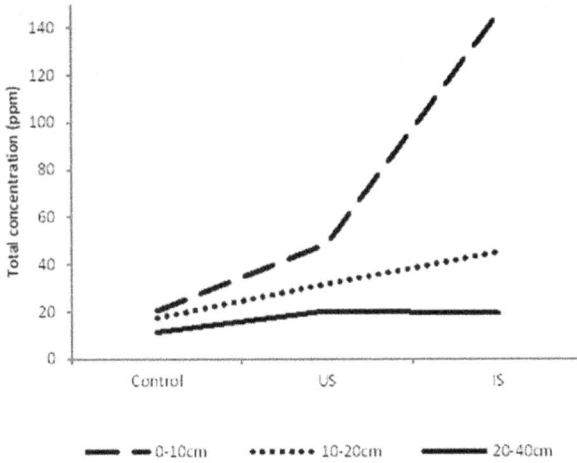

Figure 10.
Accumulation of heavy metals in the different horizons of the soil.4.

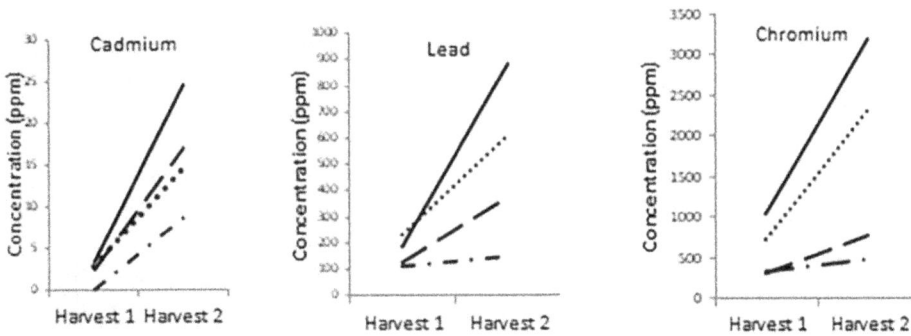

Figure 11.
Comparison of heavy metals in the different horizons of the soil (between 0 and 10 cm and 10 and 20 cm) after the first and second harvest.

Taking into account all our data, we tried to make a balance in order to assess the risk of saturation of the soil by heavy metals provided by the different types of sludge in the case of our experiment. The results obtained are shown in **Table 2**. It emerges from this table that for all the heavy metals, the soil reaches saturation much more quickly with the input of industrial sludge than for urban sludge and this is easily explained by the respective quality of sludge. For urban sludge, the probable duration of saturation varies between 361 and more than 160,000 years depending on the sequence: Cd >> > Ni > Pb > Cr > > Cu > Zn. This is in favor of the use of urban sludge given that the risk of soil contamination is low, in particular for toxic metals (Cadmium, Chromium and Lead) and that the elements which arrive first at the thresholds are copper and zinc which are trace elements essential to the plant. The problem arises differently with the spreading of industrial sludge highly loaded with heavy metals. Indeed, in this case the risk of contamination is present since the saturation time drops to very low levels for toxic heavy metals to reach less than 10 years for cadmium and chromium and 30 years for lead. The sequence found is: Ni > Cu > Zn > Pb > Cd > Cr.

	Cd	Cu	Cr	Ni	Pb	Zn
Heavy metals soil standard (ppm)	3	140	150	75	150	300
Wheat and rapeseed soil (ppm)	0,66	28,03	27,84	6,35	19,06	56,84
Average annual contribution per BU of 5 t ha^{-1}year^{-1} (g ha^{-1} year^{-1})	0.05	998	439,25	92,75	295,5	2425,25
Increase due to sludge	1.38 10-5	0,277	0,122	0,026	0,082	0,674
Number of years before saturation for Urban sludge	168,480	404	1001	2640	1597	361
Average annual contribution per Industrial Sludge 5 t Ha^{-1}(g ha^{-1} year^{-1})	909,5	766	63,380	312,5	15,075	3800,75
Increase due to sludge	0,253	0,213	17,606	0,087	4188	1056
Number of years before saturation for Industrial Sludge	10	526	7	789	31	230

Table 2.
Assessment of the risk of soil saturation by heavy metals provided by the different types of sludge.

Once in the soil, some of these metals persist due to their immobile nature and the risk of crossing plants is low. Their in-depth migration is unlikely. On the other hand, the most mobile elements can transfer through the soil to the aquifer or be absorbed by the plant. Besides the intrinsic criteria of metals, their bioavailability and their transfer are more or less modified, in particular by edaphic factors such as pH, temperature and organic matter contents, hence the need to take into account all factors for success of a spreading project.

The decontamination of polluted soils can be considered but remains a rather long process requiring a lot of time [80] calculated that more than 1000 years would be needed to clean up a contaminated site. These results have been confirmed by [83]. Likewise, [84] studying the phytoextraction of *Brassica juncea* on a site multi-contaminated mainly by Zn, Cu and Pb showed that the time required for decontamination is between a minimum of 1150 years and a maximum of 360,000 years.

When metals migrate through plants and the food chain is involved, account must be taken of the amounts of the metals that can be transferred to consumers. As *B. napus* is an edible oil crop [85], we were concerned with the quality of the seeds and the oil extracted in terms of its fatty acid composition.

An increase in the levels of metallic trace elements was indeed mentioned at seed level during the second campaign. We also detected a decrease in the oil content with the addition of industrial sludge. The composition of total lipids in fatty acids shows an increase in the percentage of oleic acid (C18: 1) at the expense of linoleic (C18: 2) and linolenic (C18: 3) acids under the effect of heavy metals provided by industrial sludge while no difference is recorded with urban sludge regardless of the dose. No significant difference in heavy metal content was observed with the contribution of different doses of urban sludge, even after two years of spraying. On the other hand, we detected increases in most of the heavy metals, in particular for the high doses of industrial sludge.

The cadmium contents increase significantly with the contribution of 100 t ha^{-1}, it reaches 25 ppb following the cumulative contribution. For lead and chromium, the increase is especially visible at the 100 t ha^{-1} dose. These increases become more important during the second year. For nickel, the levels are higher compared to the control from the addition of 50 t ha^{-1} of industrial sludge for the two spreading operations. For wheat, the composition of the seeds in metallic trace elements is little or not affected by the contribution of sludge. The observed increases concerned only a few metals and for the excessive doses of industrial sludge. In order to assess the health risk linked to heavy metals via the consumption of products amended with sewage sludge, we tried to theoretically determine the daily exposure of consumers of these products to heavy metals (EJE) and to compare it with tolerable toxicological doses (TDI). The TDI is defined as the amount of contaminants that can be ingested daily without adverse health effects [86, 87]. The harmful potential of a product is the greater the lower the TDI value. In our calculations, we considered the seeds of wheat and rapeseed oil that are suitable for human consumption.

The daily exposure dose attributable to the consumption of these EJE products is calculated according to the formula [88]:

$$\text{EJE } (\mu g\, kg^{-1} day^{-1}) = [\text{Cproduct}] \times \text{Qty of product} \qquad (1)$$

With: [C product]: Concentration of the metal in wheat grains or rapeseed oil; Product qty.: consumption of the product at the 95th percentile (g person^{-1} day^{-1}) which is equal to 382 g day^{-1} for adult person weighing 60 kg for wheat [88] and 25 g day^{-1} for rapeseed oil [87]. For the trace and zero contents, we assimilated them to the detection limits of the dosing device.

The intake of sludge in its two forms has no effect on the theoretical exposure to Cd of high consumers of wheat, which is estimated at 3.8 μg person^{-1} day^{-1} for all the treatments (**Table 3**). These values are clearly lower than the TDI which is 60 μg person^{-1} day^{-1}. The problem arises differently for Pb where the dose of sludge (urban and industrial) influences by increasing the level of exposure of consumers but the values obtained are all below the TDI (216 μg person^{-1} day^{-1}). It should be noted that the higher the dose, the closer one gets to the TDI. For Ni, the intake of both types of sludge increases consumer exposure to this element but without a net dose effect. In addition, the values obtained remain below the TDI (720 μg person^{-1} day^{-1}). Exposure to Cr is increased with the addition of industrial sludge. Nevertheless, the comparison with the TDI could not be made because the available TDI is fixed for Cr III and Cr IV while our data relate to total Cr.

The hazard quotient (QD) is defined by the ratio between the calculated EJE and the corresponding TDI, according to the formula of [89] i.e. QD = EJE/DJT If the hazard quotient is greater than 1, the occurrence of Adverse effects related to toxicants are potentially possible. Otherwise, the risk can be considered as theoretically non-existent. All the ratios calculated for wheat seed are less than 1, however it is important to note that the high doses, especially of industrial sludge, increase the QD which rapidly approaches 1 for Ni and in particular Pb. It is probable that 'it is the same for Cr. Thus, it is imperative to note that whatever the quality of the sludge, the spreading must be done at suitable doses and must be controlled. For rapeseed oil, the estimated daily exposure is extremely low since the values are infinitely low and are much lower than the respective TDI. It should be noted that for Cd and Pb, the 100 t ha^{-1} BI increase EJE. From a metals point of view, the addition of sludge, especially urban sludge and at low doses, does not generate a health risk, however, it should not be forgotten that the metals can be introduced by other products which must be taken into account in the process risk assessment hence the need to monitor these situations.

	Estimated exposure μg person^{-1} day^{-1}				EJE/DJT			
wheat seed	**Cd**	**Pb**	**Cr**	**Ni**	**Cd**	**Pb**	**Cr**	**Ni**
Control	3,8	19,1	525,3	261,7	0,064	0,088	—	0,363
5BU	3,8	19,1	540,5	355,3	0,064	0,088	—	0,493
25BU	3,8	19,1	498,5	395,4	0,064	0,088	—	0,549
50BU	3,8	143,3	548,2	336,2	0,064	0,663	—	0,467
100BU	3,8	191,0	531,0	269,3	0,064	0,884	—	0,374
5BI	3,8	19,1	624,6	382,0	0,064	0,088	—	0,531
25BI	3,8	59,2	618,8	368,6	0,064	0,274	—	0,512
50BI	3,8	210,1	601,7	389,6	0,064	0,973	—	0,541
100BI	3,8	194,8	685,7	475,6	0,064	0,902	—	0,661
Rapeseed oil								
Control	0,003	0,005	0,0002	0,014	4,3E-05	2,52E-05	—	1,98E-05
5BU	0,003	0,005	0,0001	0,016	4,3E-05	2,44E-05	—	2,27E-05
25BU	0,003	0,006	0,0002	0,016	4,3E-05	2,60E-05	—	2,20E-05
50BU	0,003	0,005	0,0002	0,016	4,3E-05	2,44E-05	—	2,22E-05
100BU	0,003	0,005	0,0002	0,017	4,3E-05	2,52E-05	—	2,31E-05
5BI	0,003	0,005	0,0002	0,014	4,5E-05	2,44E-05	—	1,91E-05
25BI	0,005	0,006	0,0002	0,014	7,9E-05	2,60E-05	—	1,89E-05
50BI	0,016	0,006	0,0003	0,017	2,6E-04	2,83E-05	—	2,38E-05
100BI	0,055	0,014	0,0008	0,016	9,2E-04	6,61E-05	—	2,22E-05

Table 3.
Estimated exposure and calculated mean Hazard quotient.

4. Conclusion

The spreading of sludge from wastewater treatment plants increased the production of durum wheat and rapeseed. Their richness in nitrogen, phosphorus and potassium gives them a beneficial effect on crops. However, the application of sludge can induce increases in the concentration of metals in plant tissues. This increase can generate disturbances at the level of the cell and organelles like mitochondria and chloroplasts which can be altered. Repeated applications of sludge at the same site tend to increase the accumulation of heavy metals in the soil which can cause toxicities to soil microorganisms, animals and humans, via the food chain. However, it is important to note that these harmful effects mainly concerned industrial sludge, but the use of this sludge is strictly prohibited. In addition, the high doses used in our field experiments are clearly higher than those authorized in agricultural practice. Finally, the risk assessment by calculating both the level of exposure for the consumer and the number of years for a soil to be saturated shows that the use of urban sludge is safe, particularly in the short and medium term. Nevertheless, the quality of the sludge to be spread must be constantly checked. Other metallic trace elements such as mercury, boron brought in by the sludge must be taken into account.

Acknowledgements

This work was supported by National Institute for Rural Engineering research, Water and Forestry, Tunis and Ecole Nationale Supérieure des Mines de Saint Etienne, France.

Conflict of interest

The authors of chapiter submitted for publication, we confirm that the results presented in this paper are real and original. The authors declare that they have no competing interests. The opinions expressed in this article are those of the authors and do not necessarily represent any agency determination or policy.

Author details

Najla Lassoued[1,2] and Bilal Essaid[3]*

1 National Agronomic Institute of Tunisia, Tunis, Tunisia

2 National Institute for Rural Engineering Research, Water and Forestry, Ariana, Tunis, Tunisia

3 National School of Mines of Saint-Etienne, PEG, Saint Etienne, France

*Address all correspondence to: bilalessaid@gmail.com

IntechOpen

References

[1] Casado-Vela J, Sellés S, Navarro J. Evaluation of composted sewage sludge as nutritional source for horticultural soils. Waste Management. 2006;**26**(9): 946-952

[2] Degaard H, Paulsrud B, Karlsson I. Wastewater sludge as a resource: Sludge disposal strategies and corresponding treatment technologies aimed at sustainable handling of wastewater sludge. Water Science and Technology. 2002;**46**:10295-10303

[3] Lake DL, Kirk PWW, Lester JN. The fractionation, characterization and speciation of heavy metals in sewage sludge and sewage sludge amended soils: A review. Journal of Environmental Quality. 1984;**13**:175-183

[4] Lombardi AT, Garcia O Jr. An evaluation into the potential of biological processing for the removal of metals from sewage sludge. Critical Reviews in Microbiology. 1999;**25**:275-288

[5] Zachara JA, Girvin DC, Schmidt RL, Resch CT. Chromate adsorption on amorphous iron oxyhydroxide in the presence of major groundwater ions. Environmental Science & Technology. 1987;**21**:589-594

[6] Petrescu L, Bilal E. Environmental impact assessment of a uranium mine, East Carpathians, Romania: Metal distribution and partitioning of U and Th. Carpathian Journal of Earth and Environmental Sciences. 2007;**2**(1):39-50

[7] Secu CV, Iancu OG, Buzgar N. Lead, zinc and copper in the bioacumulative horizon of soils from Iaşi and the surrounding areas. Carpathian Journal of Earth and Environmental Sciences. 2008;**3**(2):131-144

[8] Lăcătuşu R. Cîtu G. Aston J. Lungu M. Lăcătuşu AR. Heavy metals soil pollution state in relation to potential future mining activities in the Roşia Montană area. Carpathian Journal of Earth and Environmental Sciences. 2009;**4**(nr. 2): 39-50

[9] Sabatini DD, Bensch K, Barrnett RJ. Cytochemistry and electron microscopy. The preservation of cellular ultrastructure and enzymatic activity by aldehyde fixation. The Journal of Cell Biology. 1963;**17**(19):19-58

[10] Spurr AR. Alow-viscosity epoxy resin embedding medium for electron microscopy. J. Ultrstruct. Res. 1969;**26**:31-43

[11] Reynolds RS. The use of lead citrate at light pH as an electron-opaque stain in electron microscopy. The Journal of Cell Biology. 1963;**17**:208-213

[12] Bilal L. Quality of rapeseed oil after different applications of sewage sludge in soils. Carpathian Journal of Earth and Environmental Sciences. 2020;**15**(2):443-452

[13] Matthews PJ, Davis RD. Control of metal application rates from sewage sludge utilization in agriculture. CRC Critical Rev. Environ. Sci. Technol. 1984;**14**:199-250

[14] Willett IR, Jakobsen P, Malafant KWJ. Fertilizer and liming value of lime-treated sewage sludge. Fertilizer research. 1986;**8**(3):313-328

[15] Bahri A. Utilization of treated wastewaters and sewage sludge in agriculture in Tunisia. Desalination. 1987;**67**:233-244

[16] Rejeb S, Bahri A. Incidence de l'apport de boues résiduaires urbaines

sur la composition minérales et la productivité de quelques espèces cultivées en Tunisie. Les Cahiers de CRGR. 1995;**24**:13-32

[17] Kchaou R, Khelil MN, Gharbi F, Rejeb S, Henchi B, Hernandez T, et al. Isotopic evaluations of dynamic and plant uptake of N in soil amended with 15N-labelled sewage sludge. Polish J. of Environ. Stud. 2010;**19**(2):363-370

[18] Su DC, Wong JWC. Chemical speciation and phytoavailability of Zn, Cu, Ni and Cd in soil amended with fly ash-stabilized sewage sludge. Environment International. 2003;**29**:895-900

[19] Vlamis J, Williams DE, Corey JE, Page AL, Ganje TJ. Zinc and cadmium uptake by barley in field plots fertilized seven years with urban and suburban sludge. Soil Science. 1985;**139**:81-87

[20] Lobo TF, Grassi FH. Níveis de lodo de esgoto na produtividade do girassol. R. C. Suelo Nutr. Veg. 2007;**7**:16-25

[21] Gukert A, Morel JL. Billon de cinq années d'utilisation de boues résiduaires urbaines Sur plantes de grande culture dans les conditions agro-climatiques lorraines. In: Alexendre Detott H, editor. Frist European Symposium of Traitemant and Use of Sewagesludge. 1979. pp. 269-282

[22] Nejmeddine A, Echab A, Fars S, Hafidi M. Accumulation of trace elements by ryegrass grown on soils amended by sewage sludge. Cahiers Agricultures. 2003;**12**:33-38

[23] Lassoued N, Khelil M, Rejeb S, Bilal E, Chaouachi M. Transfert heavy metal sewage sludge as fertilizer from soil to ray grass. Banat's Journal of Biotechnology. 2014;**9**:86

[24] Zaier H, Ghnaya T, Ben RK, Lakhdar A, Rejeb S, Jemal F. Effects of EDTA on phytoextraction of heavy metals (Zn, Mn and Pb) from sludge-amended soil with Brassica napus. Bioresource Technology. 2010;**101**(2010):3978-3983

[25] Lakhdar A, Iannelli MA, Debez A, Massacci A, Jedidi N, Abdelly C. Effect of municipal solid waste compost and sewage sludge use on wheat (Triticum durum): Growth, heavy metal accumulation, and antioxidant activity. Journal of the Science of Food and Agriculture. 2010;**90**(6):965-971

[26] Pasqualone A, Laura ND, Giovanni L, Luciana P, Rosanna S, Giovanna C. Effect of composted sewage sludge on durum wheat: Productivity, phenolic compounds, antioxidant activity, and technological quality, food. Agriculture and Environment (JFAE). 2014;**12**(3-4):276-280

[27] Boudjabi S, Kribaa M, Tamrabet L. Contribution of sewage sludge to the fertility of the soil and the growth of barley (Hordium vulgare L) variety Jaidor. In: Efficient Management of Wastewater. Berlin, Heidelberg: Springer; 2008. pp. 227-235

[28] Larry DK. Interaction chimique pour l'élévation de la biodisponibilité des métaux en traces du sol. Agronomie. 1981;**16**:201-215

[29] Rejeb S, Bahri A. Impact of the contribution of urban sewage sludge on the mineral composition and productivity of some crops grown in Tunisia. Books of the RMAF. 1995;**24**:13-32

[30] Rejeb S, Khelil MN, Gharbi F, Ghorbal MH. Effet des boues urbaines sur la production dela pomme de terre. France. 2003;**12**:39-42

[31] Boudjabi S, Kribaa M, Chenchouni H. Growth, physiology and

yield of durum wheat (Triticum durum) treated with sewage sludge under water stress conditions. EXCLI Journal. 2015;**14**:320-334

[32] Zantopoulos N, Antoniou V, Nikolaidis E. Copper, zinc, cadmium, and lead in sheep grazing in North Greece. Bulletin of Environmental Contamination and Toxicology. 1999;**62**:691-699

[33] Brun LA, Maillet J, Hinsinger P, Pepin M. Evaluation of copper availability to plants in copper-contaminated vineyard soils. Environmental Pollution. 2001;**111**:293-302

[34] Gilmour JT, Skinner V. Predicting plant available nitrogen in land-applied biosolids. Journal of Environmental Quality. 1999;**28**:1122-1126

[35] Marchiol L, Assolari S, Sacco P, Zerbi G. Phytoextraction of heavy metals by canola (Brassica napus) and radish (Raphanus sativus) grown on multicontaminated soil. Environmental Pollution. 2004;**132**:21-27

[36] Johnson P. Foliar phosphate applications to potatoes. In: Symposium on Plant Health and the European Single Market. UK: Reading, ROYAUME-UNI; 1993. pp. 575-586

[37] Huang JW, Chen J, Berti WR, Cunningham SD. Phytoremediation of lead-contaminated soils: Role of synthetic chelates in lead phytoextraction. Environmental Science and Technology. 1997;**31**:800-805

[38] Wollan E, Davis RD, Sally J. Effects of sewage sludge on seed germination. Environmental Pollution. 1978;**17**(3): 195-205

[39] Ramírez W, Domene X, Andrés P, Alcañiz JM. Phytotoxic effects of sewage

sludge extracts on the germination of three plant species. Ecotoxicology. 2008;**17**(8):834-844

[40] Day A, Sr S, Taylor B, Pepper I, Minnich M. Effects of Sewage Sludge on Wheat Forage Production. Forage and Grain: A College of Agriculture Report 370071. Series 71p. Tucson, AZ: College of Agriculture, University of Arizona; 1987

[41] Chino M, Goto S, Youssef RA. Behaviour of zinc and copper in soil with long term application of sewage sludge. Soil Science & Plant Nutrition. 1992;**38**(1):159-167

[42] Chaudri AM, Allain CM, Badawy SH, Adams ML, McGrath SP, Chambers BJ. Cadmium content of wheat grain from a long-term field experiment with sewage sludge. Journal of Environmental Quality. 2001;**30**:1575-1580

[43] Misra SG, Mani D. Soil Pollution. New Delhi, India: Ashish Publishing House; 1991. pp. 43-80

[44] Kumar PBAN, Dushenkov V, Motto H, Raskin I. Phytoextraction: The use of plants to remove heavy metals from soils. Environmental Science and Technology. 1995;**29**:1232-1238

[45] Liphadzi MS, Kirkham MB. Phytoremediation of soil contaminated with heavy metals: A technology for rehabilitation of the environment. South African Journal of Botany. 2005;**71**:24-37

[46] Gripsen VMJ, Nelissen HJM, Verklei JA. Phytoexttraction with Brassica napus: A tool for sustainable management of heavy metal contamianted soils. Environmental. Pollution. 2006;**144**:77-83

[47] Ebbs SD, Kochian LV. Toxicity of zinc and copper to brassica species:

Implications for phytoremediation. Journal of Environmental Quality. 1997;**26**:776-781

[48] Tariq M, Islam KR, Muhammad S. Toxic effects of heavy metals on early growth and tolerance of cereal crops. Pakistan Journal of Botany. 2007;**39**(2):451-462

[49] Do Nascimento CW, Siriwardena AD, Xing B. Comparison of natural organic acids and synthetic chelates at enhancing phytoextraction of metals from a multi-metal contaminated soil. Environmental Pollution. 2006;**140**:114-123

[50] Rascio N. Metal accumulation and damage in rice (Vialone nano) seedlings exposed to cadmium. Environmental and Experimental Botany. 2008;**62**:267-278

[51] Hall JL. Cellular mechanisms for heavy metal detoxification and tolerance. Journal of Experimental Botany. 2002;**53**:1-11

[52] Navari-Izzo F, Quartacci MF, Pinzino C, Dalla VF, Sgherri C. Thylakoid-bound and stromal antioxidative enzymes in wheat treated with excess of copper. Physiologia Plantarum. 1998;**104**:630-638

[53] Sgherri C, Cosi E, Navari-Izzo F. Phenols and antioxidative status of Raphanus sativus grown in copper excess. Physiologia Plantarum. 2003;**118**:21-28

[54] Navari-Izzo F, Quartacci MF, Pinzino C, Dalla Vecchia F, Sgherri C. Thylakoid-bound and stromal antioxidative enzymes in wheat treated with excess of copper. Physiol. Plant. 1998;**104**:630-638

[55] Sandalio LM, Dalurzo HC, Gómez M, Romero-Puertas MC, del Rio LA. Cadmium-induced changes in the growth and oxidative metabolism of pea plants. Journal of Experimental Botany. 2001;**52**:2115-2126

[56] Elloumi N, Ben F, Rhouma A, Ben B, Mezghani I, Boukhris M. Cadmium induced growth inhibition and alteration of biochemical parameters in almond seedlings grown in solution culture. Acta Physiologiae Plantarum. 2007;**29**:57-62

[57] Cataldo DA, Garland TR, Wildung RE. Cadmium uptake kinetics in intact soybean plants. Plant Physiology. 1983;**73**:844-848

[58] Srivastava A, Jaiswal VS. Biochemical changes in duck weed after cadmium treatment. Water, Air, and Soil Pollution. 1990;**50**:163-170

[59] Costa G, Morel J. Water relations, gas exchange and amino acid content in Cd-treated lettuce. Plant Physiology and Biochemistry. 1994;**32**(4):561-570

[60] Ouariti O, Boussama N, Zarrouk M, Cherif A, Ghorbal MH. Cadmium-and copper-induced changes in tomato membrane lipids. Phytochemistry. 1997;**45**:1343-1350

[61] Haghiri F. Cadmium uptake by plants. Journal of Environmental Quality. 1973;**2**:93-96

[62] Watanabe T, Osaki M. Mechanism of adaptation to high aluminium condition in native plant species growing in acid soils: A review. Communications in Soil Science and Plant Analysis. 2002;**33**(2002):1247-1260

[63] Vecchia F, Dalla. Morphogenetic, ultrastructural and physiological damages suffered by submerged leaves of Elodea canadensis exposed to cadmium. Plant Science. 2005;**168**:329-338

[64] Ebbs S, Uchil S. Cadmium and zinc induced chlorosis in Indian mustard

Brassica juncea (L.) involves preferential loss of chlorophyll b. Photosynthetica. 2008;**46**:49-55

[65] Drażkiewicz M, Baszyński T. Growth parameters and photosynthetic pigments in leaf segments of Zea mays exposed to cadmium, as related to protection mechanisms. Journal of Plant Physiology. 2005;**162**:1013-1021

[66] Baszyński T, Tukendorf A, Ruszkowska M, Skorzyńska E, Maksymiec W. Characteristics of the photosynthetic apparatus of copper tolerant spinach expose to excess copper. Journal of Plant Physiology. 1988;**132**:708-713

[67] Mishra S, Srivastava S, Tripathi RD, Govindarajan R, Kuriakose SV, Prasad MNV. Phytochelatin synthesis and response of antioxidants during cadmium stress in Bacopa monniera. Plant Physiology and Biochemistry. 2006;**44**:25-37

[68] Bertrand M, Poirier I. Photosynthetic organisms and excess of metals. Photosynthetica. 2005;**43**:345-353

[69] Briat JF, Curie C, Gaymard F. Iron utilization and metabolism in plants. Current Opinion in Plant Biology. 2007;**10**:276-282

[70] Bączek-Kwinta R, Bartoszek A, Kusznierewicz B, Antonkiewicz J. Physiological response of plants and cadmium accumulation in heads of two cultivars of white cabbage. Journal of Elementology. 2011;**16**(3):355-364

[71] Mysliwa-Kurdziel B, Strzalka K. Influence of metal on biosynthesis of photosynthetic pigments. In: Prasad MNV, Strzalka K, editors. Physiology and Biochemistry of Metal Toxicity and Tolerance in Plants.

Dordrecht, Netherlands: Kluwer Academic Publishers; 2002. pp. 201-227

[72] Mysliva-Kurdziel B, Prasad MNV, Strzalka K. Photosynthesis in heavy metal stressed plants. In: Prasad MNV, editor. Heavy Metal Stress in Plants: From Biomolecules to Ecosystems. Berlin: Springer Verlag; 2004. pp. 146-181

[73] Baszinsky T, Wajda L, Krol M, Wolinska D, Krupa Z, Tukendrof A. Photosynthetic activates of cadmium-treated tomato plants. Physiologia Plantarum. 1980;**48**:365-370

[74] Küpper H, Šetlik I, Spiller M, Küpper FC, Prášil O. Heavy metal-induced inhibition of photosynthesis: Targets of in vivoheavy metal chlorophyll formation. Journal of Phycology. 2002;**38**:429-441

[75] Küpper H, Setlík I, Setliková E, Ferimazova N, Spiller M, Küpper FC. Copper-induced inhibition of photosynthesis: Limitingsteps of in vivo copper chlorophyll formation in Scenedesmusquadricauda. Functional Plant Biology. 2003;**30**:1187-1196

[76] Stefanov S, Pandev SD, Seizova K, Tyankova LA, Popov S. Effect of lead on the lipid metabolism in spinach leaves and thylakoid membranes. Biologia Plantarum. 1995;**37**:251-256

[77] Kucera T, Horáková H, Sonská A. Toxic metal ions in photoautotrophic organisms. Photosynthetica. 2008;**46**: 481-489

[78] Zhang J, Sun W, Li Z, Liang Y, Song A. Cadmium fate and tolerance in rice cultivars. Agronomy for Sustainable Development. 2009;**29**:483-490

[79] Haydon MJ, Cobbett CS. Transporters of ligands for essential metal ions in plants. The New Phytologist. 2007;**174**:499-506

[80] Brunetti G, Farrag K, Rovira PS, Nigro F, Senesi N. Greenhouse and field studies 20 on Cr, Cu, Pb and Zn phytoextraction by Brassica napus from contaminated soils in the 21 Apulia region, southern Italy. Geoderma. 2011;**160**:517-523

[81] Chang PY, Hao E, Patt Y. Alternative implementations of hybrid branch predictors. In: Proceedings of the 28th Annual International Symposium on Microarchitecture. Michigan; 1995;**28**:252-257

[82] Juste C, Solda P. Effet de l'application de boues de stations d'épuration urbaines en monoculture de mais. Action sur le rendement et la composition des plantes et sur quelques caractéristiques du sol. Sciences du Sol. 1977;**3**:147-155

[83] Yu R, Ji J, Yua X, Song Y, Wang C. Accumulation and translocation of heavy metals in the canola (Brassica napus L.)-soil system in Yangtze River delta, China. Plant and Soil. 2012;**353**:33-45

[84] Clemente R, Walker DJ, Bernal MP. Uptake of heavy metals and as by Brassica juncea grown in a contaminated soil in aznalcollar (Spain): The effect of soil amendments. Environmental Pollution. 2005;**138**:46-58

[85] Hayward A. 2012. In: Edwards D, Batley J, Parkin I, Kole C, editors. Genetics, Genomics and Breeding of Oilseed Brassicas. Boca Raton, FL, USA: CRC Press; 2012. pp. 1-13

[86] WHO. Report on Risk to Health from Microbes in Sewage Sludge Applied to Land. Regional Office for Europe, Reports and Studies of World Health Organization; 1981;**54**:27

[87] FAO, 2014. Organisation des Nations unies pour l'alimentation et l'agriculture, la situation mondiale de l'alimentation et de l'agriculture. Available from: http://www.fao.org/home/fr/

[88] Anses 2011. National food safety agency, environment and work. Anses Saisine n° 2011-SA-0136. p. 23

[89] Ricoux C, Gasztowtt B. Assessment of the health risks linked to the exposure of heavy consumers of river fishing products contaminated by environmental toxins. Report of the National Institute for Health Surveillance; 2005. p. 65

Section 4

Membrane Technologies for Wastewater Treatment and Resource Recovery

Chapter 9

Technologies for Removal of Emerging Contaminants from Wastewater

Tahira Mahmood, Saima Momin, Rahmat Ali, Abdul Naeem and Afsar Khan

Abstract

Emerging contaminants (ECs) include both natural and man-made compounds that have recently been found to be present in wastewater and have a harmful effect on human health and aquatic environment. Several ECs such as pharmaceuticals, antibacterial, hormones, synthetic dyes, flame retardants are directly or indirectly discharged from hospitals, agricultural, industrial and other sources to the environment. Strategies have been developed to overcome the challenges faced by contaminated water treatment technologists. Advanced treatment technologies such as physical, chemical, and biological methods have been studied for ECs removal as well as for reduction of effluents levels in discharged water. Techniques such as membrane filtration, adsorption, coagulation-flocculation, solvent extraction, ion exchange, photodegradation, catalytic oxidation, electrochemical oxidation, ozonation and precipitation, etc., have been investigated. Based on past research, these techniques significantly remove one or more pollutants but are insufficient to remove most of the toxic contaminants efficiently from wastewater. Nanomaterial incorporated technologies may be a proficient approach for removing different contaminants from wastewater. These technologies are costly because of high-energy consumption during the treatment of wastewater for reuse on large scale. Consequently, comprehensive research for the improvement of wastewater treatment techniques is required to obtain complete and enhanced EC removal by wastewater treatment plants.

Keywords: wastewater treatment, membrane filtration, available technologies, effluents, emerging contaminants, nanotechnology, adsorption, personal care products, pharmaceuticals, aquatic environment

1. Introduction

Wastewater is the water having surplus substances that may be dissolved or suspended solid particles or organic and inorganic substances or other impurities that critically influence its quality and make it unsuitable for use [1]. Wastewater composition varies and is highly dependent on major sources of generation as industries, commercial and residential areas, agricultural sources, etc. [2]. In developing countries, the risk of consumption of contaminated water and its sanitation problem is increasing day by day.

Water covers about 70% of the earth's shells and is essential for all living organisms to survive and also for various manufacturing industries. About 3% of the total water on earth is fresh water of 0.01% is available for human use. The discharge of untreated contaminants from various industries directly to groundwater hinders the favorable use of water in normal operations of the ecosystem and causes water scarcity. Water deficiency is considered one of the most significant alarms for humanity and sustainable development [3]. According to the UNO report, about 1.2 billion people are affected by severe water scarcity due to the increasing world population and in future, 1.8 billion citizens are predicted to be affected by water insufficiency. Beyond water scarcity, water pollution also poses a greater threat to human health and aquatic life as well as the environment. Several new compounds recently detected in drinking, ground, and surface water have a major effect on water parameters. Water is a universal solvent and water quality is affected due to contamination by toxic substances dissolved in it which causes water pollution [4]. Water requirement is increasing due to adaptation in atmosphere, industrialization, increase in population, and obliteration of the surroundings [5]. The occurrence of organic and inorganic pollutants in wastewater is a major challenge to recycle water sources. To determine small amounts of unknown pollutants in the evaluation of emerging contaminants, the latest modern treatment techniques are still limited [6].

Currently, various analytical methods have been developed for different kinds of emerging pollutants. The separation of these toxic pollutants from water becomes important before the discharge of industrial wastewater into the aquatic environment. For this purpose, the development of proficient techniques has been a major area of environmental research. In general, traditional clean-up methods are classified as biological, physical, and chemical. Biological treatment is of low cost and simple, but not effective for synthetic dyes as they are resistant to aerobic biodegradation. Chemical treatments produce toxic by-products and are low efficient, while physical treatment is usually effective. For treating these organic pollutants present in water, several techniques such as membrane filtration, coagulation-flocculation, solvent extraction, ion exchange, catalytic oxidation, electrochemical oxidation, precipitation, etc. have been tested. However, these techniques are less effective, very expensive, and do not eliminate the contaminants from polluted water which makes this issue more challenging for the researchers. Besides these techniques, adsorption and photocatalytic degradation are considered the most potential approaches to removing wastewater contaminants [7].

The current chapter focuses on classification, potential sources, occurrence, prevention, control, and elimination of emerging contaminants. The major objective of this section is to study available technologies currently used for the removal of ECs from wastewater. This chapter also focuses on selecting the best available technology for removing emerging contaminants from wastewater. A schematic representation of treatment technologies, their principal advantages, performance efficiency, and limitations are discussed in the present study. Furthermore, future research opportunities are examined to provide more suitable and strategic recommendations for ECs removal from the aquatic environment.

2. Emerging contaminants

Emerging contaminants (ECs), termed contaminants of emerging concern, emerging pollutants (EPs), micro-pollutants, or trace organic compounds (TrOCs) are derived from different natural as well as anthropogenic sources that extensively influence water quality [8]. They are termed as emerging not because they are new but due to enhancement in the level of concern. These contaminants are generally in small concentrations, ranging from nano-gram per liter (ng L^{-1}) to micrograms per liter (µg L^{-1}) in the atmosphere. United States Environmental Protection Agency (USEPA) describes ECs as new chemical compounds that have the potential to cause harmful effects on individual health and the surroundings [9]. It is essential to treat and recycle wastewater to an acceptable standard to fulfill water demands.

2.1 Classification of ECs

ECs are classified into organic, inorganic micro-pollutants like pesticides, personal care products (PCPs), pharmaceuticals, synthetic organic dyes, polycyclic aromatic hydrocarbons (PAHs), heavy metals ions, plasticizers, per-fluorinated compounds, flame retardants, surfactants, etc. (**Figure 1**) generated by human activities such as domestic, health care units, agricultural and industrial pathways [10]. These compounds are a source of concern due to their physical and chemical properties because they are widely distributed in the environment which is harmful to humans and wildlife. These pollutants are difficult to detect and have varied activities and miscellaneous sources of production. Their presence in small concentrations causes chronic toxicity, endocrine disruption, and the expansion of pathogen resistance [11].

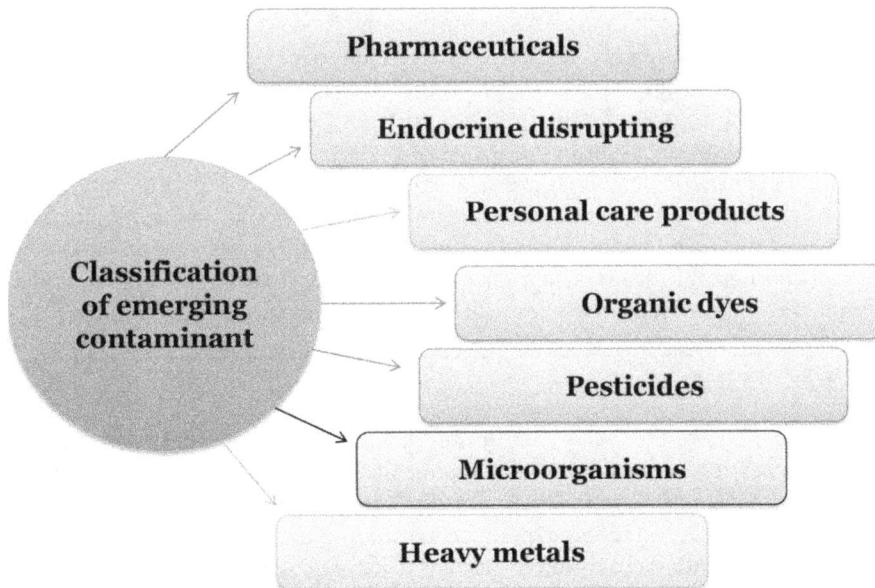

Figure 1.
Classification of emerging contaminants.

2.1.1 Pesticides

Pesticides, a class of organic contaminants, based on their physical and chemical properties are categorized as fungicides, herbicides, bactericides, and insecticides which are used in the agricultural sector to control dangerous insects, weeds, and microorganisms, etc. Based on their application sites, pesticides are frequently detected in groundwater causing toxicity and may bio-accumulate in humans and plants, or sediments depending on solubility, reactivity, and characteristics of soil and environment. Among the pesticide contaminants, dichlorodiphenyltrichloroethane (DDT) and hexachlorocyclohexane are commonly used pesticides (about 67%) as compared to other compounds such as phorate, chlorpyriphos, Atrazine, methyl parathione, Bentazone, Diazinon, Cyanazine, Simazine, phosphamidone, Terbuthylazine, Alachlor and Dimethoate [12].

2.1.2 Pharmaceutical industry

Pharmaceuticals are major emerging organic contaminants occurring in small amounts in water resources worldwide [13]. Pharmaceuticals are extensively used on daily basis in human healthcare as well as veterinary medicine such as nutrition, investigative aids, therapy and preventive medicine. Many pharmaceutical products such as drugs (both prescribed and non-prescribed), hormones and antibiotics are extensively detected in the aquatic environment, surface and groundwater and have adverse effects on humans, poultry, livestock and fish farming, etc. Generally, livestock is given medications to reduce diseases and infections. Researchers have examined more than 3000 chemicals used in therapeutic products but only small proportion (ng L^{-1} doses) has been studied in the field, which possibly will lead to negative effects on human and wildlife. To enhance animal farming, organic fertilizer such as manure and purines as medicines are used which indirectly affect the atmosphere and can reach living organisms through food stuff. Commonly reported pharmaceuticals in wastewater are antibiotics, diclofenac, antacids, clofibric acid, steroids, antidepressants, ciprofloxacin, propranolol, beta blockers, analgesics, salicylic acid, fluoxetine, antipyretics, anti-inflammatory drugs, nitroglycerin, tranquilizers, lipid-lowering drugs and stimulants [14].

Natural or synthetic hormones are also essential ecological contaminants, because of their estrogenic and androgenic impacts on wildlife. Organic and inorganic hormones consist of 17α-estradiol, 17β-estradiol, estrone, equiline, equilenin, estriol, mestranol and norethindrone which can enter atmosphere through farming, and are not completely eliminated from wastewater and harm aquatic life and humans.

2.1.3 Personal care products (PCPs)

Personal care products (PCPs) are household chemicals commonly used for health, odor, beauty, or cleaning. These chemicals are used in personal care products like ornamental cosmetics, soaps, hair and skin care products, lotions, fragrances and sunscreens. PCPs are used in large quantities throughout the world due to which the release of these pollutants in the environment is increasing day by day [15]. Mostly these substances are bioactive and bioaccumulative and harm the environment and humans [16]. The most probable emerging contaminants in PCPs are antiseptics, perfumes pollutants like galaxolide, pest repellants, preservatives diethyl phthalate ultraviolet (UV) filters and Triclosan (TCS) and triclocarban as disinfectant pollutant. Parabens are antimicrobial preservatives used in cosmetic items, pharmaceuticals, and some food stuffs such as benzyl, butyl, ethyl, isobutyl, isopropyl, methyl, and propyl hydroxybenzoates. Polycyclic musks are used in numerous products such as clean-up products, shampoos, hair care and washing products and cosmetic products. Their use on the outside of human skin increases its discharge in environment without any metabolic changes. Among all these products, cosmetics are frequently used, thus its occurrence in air at low quantity may be a source of damage to human beings, wildlife and environment.

2.1.4 Surfactants

Surfactants are synthetic organic compounds used all over the world in making of household products such as emulsifiers, detergents, paints, and pesticides, in addition to personal care products and are harmful to aquatic species [17]. They are classified as cationic, anionic and zwitterionic surfactants. Frequently used surfactants such as fatty alcohol ethoxylates, linear alkyl benzene sulfonates, lignin sulfonates, and alkyl phenol ethoxylates are produced on a large scale. Furthermore, octylphenol and nonyl-phenol ethoxylates, are highly toxic even at low concentrations and must be substituted in all their uses.

2.1.5 Food additives

Numerous artificial sweeteners such as acesulfame, saccharin and sucralose which are extensively used in foodstuff, pharmaceuticals and hygiene products find their way to domestic wastewater via human excretion. These moderately metabolized sweeteners which pollute the environment are usually hard to remove. Though, latest calculated ratio for predicted environmental concentration (PEC) and predicted no effect concentration (PNEC) of compound sucralose for marine system is below 1 indicating limited threat to aquatic system (plants, algae and fish) [18].

2.1.6 Flame retardants (FRs)

Among all flame retardant compounds, organophosphate ester flame retardants (OPEFRs) class of phosphorus-containing flame retardants (FRs) and halogenated FRs such as polybrominated diphenyl ethers (PBDEs)) are known FR groups that decrease the flammability of industrial and consumer products. Organophosphate flame retardants (OPFRs) are used in furnishings, textiles, construction materials, electronics and as plasticizers in floor polishes and coatings. The discharge of OPFRs from wastewater treatment plants (WWTPs) into the surface water polluted marine environment causes toxicity. PBDEs are hydrophobic in nature and are mostly used as FRs in the manufacturing of carpets, computers, polyurethane foams, electronic cables, etc. [19].

2.2 Potential sources of emerging pollutants

Emerging contaminants sources are the same as those of traditionally known contaminants and they are released to environment by agricultural, domestic, mining and industrial activities and hospitals. These sources are categorized as: point sources and non-point sources [20]. Contaminants from point sources are discharged from a particular site in high concentration and enter the ecological system in a spatially distinctive way. Examples are discharges from industrial activities, mineral extraction and sewage treatment plants. While non-point sources also termed as diffuse sources release pollutants from indistinguishable disperse sources usually over large areas in low quantity. Examples are runoff of bio-solids or fertilizer applied to soils and rain overflow in urban or industrial areas (**Figure 2**) [21].

Water resources contamination by ECs from Wastewater is taking place all over the world particularly in those areas where wastewater treatment is not properly organized. Frequent use of drugs and personal care products lead to discharge of low quantity of different by-products. For example, triclosan, bisphenol-A and phthalates are significant industrial compounds integrated into several commercial household products. Their existence in water and environment affects their physical and chemical properties. PPCPs and other ECs metabolites are complex and hydrophobic in nature when released in water and settle at water surface. Thousands of these ECs and their metabolites have been discovered in the marine environment and are more noxious and harmful. Basically, wastewater treatment plants are not specifically designed for the effective removal of emerging contaminants [22].

2.3 Toxicological effects of emerging contaminants

The adverse effects of ECs on living bodies have been widely reported which confirm that even small amount of ECs pose negative effects such as chronic toxicity and endocrine disruption in humans and animals. The major route of human contact with endocrine-disrupting chemicals (EDCs) is taking of foods and drinks connected to contaminated soil, water and microorganisms leading to bio magnification and

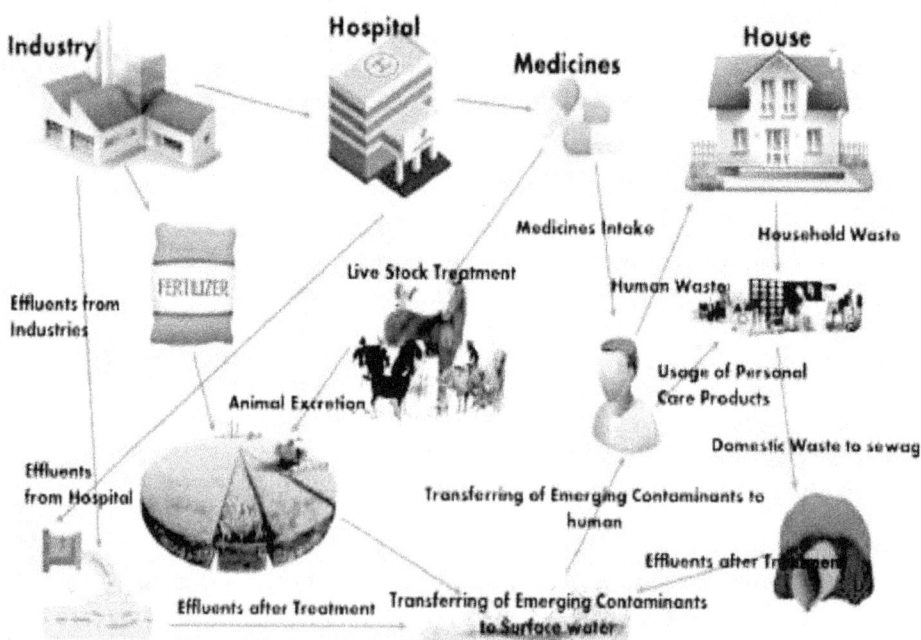

Figure 2.
Sources and their pathways of emerging contaminants.

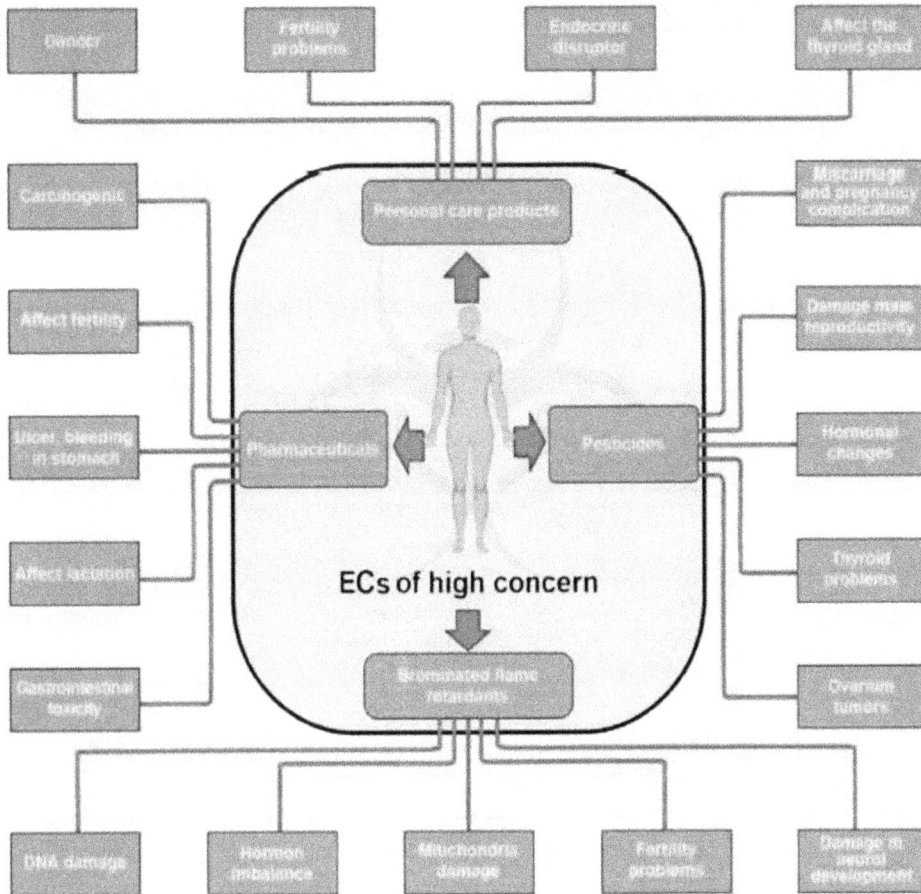

Figure 3.
Harmful effects on human health.

bioaccumulation in human body (**Figure 3**). Currently, researchers are focusing on ECs present in surface waters for many reasons: firstly, surface waters commonly contain high quantity and a diverse range of contaminants particularly when surface water is directly associated with industrial discharges and secondly it is easily monitored as compared to groundwater [23].

3. Traditional wastewater treatment methods

Conventional techniques for the treatment of wastewater consist of physical, chemical and biological techniques for the removal of soluble and insoluble pollutants. Benefits and challenges of wastewater treatment technologies are given in **Table 1**. Biological treatment is of low cost and simple, but not effective for synthetic pollutants such as dyes as they are resistant to aerobic bio-degradation. Chemical treatments produce toxic by-products and are less efficient, while physical treatment is usually effective. Different phases included in wastewater treatment preliminary, primary, secondary and tertiary [24].

Treatment	Basic methodologies	Benefits	Challenges
Screening	Coarse Screening: removes solid materials with a size below 6 mm.	Minimizes interruption and blockage of the treatment technologies	Not effective in the removal of the ECs
	Fine Screening: removal of contaminants in between 0.001 and 6 mm	Efficiency regulated by altering the fineness of the screen openings.	The screen must be cleaned due to the blockage in the small openings
		Highly recommended to regulate the temperature of the process	
		Less expensive and less complex process	
Adsorption	Process of removing soluble substances by solid substrates of very specific surface area	Capable of very specific removal of the ECs	Accumulation of cyanotoxin in the adsorbent
		Accurate and efficient removal	Difficult to remove the unknown type of contaminants, since adsorbents are highly specific
		Can assist other treatment processes	
		Less complex and less expensive, adopted easy	Regulation of selectivity of the membranous system is difficult
			Reverse osmosis requires external energy;
Biosorption	Immobilization of the microbes on absorbents	Efficient treatment	Absorbents need to be cleaned at a certain interval of time
		Specific removal of certain ECs	

Table 1.
Benefits and challenges of wastewater treatment technologies.

3.1 Preliminary treatment

Preliminary treatment helps in the removal of suspended materials like dead animals, papers, oils, grease, etc., from wastewater. Different components such as screening, accumulation and floatation tanks and skimming reservoir are used in preliminary treatment. The accumulation tank is used for the elimination of sand and grit while oils and greases are removed by floatation units and skimming tanks.

3.2 Primary treatment

In primary treatment, organic and inorganic components are removed by floatation and sedimentation processes. Throughout this treatment, untreated nitrogen, unrefined phosphorus, and heavy metals related with suspended impurities are drained off. This method reduces biochemical oxygen demand (BOD) ranges by 5–40%, 50–70% of entire floating particles and oil and grease up to 65% from wastewater. In various developed countries, primary treatment is required for the reuse of wastewater irrigation, i.e., for crops not used by humans.

3.3 Secondary treatment

The secondary or biological treatment is used to eliminate organic effluent that escapes from primary treatment. This method modifies organic matter and transforms it into stabilized form by oxidation or nitrification. This treatment method for sewage is divided into two groups known as filtration and activated sludge methods. Different filters such as contact beds, irregular sand and trickling filters are used in this treatment [25].

3.4 Tertiary treatment

Tertiary treatment is employed for the removal of specific effluents which cannot be completely removed by secondary method. During this process, around 99% of all contaminants are eliminated. This process removes inorganic substances such as nitrogen and phosphorous and recovers wastewater quality which can be reused for irrigation and drinking and have no harmful effect when discharged to the environment [25].

4. Available technologies for wastewater treatment

Generally, conventional wastewater treatment plants are not constructed to eliminate emerging contaminants. The occurrence of ECs in the environment affects public health, marine life and produces resistant bacteria, neurotoxin effects, endocrine interruption, and tumors. To eliminate these organic pollutants from water, several techniques such as membrane filtration, coagulation-flocculation, solvent extraction, ion exchange, catalytic oxidation, electrochemical oxidation, and precipitation, etc. have been tested (**Figure 4**). However, these techniques are less effective, very expensive and do not eliminate the contaminants completely from polluted water which makes this issue more challenging for the researchers. Besides these techniques, adsorption and photocatalytic degradation are considered potential approaches to remove wastewater contaminants [26].

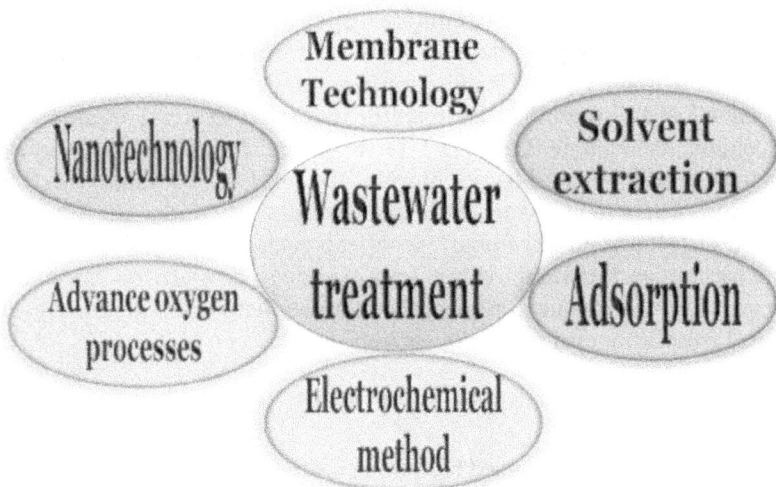

Figure 4.
Various treatment methods used for wastewater.

4.1 Membrane filtration

Membrane technology is a physical method implemented to eliminate emerging contaminants from aquatic system. Membranes are formed from substances having filtering properties such as specific surface charge, pore size and hydrophobicity to remove suspended contaminants. Membrane filtration is categorized as ultra-filtration (UF), nano-filtration (NF), microfiltration (MF), forward osmosis (FO) and reverse osmosis (RO). Major membrane processes including forward osmosis, membrane refinement and electro-dialysis of the membrane have the ability to reduce emerging contaminants upto greater than 99% but still have not been executed on large scale [27].

The ultrafiltration technique works at low pressure for the removal of colloidal, suspended or dissolved pollutants depending on the membrane and pollutant type. UF has pore size in range of 0.001–0.1 µm which is larger than dissolved hydrated metals ions, thus easily pass through it. Polymer enhanced ultrafiltration (PEUF) and Micellar enhanced ultrafiltration (MEUF) processes were studied to enhance the removal efficiency of metal ions such as copper, zinc, chromate, arsenate, cadmium, nickel, serinium, and organics like phenol, o-cresol, etc.

Microfiltration has pore size ranges from 0.1 to 10 µm and is commonly operated at atmospheric pressure but cannot effectively remove contaminants of size greater than 1 µm. Reverse osmosis and forward osmosis depend on the osmotic pressure gradients and use semi-permeable membrane to efficiently remove dissolved particles up to 1 nm from water. Nanofiltration membrane possess small pore size ranges from 1 to 10 nm and have high competency for removal of ECs based on type of membrane and contaminant. NF can be used for removal of pharmaceuticals and natural hormones such as anti-inflammatory drugs, sulfonamide and fluoroquinolone antibiotics, testosterone, estradiol, and progesterone [28].

4.2 Coagulation-flocculation

Coagulation-flocculation process is effective for the elimination of larger colloidal or suspended particles of disperse dyes colored wastewater. Coagulation is a procedure in which dye solution systems are dispersed to form flocs and agglomerates while in flocculation aggregated flocs are joined to form larger agglomerates which settle down due to gravity [29]. Coagulation/flocculation is economically feasible and simply operated and commonly used in textile industries to purify wastewater. In this method, coagulants like lime ($Ca(OH)_2$), ferric sulfate ($Fe_2(SO_4)_3 \cdot 7H_2O$), aluminum sulfate ($Al_2(SO_4)_3 \cdot 18H_2O$), and ferric chloride ($FeCl_3 \cdot 7H_2O$), combine with the pollutants and remove them by electrostatic interactions or sorption. Use of aluminum sulfate ($Al_2(SO_4)_3$) for removal of pharmaceuticals such as betaxolol, chlordiazepoxide, bromazepam, warfarin and hydrochlorothiazide by coagulation-flocculation has been reported. This technique diminishes suspended matter, soluble dyes, colloidal particles and coloring agents from wastewater [30].

4.3 Solvent extraction

Solvent extraction is widely used technique for the elimination of organic and inorganic pollutants discharged into wastewater from various industries. It is based on three major operations. First is the extraction/transferring of solute particles to solvent from water. Secondly, the separation of solute from solvent and the third stage is the solvent recovery stage. Solvent extraction is mostly operated for exclusion of phenols, creosols and other phenolic acids from contaminated water containing low quantity of solute arising from petroleum processing plant, coke-oven plants in the steel and plastics manufacturing [31].

4.4 Adsorption

Adsorption is one of the most efficient techniques used for treating wastewater due to its simple design, high competence and ease of operation, capital cost, easy recovery, adaptability and technical feasibility without producing harmful by-products. This technique is not new but is recognized throughout the world because of removal capacity and regeneration of adsorbents. This technique has been broadly applied for both organic inorganic toxins from household and industrial wastewater [32]. Various research efforts have been devoted to discover low-cost adsorbents having large surface area and excellent binding capacity to enhance their adsorption efficiency. Different types of adsorbents, e.g., peat, bamboo dust, chitosan, silica gel, activated carbon,, fly ash, zeolites, metal organic frameworks nano-adsorbents for example carbon nanotubes and graphene have been applied for elimination of emerging contaminants [33, 34]. Activated carbon is widely used as traditional adsorbent because of highly porous surface area, convenient pore composition and thermo stability for removal of dyes and pharmaceutical products, e.g., 17β-estradiol, 17α-ethynylestradiol, bisphenol A, and fluoroquinolonic Caffeine from wastewater [35].

4.5 Advanced oxidation process

Advanced oxidation processes (AOPs) have been introduced as proficient technology in wastewater treatment. AOPs are based on the generation of hydroxyl (OH) or sulfate radicals for oxidation of ECs while sometimes ozone and UV irradiation are used for enhanced removal efficiency. AOPs methods efficiently remove biologically injurious or non-degradable compounds such as pesticides, aromatics, petroleum essentials and volatile organic compounds (VOCs) rather than transferring these to another phase. AOPs are applicable for the removal of many organic contaminants at the same time without producing any hazardous substance in water, as OH˙ is reduced to form H_2O as byproduct. AOPs include ozonation (O_3), hydrogen peroxide (H_2O_2), electrochemical oxidation, Fenton process, UV light and photocatalytic process [36].

4.5.1 Non-photochemical processes

Ozonation: Ozone is an extremely efficient oxidizing agent and has the potential for elimination of organic and inorganic compounds from industrial effluents. It is a complex oxidation method. Pre-oxidation processes give significant development in biological degradation while post-oxidation process improves effluent quality. The limitation of ozonation is low solubility, stability and short half-life. O_3/H_2O_2 and catalytic ozonation have been investigated for generation of hydroxyl radical which efficiently removed organic pollutants such as antibiotics, antiphlogistics, beta blockers, lipid regulators and their metabolites, natural estrogen estrone, antiepileptic drug carbamazepine and musk fragrances in wastewaters effluents.

Electrochemical method: Electrochemical procedure is generally applied for removal of toxic contaminants from textile effluents by direct or indirect oxidation. This procedure is commonly applied for elimination of ECs like dye by using either mercury electrode, graphite rod, boron doped diamond electrode, platinum foil or titanium/platinum as anode while SS304 is used as cathode in textile sewage treatment. This process is cost effective as minute amount of chemical is required and stability is easily attained by manipulating the electric current.

Fenton process: Reaction between ferrous iron and hydrogen peroxide is termed as Fenton's reaction. Fenton method is used for removal of organic pollutants like phenols, reactive dyes and pesticides. Fenton process is of low cost as no energy is required for activating H_2O_2, environmental friendly, easy to control and efficient for elimination of organic pollutants.

4.5.2 Photolytic chemical process

4.5.2.1 Homogeneous photolytic chemical process

Ultraviolet lamp (UV): In this process oxidizing agent like H_2O_2 is initiated by UV process to produce OH˙ and can degrade micropollutants efficiently which can be affected by various parameters such as pH, structure of dye, composition of effluent and intensity of UV radiation. Generally, UV process occurs at standard wavelength of 254 nm at low pressure. A pilot plant with UV/H_2O_2 produced hydroxyl radical to treat effluent, achieved 98% removal of mecoprop and diclofenac. O_3/H_2O_2/UV processes were examined in treatment of textile effluent to achieve complete degradation [36].

Photo-fenton process: In this process, formation of hydroxyl radical is improved by UV light in the presence of Fe and competently degrades wastewater effluents. Fenton process and photo-fenton process are similar but in the later process mineralization is much better. Removal of numerous ECs like pharmaceuticals, beta blockers and pesticides excluding triclosan by photo-fenton process is enhanced significantly (95–100%).

4.5.2.2 Heterogeneous photolytic chemical process

Mostly semiconductor consists of two energy bands, high conduction band and low energy valence band and these two are separated by band gap. In heterogeneous processes, semiconductor sensitized photolytic chemical oxidation produces OH radical. Adsorption of photon having energy ($\geq$band gap energy of catalyst) is needed for photocatalytic reaction to occur. ZnO, strontium titanium trioxide and TiO_2 have been utilized extensively as photocatalysts for commercial application. Photo-catalysis is commonly used for dyestuff degradation from textile wastewater. Photocatalytic process enhances efficiency in the presence of H_2O_2 up to 100% for numerous pollutants such as bisphenol A, pesticides, pharmaceuticals [37].

4.6 Application of nanotechnology for ECs removal

Nanomaterials are generally defined as the materials having at least one dimension smaller than 100 nm. Nanomaterials have higher density and larger surface area resulting in increasing adsorption efficiency, surface reactivity, and resolution mobility. Current investigation in the exploitation of nanomaterials has facilitated the application of nanotechnology in wastewater treatment via adsorption, AOPs and filtration. Nanomaterials have been reported to effectively eliminate emerging contaminants from wastewater. A variety of nanomaterials have been reported for wastewater treatment (**Figure 5**) such as zerovalent metal nanoparticles, metal-oxide nanoparticles, carbon nanomaterials and nanocomposites [38].

4.6.1 Zerovalent metal nanomaterials

Zerovalent metal is a significant wastewater treatment nanomaterial which is highly reactive because of small size and high surface area. Recently, several zerovalent metal nanoparticles for example silver, zinc, iron, aluminum and nickel received attention of researchers for contaminant removal. Silver nanoparticles have potential antimicrobial properties and are generally used as disinfectant to eliminate a large amount of microorganisms, like viruses and bacteria, as well as fungi [39]. It is extremely reactive, cost effective, environment friendly and has multiple pathways for wastewater treatment. Iron nanomaterial can proficiently remove contaminants such as cadmium nitrates, colorant and antibiotics from wastewater by adsorption, redox reaction, and co-precipitation technique. Li et al. [40] reported two-step technique to form zero-valent metal nanomaterials covered with silica and polydopamine ($nZVI/SiO_2/PDA$) for use as sorbent which shows high capacity, selectivity and reusability up to 10 cycles.

4.6.2 Metal-oxide nanomaterials

Metal-oxide nanomaterials like ferric oxides, manganese oxides, aluminum oxides and titanium oxides have been effectively utilized in removing noxious waste such as arsenic, uranium, phosphate, and organics. Titanium oxide nanomaterial is a capable photocatalyst having band gap of 3.2 eV with high photostability, low price and outstanding photocatalytic behavior. TiO_2 nanomaterials are suitable for degradation of pollutants like organic chlorine, polycyclic aromatic compounds, pigments, phenols, pesticides, and heavy metals [41].

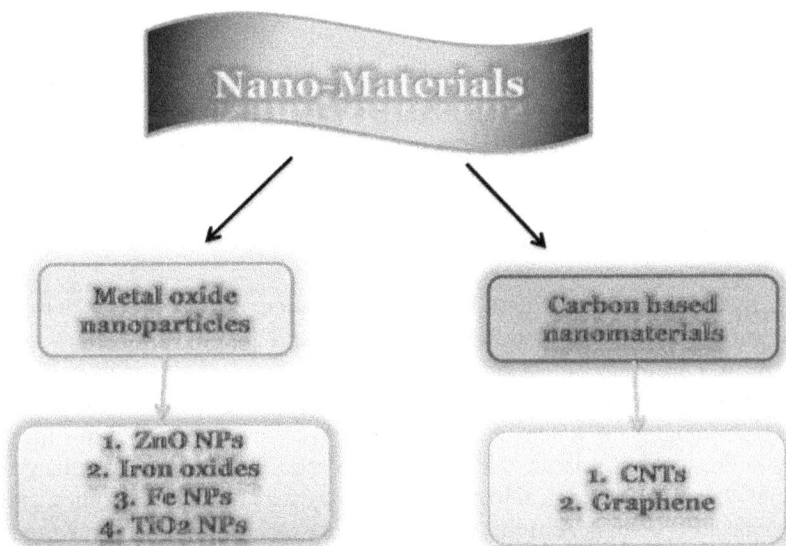

Figure 5.
Various groups of nanomaterials.

Zinc oxide (ZnO) nanomaterial is competent material for purifying wastewater having a strong oxidizing capacity, wide wavelength and admirable photocatalytic properties. ZnO nanomaterial is environment friendly and captures more light as compared to other metal oxides possessing semiconducting properties. Iron oxide nanoparticles have versatility and are available as potent sorbent material-removing heavy metals from wastewater [42].

4.6.3 Carbon-based nanomaterials

Carbon nanomaterials comprise distinctive structural and electronic properties duet to which they perform complex applications particularly in adsorption [43]. They have high adsorption capacity for removal of various pollutants, high surface area and aromatic selectivity. These nanomaterials are categorized as carbon beads, nonporous carbon, carbon nanotubes (CNTs) and carbon fibers. CNTs have well-defined cylindrical structures, stronger physicochemical interactions, porosity, large surface area, adaptable hydrophobic side and high adsorption capacity for dichloro-benzene, ethylbenzene, dyes, Pb^{2+}, Zn^{2+}, Cd^{2+} and Cu^{2+} [44].

Another class is graphene-based nanomaterial which is a single carbon atom layer having honeycomb like structure [45]. Graphene oxide is a graphene layer consisting of hydroxyl, epoxy, carboxyl, and carbonyl groups and is identified for eradicating heavy metals such as lead, zinc, copper, cadmium, mercury and arsenic. Graphene hybrid with nanoparticles of manganese ferrite can be exploited to proficiently remove Pb(II), As(III), and As(V) from contaminated water. Rajabi et al. [46] compared the adsorption efficiency of MWCNTs and functionalized CNTs by varying experimental conditions including pH, times, and temperatures. From results it was clear that f-CNTs possess a higher removal capacity than pristine CNTs. The maximum removal capacity ($166.7 \, mg \, g^{-1}$) of methylene blue (MB) with functionalized multi-walled carbon nanotubes (f-MWCNTs) was higher as compared to MWCNTs, which was $100 \, mg \, g^{-1}$.

5. Future challenges

Management of ECs in water sources is extremely challenging for humans. World population is increasing considerably every year, which leads to increase of freshwater demand for domestic use and also generate wastewater causing water deficiency. Besides, technological advancement, demand of water from cultivation, urban areas and industries, are main causes of water scarcity, resulting in adverse effects on environment. That is why, a highly efficient and low cost waste water management methods and society alertness is necessary.

Currently wastewater treatment is a difficult challenge as it has considerable effect on bio-physical environment and living organisms and depends on socioeconomic circumstances. Discovery of a general technique for complete elimination of all pollutants from wastewaters is complicated. A number of biological, physical, and chemical technologies for wastewater management have been studied to eliminate emerging pollutants but unable to identify best method to overcome challenges in operational obscurity, ecological impact, efficiency, feasibility, probability, and cost-efficiency. For enhanced removal, two or more techniques are merged to reach favorable water quality at low cost.

To overcome these challenges, some proposed potential directions in future are required as follows:

- To incorporate new concepts such as nano-technology and genetic engineering for production of environment friendly and non-hazardous techniques for synthesis of nanoparticles for pollution degradation.

- Effective assessment of treatment to select most suitable treatment technique depending on numerous parameters like water quality, environmental compatibility, consistency, elasticity, working and effective costs technique.

- Apply green technologies on an industrialized scale like membrane filtration nanotechnology, and microbial fuel cells as competent and cost maintenance solution.

- The exploration of cross treatment systems, e.g., combination of photo Vs electro-Fenton, UV photolysis, ozonation and biological treatment technologies is required for the development of the appropriate model.

6. Conclusion

Emerging contaminants are man-made toxic compounds discharged into wastewater. This chapter includes sources of emerging contaminants, their toxicity and treatment techniques. Pharmaceuticals, personal care products and fertilizers are the main sources of ECs. Their presence, even in small concentrations, cause toxic impacts on human health as well as marine organisms. They cannot be successfully eliminated by conventional wastewater treatment methods. Various treatment methods like membrane technology, coagulation-flocculation, solvent extraction, adsorption, advanced oxidation processes and nanotechnology have been discussed. These techniques have their advantages and limitations. Hybrid systems have been found more effective for EC elimination than individual techniques however they have issues regarding time, energy and cost. To overcome these limitations nanotechnology is a promising approach. Thus, comprehensive research on waste water treatment technologies which are technically and economically feasible is required to attain complete and efficient removal of ECs from contaminated water.

Acknowledgements

The authors would like to convey their gratitude to National Centre of Excellence in Physical Chemistry, University of Peshawar for providing us necessary support.

Author details

Tahira Mahmood*, Saima Momin, Rahmat Ali, Abdul Naeem and Afsar Khan
National Centre of Excellence in Physical Chemistry, University of Peshawar,
Peshawar, Pakistan

*Address all correspondence to: tahiramahmood@uop.edu.pk

IntechOpen

References

[1] Lee CS, Robinson J, Chong MF. A review on application of flocculants in wastewater treatment. Process Safety and Environmental Protection. 2014;**92**(6):489-508. DOI: 10.1016/j.psep.2014.04.010

[2] Abdelbasir SM, Shalan AE. An overview of nanomaterials for industrial wastewater treatment. Korean Journal of Chemical Engineering. 2019;**36**(8):1209-1225. DOI: 10.1007/s11814-019-0306-y

[3] Raouf MEA, Maysour NE, Farag RK. Wastewater treatment methodologies, review article. International Journal of Environment and Agricultural Science. 2019;**3**(1):18

[4] Karimi-Maleh H, Ranjbari S, Tanhaei B, Ayati A, Orooji Y, Alizadeh M, et al. Novel 1-butyl-3-methylimidazolium bromide impregnated chitosan hydrogel beads nanostructure as an efficient nanobio-adsorbent for cationic dye removal: Kinetic study. Environmental Research. 2021;**195**:110809. DOI: 10.1016/j.envres.2021.110809

[5] Taheran M, Naghdi M, Brar SK, Verma M, Surampalli RY. Emerging contaminants: Here today, there tomorrow! Environmental Nanotechnology, Monitoring & Management. 2018;**10**:122-126. DOI: 10.1016/j.enmm.2018.05.010

[6] Petrie B, Barden R, Kasprzyk-Hordern B. A review on emerging contaminants in wastewaters and the environment: Current knowledge, understudied areas and recommendations for future monitoring. Water Research. 2015;**72**:3-27. DOI: 10.1016/j.watres.2014.08.053

[7] Hemavathy RV, Kumar PS, Kanmani K, Jahnavi N. Adsorptive

separation of Cu (II) ions from aqueous medium using thermally/chemically treated Cassia fistula based biochar. Journal of Cleaner Production. 2020;**249**:119390. DOI: 10.1016/j.jclepro.2019.119390

[8] Varsha M, Senthil Kumar P, Senthil Rathi B. A review on recent trends in the removal of emerging contaminants from aquatic environment using low-cost adsorbents. Chemosphere. 2022;**287**:132270. DOI: 10.1016/j.chemosphere.2021.132270

[9] de Oliveira M, Frihling BEF, Velasques J, Magalhães Filho FJC, Cavalheri PS, Migliolo L. Pharmaceuticals residues and xenobiotics contaminants: Occurrence, analytical techniques and sustainable alternatives for wastewater treatment. Science of the Total Environment. 2020;**705**:135568. DOI: 10.1016/j.scitotenv.2019.135568

[10] Rout PR, Zhang TC, Bhunia P, Surampalli RY. Treatment technologies for emerging contaminants in wastewater treatment plants: A review. Science of the Total Environment. 2021;**753**:141990. DOI: 10.1016/j.scitotenv.2020.141990

[11] Calderon AG, Duan H, Seo KY, Macintosh C, Astals S, Li K, et al. The origin of waste activated sludge affects the enhancement of anaerobic digestion by free nitrous acid pre-treatment. Science of the Total Environment. 2021;**795**:148831. DOI: 10.1016/j.scitotenv.2021.148831

[12] Poonia T, Singh N, Garg MC. Contamination of arsenic, chromium and fluoride in the Indian groundwater: A review, meta-analysis and cancer risk assessment. International journal of Environmental Science and Technology.

2021;**18**(9):1-12. DOI: 10.1007/s13762-020-03043-x

[13] Chinnaiyan P, Thampi SG, Kumar M, Mini K. Pharmaceutical products as emerging contaminant in water: Relevance for developing nations and identification of critical compounds for Indian environment. Environmental Monitoring and Assessment. 2018;**190**(5):1-13. DOI: 10.1007/s10661-018-6672-9

[14] Richardson SD, Kimura SY. Emerging environmental contaminants: Challenges facing our next generation and potential engineering solutions. Environmental Technology & Innovation. 2017;**8**:40-56. DOI: 10.1016/j.eti.2017.04.002

[15] Kim E, Jung C, Han J, Her N, Park CM, Jang M, et al. Sorptive removal of selected emerging contaminants using biochar in aqueous solution. Journal of Industrial and Engineering Chemistry. 2016;**36**:364-371. DOI: 10.1016/j.jiec.2016.03.004

[16] Juliano C, Magrini GA. Cosmetic ingredients as emerging pollutants of environmental and health concern. A mini-review. Cosmetics. 2017;**4**(2):11. DOI: 10.3390/cosmetics4020011

[17] Mandaric L, Celic M, Marcé R, Petrovic M. Introduction on emerging contaminants in rivers and their environmental risk. In: Emerging Contaminants in River Ecosystems. Cham: Springer; 2015. pp. 3-25. DOI: 10.1007/698_2015_5012

[18] Tollefsen KE, Nizzetto L, Huggett DB. Presence, fate and effects of the intense sweetener sucralose in the aquatic environment. Science of the Total Environment. 2012;**438**:510-516. DOI: 10.1016/j.scitotenv.2012.08.060

[19] Venier M, Dove A, Romanak K, Backus S, Hites R. Flame retardants and legacy chemicals in Great Lakes' water. Environmental Science & Technology. 2014;**48**(16):9563-9572. DOI: 10.1021/es501509r

[20] Tijani JO, Fatoba OO, Babajide OO, Petrik LF. Pharmaceuticals, endocrine disruptors, personal care products, nanomaterials and perfluorinated pollutants: A review. Environmental Chemistry Letters. 2016;**14**(1):27-49. DOI: 10.1007/s10311-015-0537-z

[21] Rathi BS, Kumar PS, Show PL. A review on effective removal of emerging contaminants from aquatic systems: Current trends and scope for further research. Journal of Hazardous Materials. 2021;**409**:124413. DOI: 10.1016/j.jhazmat.2020.124413

[22] Munthe J, Brorström-Lundén E, Rahmberg M, Posthuma L, Altenburger R, Brack W, et al. An expanded conceptual framework for solution-focused management of chemical pollution in European waters. Environmental Sciences Europe. 2017;**29**(1):1-16

[23] Stuart M, Lapworth D, Crane E, Hart A. Review of risk from potential emerging contaminants in UK groundwater. Science of the Total Environment. 2012;**416**:1-21. DOI: 10.1016/j.scitotenv.2011.11.072

[24] Crini G, Lichtfouse E. Advantages and disadvantages of techniques used for wastewater treatment. Environmental Chemistry Letters. 2019;**17**(1):145-155. DOI: 10.1007/s10311-018-0785-9

[25] Ungureanu N, Vlăduț V, Voicu G. Water scarcity and wastewater reuse in crop irrigation. Sustainability. 2020;**12**(21):9055. DOI: 10.3390/su12219055

[26] Vieira WT, de Farias MB, Spaolonzi MP, da Silva MGC,

Vieira MGA. Removal of endocrine disruptors in waters by adsorption, membrane filtration and biodegradation. A review. Environmental Chemistry Letters. 2020;**18**(4):1113-1143. DOI: 10.1007/s10311-020-01000-1

[27] Nghiem LD, Fujioka T. Removal of emerging contaminants for water reuse by membrane technology. In: Emerging Membrane Technology for Sustainable Water Treatment. Amsterdam, Netherlands: Elsevier; 2016. pp. 217-247. DOI: 10.1016/B978-0-444-63312-5.00009-7

[28] Dhangar K, Kumar M. Tricks and tracks in removal of emerging contaminants from the wastewater through hybrid treatment systems: A review. Science of the Total Environment. 2020;**738**:140320. DOI: 10.1016/j.scitotenv.2020.140320

[29] Teh CY, Budiman PM, Shak KPY, Wu TY. Recent advancement of coagulation–flocculation and its application in wastewater treatment. Industrial & Engineering Chemistry Research. 2016;**55**(16):4363-4389. DOI: 10.1021/acs.iecr.5b04703

[30] Mohamed Noor MH, Wong S, Ngadi N, Mohammed Inuwa I, Opotu LA. Assessing the effectiveness of magnetic nanoparticles coagulation/flocculation in water treatment: A systematic literature review. International journal of Environmental Science and Technology. 2021;**28**(22):1-22. DOI: 10.1007/s13762-021-03369-0

[31] Rajan A, Sreedharan S, Babu V. Solvent extraction and adsorption technique for the treatment of pesticide effluent. Civil Engineering and Urban Planning: An International Journal (CiVEJ). 2016;**3**(2):155-165. DOI: 10.5121/civej.2016.3214

[32] Ngueagni PT, Woumfo ED, Kumar PS, Siéwé M, Vieillard J, Brun N, et al. Adsorption of Cu (II) ions by modified horn core: Effect of temperature on adsorbent preparation and extended application in river water. Journal of Molecular Liquids. 2020;**298**:112023. DOI: 10.1016/j.molliq.2019.112023

[33] Yaashikaa PR, Kumar PS, Varjani SJ, Saravanan A. Advances in production and application of biochar from lignocellulosic feedstocks for remediation of environmental pollutants. Bioresource Technology. 2019;**292**:122030. DOI: 10.1016/j.biortech.2019.122030

[34] Tran NH, Reinhard M, Gin KYH. Occurrence and fate of emerging contaminants in municipal wastewater treatment plants from different geographical regions-a review. Water Research. 2018;**133**:182-207. DOI: 10.1016/j.watres.2017.12.029

[35] Gogoi A, Mazumder P, Tyagi VK, Chaminda GT, An AK, Kumar M. Occurrence and fate of emerging contaminants in water environment: A review. Groundwater for Sustainable Development. 2018;**6**:169-180. DOI: 10.1016/j.gsd.2017.12.009

[36] Deng Y, Zhao R. Advanced oxidation processes (AOPs) in wastewater treatment. Current Pollution Reports. 2015;**1**(3):167-176. DOI: 10.1007/s40726-015-0015-z

[37] Donkadokula NY, Kola AK, Naz I, Saroj D. A review on advanced physico-chemical and biological textile dye wastewater treatment techniques. Reviews in Environmental Science and Bio/technology. 2020;**19**(3):1-18. DOI: 10.1007/s11157-020-09543-z

[38] Nasrollahzadeh M, Baran T, Baran NY, Sajjadi M, Tahsili MR,

Shokouhimehr M. Pd nanocatalyst stabilized on amine-modified zeolite: Antibacterial and catalytic activities for environmental pollution remediation in aqueous medium. Separation and Purification Technology. 2020;**239**:116542. DOI: 10.1016/j. seppur.2020.116542

[39] Borrego B, Lorenzo G, Mota-Morales JD, Almanza-Reyes H, Mateos F, López-Gil E, et al. Potential application of silver nanoparticles to control the infectivity of Rift Valley fever virus in vitro and in vivo. Nanomedicine: Nanotechnology, Biology and Medicine. 2016;**12**(5):1185-1192. DOI: 10.1016/j. nano.2016.01.021

[40] Li J, Zhou Q, Liu Y, Lei M. Recyclable nanoscale zero-valent iron-based magnetic polydopamine coated nanomaterials for the adsorption and removal of phenanthrene and anthracene. Science and Technology of Advanced Materials. 2017;**18**(1):3-16. DOI: 10.1080/14686996.2016.1246941

[41] Fagan R, McCormack DE, Dionysiou DD, Pillai SC. A review of solar and visible light active TiO_2 photocatalysis for treating bacteria, cyanotoxins and contaminants of emerging concern. Materials Science in Semiconductor Processing. 2016;**42**:2-14. DOI: 10.1016/j.mssp.2015.07.052

[42] Lu H, Wang J, Stoller M, Wang T, Bao Y, Hao H. An overview of nanomaterials for water and wastewater treatment. Advances in Materials Science and Engineering. 2016;**2016**:4964828. DOI: 10.1155/2016/4964828

[43] Nasrollahzadeh M, Sajjadi M, Iravani S, Varma RS. Carbon-based sustainable nanomaterials for water treatment: State-of-art and future perspectives. Chemosphere. 2021;**263**:128005. DOI: 10.1016/j. chemosphere.2020.128005

[44] Ouni L, Ramazani A, Fardood ST. An overview of carbon nanotubes role in heavy metals removal from wastewater. Frontiers of Chemical Science and Engineering. 2019;**13**(2):1-22. DOI: 10.1007/s11705-018-1765-0

[45] Al-Wafi R, Ahmed MK, Mansour SF. Tuning the synthetic conditions of graphene oxide/magnetite/ hydroxyapatite/cellulose acetate nanofibrous membranes for removing Cr (VI), Se (IV) and methylene blue from aqueous solutions. Journal of Water Process Engineering. 2020;**38**:101543. DOI: 10.1016/j.jwpe.2020.101543

[46] Rajabi M, Mahanpoor K, Moradi O. Removal of dye molecules from aqueous solution by carbon nanotubes and carbon nanotube functional groups: Critical review. RSC Advances. 2017;**7**(74):47083-47090. DOI: 10.1039/c7ra09377b

Chapter 10

Preparation and Evaluation of Hydrophobic Grafted Ceramic Membrane: For Application in Water Desalination

Sabeur Khemakhem

Abstract

A new inorganic hydrophobic porous membrane was prepared and applied in desalination with the air-gap membrane distillation process. Ceramic supports from low-cost natural Tunisian sand have been elaborated by the extrusion method. The microfiltration layer has been elaborated from ZrO_2 powder by slip casting technical using a solution of water, sand powder, and polyvinyl alcohol solution. The hydrophobic surface of the active layer was elaborated by grafting 1H,1H,2H,2H-perfluorodecyltriethoxysilane on the ceramic microfiltration membrane surface (Tunisian Sand/Zirconia), to prepare a hydrophobic surface. The contact angle method allows showing the hydrophobic nature on the grafted membrane surface since it increases from 25° before grafting to values exceeding 140° after grafting. The efficiency of the grafting process was characterized by scanning electron microscopy (SEM). The membrane permeability varies from 700 $l.h^{-1}.m^{-2}$ before grafting to 10 $l.h^{-1}.m^{-2}$ after grafting. The new hydrophobic membrane seems to be promising in the field of membrane distillation. Salt retention higher than 98% was obtained using a modified microfiltration ceramic membrane.

Keywords: membrane, ceramic, hydrophobic, Tunisian sand, desalination

1. Introduction

Currently, several searches are focusing on the different applications of ceramic membranes [1–3]. The major application fields for ceramic membranes are water purification, pharmaceutical industries, and biotechnology. In this research work, we have prepared a ceramic microfiltration membrane that used the mud of hydrocyclone laundries of phosphates as support, which has very interesting properties in terms of mechanics, chemical resistance, and thermal compared with commercial inorganic ceramic membranes based on pure oxide mineral material such as zirconia, titania, and alumina [4]. Because of the presence of surface hydroxyl groups, these ceramic materials originally have a hydrophilic appearance that can form bonds easily with water molecules [1, 5, 6]. Modifying the hydrophilic character into hydrophobic ceramic membrane is very interesting for several applications. The grafting process can be performed by reaction between the -OH surface groups of the microfiltration layer and the ethoxy (O-Et) groups of the organosilane, leading to the increase of the hydrophobic properties [5, 7, 8]. Changing the hydrophobicity of the microfiltration layer surface would expand the potential applications of this material. The modification of the microfiltration layer surface allows forming a monomolecular layer of organosilane nature on the ceramic microfiltration layer [9, 10]. Fluoroalkylsilanes are organic groups used to modify the nature of different surfaces and create hydrophobic properties [11–13]. Several researchers have used the transplant process of different fluoroalkylsilanes on the surface of zirconia, silica, titania, and alumina membranes and their applications in different fields [5, 8, 14–16]. The modified inorganic ceramic membranes will be used, for example, in pervaporation and membrane distillation processes. Due to its hydrophobic character, the membrane will not be wet by the aqueous supply solution; then filling of the pores will occur due to capillary forces. The membrane distillation is a thermally driven separation process; in this case, only vapor is transported through a membrane. The membrane distillation technique allows potential applications in several fields of great scientific and industrial interest. It has been used in the treatment and purification of thermally sensitive industrial products such as the concentration of aqueous solution for wastewater treatment [17], water desalination [18–21], fruit juices [22] and in the pharmaceutical industry [23]. The technique of air-gap membrane distillation configuration was used in this work because it is simple to be performed in the laboratory, and it generally produces high water permeate flux. Inorganic ceramic membranes have several advantages in the application of oil-water separation, such as high mechanical strength, chemical inactivity, and thermal stability. Currently, a large amount of oily wastewater is discarded by food industries, petrochemical, and pharmaceutical, leading to further pollution of the environment. Widely used methods of purification of oily wastewater, such as air flotation, gravity separation, coagulation, skimming, and flocculation, have many disadvantages such as high operating costs, low efficiency, corrosion, and recontamination problems [24, 25]. The most widely used techniques are based on the membrane separation process such as dehydration of oil emulsion by pervaporation [26], by reverse osmosis [27], flocculation followed by microfiltration [28], microfiltration alone [29], membrane distillation [30], and ultrafiltration [31].

In this work, a microfiltration membrane prepared of sand/zirconia was surface modified with grafted perfluoroalkyl-silane. Ceramic membranes surface grafted with perfluoroalkyl-silane changed the hydrophilic character into a hydrophobic one. The prepared hydrophobic ceramic membrane was used for water desalination for the air-gap membrane distillation process. It is noted that fresh water can be produced from NaCl solutions as well as seawater.

2. Membrane distillation process

2.1 Principle

Membrane distillation is a thermal membrane process in which a hydrophobic membrane separates a cold and hot stream of water [32, 33]. The hydrophobic propriety of the membrane stops the passage of liquid water through the pores while allowing the passage of water vapor. The temperature shift between the two membrane surfaces produces a vapor pressure gradient that allows water vapor to cross through the membrane, and finally, the water vapor is condensed on the colder surface.

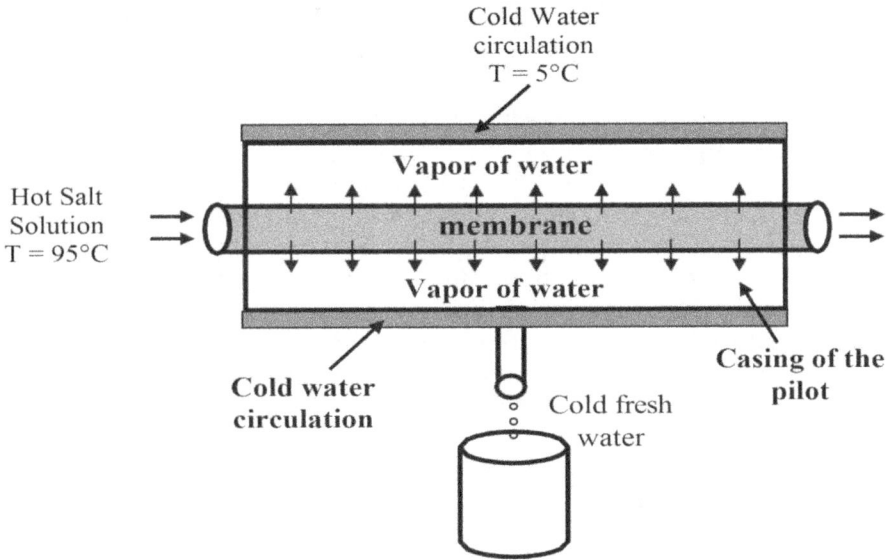

Figure 1.
Schematic presentation of AGMD in counter current flow configuration.

Finally, we obtained a distillate of very high purity. Membrane distillation may be used in various modes differing, for example, in a protocol of permeates collection [34].

2.2 Air-gap membrane distillation process

In the technical of air-gap membrane distillation (**Figure 1**), only the effluent to be treated is in contact with the active membrane layer. After crossing the membrane, the vapor permeates passing through the membrane and is condensed on the cold outer surface. There is an air distance between the outer surface of the membrane and the cold surface to reduce the loss of energy by thermal conduction through the membrane. The main male of the air space is that there is an additional resistance to mass transfer. In this technique of distillation, permeate is not in contact with the outer surface of the membrane, there is no fear of wetting the membrane on the permeate side in this process.

3. Experimental

3.1 Reagents

A composite microfiltration membrane mud of Tunisian sand/zirconia with pore diameters in the range of 0.2 μm entirely prepared in our laboratory [35] was used. Every grafting process was performed using 10^{-2} mol.l^{-1} solution of 1H,1H,2H,2H-perfluorodecyltriethoxysilane (97%) supplied from Sigma-Aldrich, it was dissolved in ethanol with an analytical grade of 95% purchased from Chemi-Pharma.

3.2 Preparation and characterizations of hydrophobic membranes

Hydrophobic membranes were prepared by grafting C8 onto the sand/zirconia ceramic membrane. Grafting is performed with several condensation reactions between the OH groups found on the membrane surface and the silane functional group. Samples of flat membranes and tubular membranes are fully immersed in fluoroalkylsilanes solutions for 1 h at room temperature. After drying the samplers at 90°C for 1 h, the grafted ceramic membranes were rinsed successively in ethanol and acetone and finally placed in a 100°C oven for 1 h. The contact angles were measured at room temperature (23°C) using a Dataphysics OCA 15 camera with a resolution of 752 to 582 square pixels. The camera is equipped with a CCD camera and operates at an acquisition rate of four frames per second. The surfaces of the samples before and after grafting were characterized by scanning electron microscopy (SEM) (Hitachi S-4500). The pore diameter of the modified microfiltration layer was determined through the nitrogen adsorption/desorption isotherm using the 2010 Asap Micro Metrics Gas Analyzer, the exact diameter of the pores was determined using the BJH (Barret–Joyner–Halenda) method [36]. The permeability was obtained for membranes grafted by a home-made pilot plant.

3.3 Membrane distillation

The setup scheme in **Figure 1** was used for the application of grafted membranes in the air-gap membrane distillation process. In this work, the solution to be treated was heated in a stainless steel feed tank and then distributed in the filtration module. The working temperature on the power side ranged from 75 to 95°C, while on the cooling side was kept constant at 5°C. Permeate vapor will be condensed on a surface of cooled stainless steel near the membrane (**Figure 1**). The temperatures were measured with two thermometers located at the feed cell frame and at the cooling plates. It is noted that each analysis was performed at least twice. The water permeates flux was determined by measuring the volume of distilled water permeate as a function of time.

3.4 Saline water

Both kinds of saline water were used: NaCl solutions prepared by using deionized water and pure NaCl (ProLabo) in the concentrations range of 0.5–3 mol $NaCl.L^{-1}$ and seawater from SIDI MANSOUR Sea located in the city of Sfax (Tunisia) with salt concentration at about 0.5 $mol.L^{-1}$. The saline water was heated in a feed tank and then circulated through the membrane module in the air-gap membrane distillation configuration process. The feed velocity of the used circulation water is 2.6 $m.s^{-1}$. The feed pan temperature ranges from 75 to 95°C, while keeping the cooling system side temperature constant at 5°C.

4. Results and discussion

4.1 Scanning electron microscopy (SEM)

The surface quality and morphology of the modified and unmodified ceramic microfiltration layer were examined by scanning electron microscopy. **Figure 2** shows pictures of the surfaces of grafted and ungrafted microfiltration layers. The obtained photo shows that the modified surface is homogeneous without defects and was completely covered with fluoroalkylsilane (**Figure 2b**). We note the same that grafting

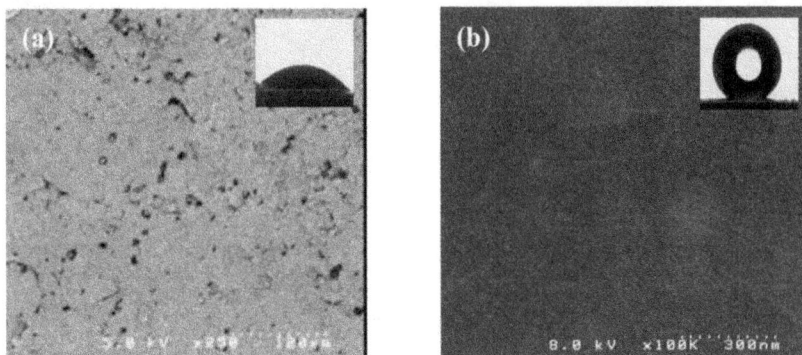

Figure 2.
SEM photo of surface of ungrafted (a) and grafted (b) membrane.

perfluorodecyltriethoxysilane on the surface of the membrane (Sand/Zirconia) led to a sharp decrease in pore size (**Figure 3a** and **b**).

4.2 Determination of pore size

The method based on N_2 adsorption and desorption was used for the determination of the pore size of the modified zirconia membrane. **Figure 3** showed a type IV isotherm with hysteresis behavior associated with capillary condensation of the adsorbate in mesopores. This is done because the smaller pores became completely filled with liquid nitrogen by reducing the saturation vapor pressure, according to the Kelvin Eq. [37]. The pore diameters measured are in order of 10 nm. The decrease in the pore size of 0.22 μm before grafting to values of 10 nm after grafting clearly confirms the densification of the membrane surface shown by SEM (**Figure 3**).

4.3 Contact angle measurement

The hydrophobic character of the ceramic membrane was determined by measuring the contact angle of the water drop. The low contact angle of the membrane, which is approximately 18° (**Figure 4**), is attributed to the very hydrophilic character of the membrane due to the high density of the hydroxyl group on the surface of the membrane. After grafting, the value of contact angle increased exceeding 170°, which confirmed that the grafted membrane acquired a very hydrophobic character (**Figure 4**). After grafting the membrane surface, the measured contact angle increased by more than 170°. This result confirms that the grafted membrane has become very hydrophobic (**Figure 5**). In addition, the C8 used in this research work has a long alkyl chain of eight carbon atoms, leading to a very high increase in surface hydrophobicity.

4.4 Cross-flow filtration experiments

Water permeability measurements of the membrane were determined to assess and demonstrate the hydrophobic character of the membrane after grafting. For this, a test with grafted and ungrafted membranes was achieved. The water permeability determination of the non-grafted membrane is of the order of 720 $L.h^{-1}.m^{-2}.bar^{-1}$.

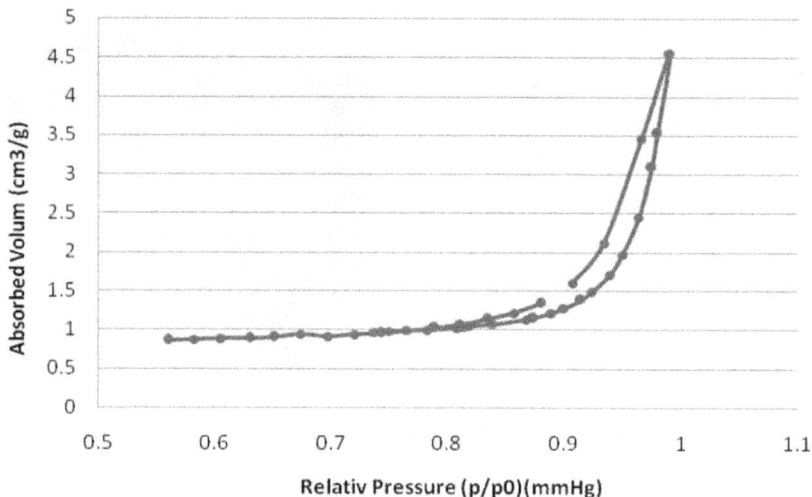

Figure 3.
Nitrogen adsorption-desorption of modified zirconia membrane.

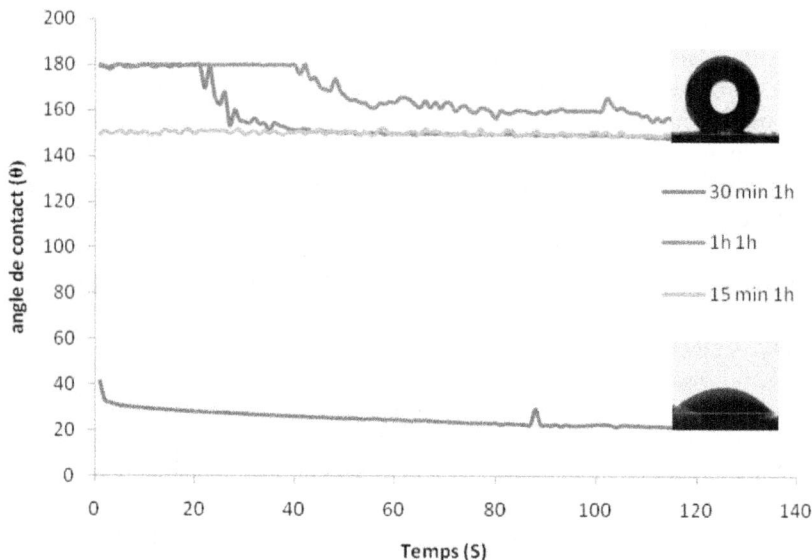

Figure 4.
Time dependence of contact angles for the modified and unmodified membranes.

After grafting, there was a very large reduction in permeability, indeed for the micro-filtration membrane grafted only 7 L.h^{-1}.m^{-2}.bar^{-1} was obtained (**Figure 6**). So, we can say that the grafted molecules were responsible for reducing the size of the pores causing the decrease in membrane permeability, which reflected the efficiency of the graft (C8) on the zirconium oxide membrane.

4.5 Membrane distillation process of saline water

Air interval membrane distillation experiments were conducted in saline water through the prepared hydrophobic membrane. The evolution of permeate flow and

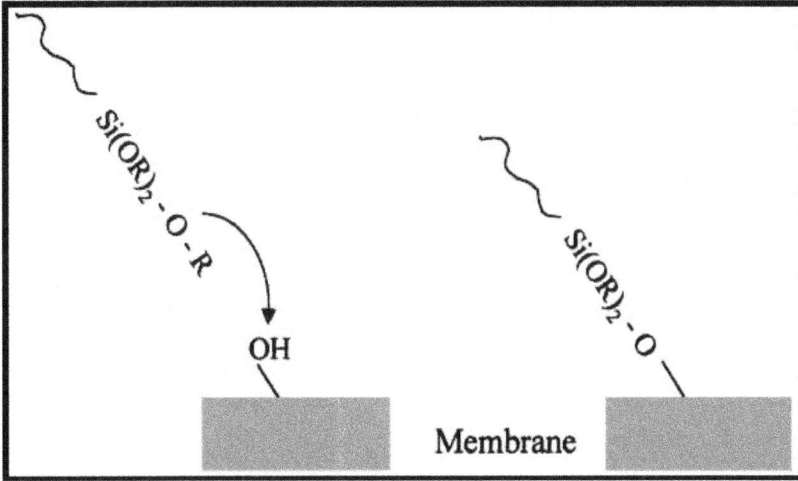

Figure 5.
Schematic representation of grafting process.

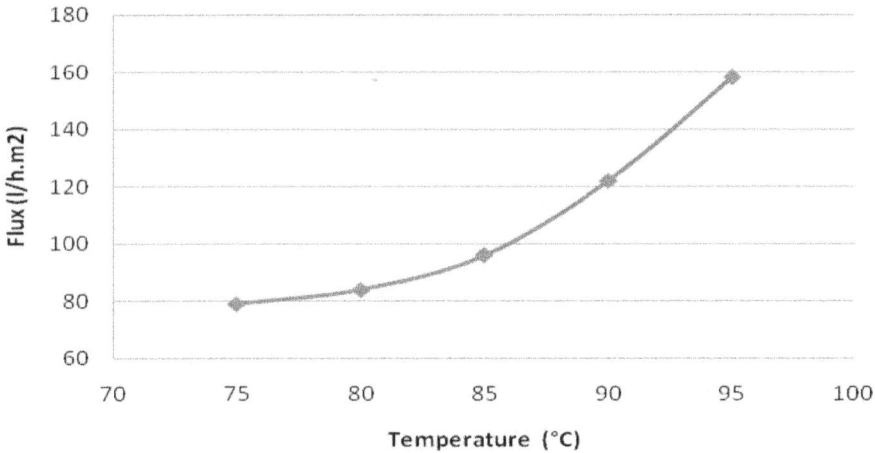

Figure 6.
Variation of the flux, with the bulk feed temperature in desalination of a 1 M NaCl solution by AGMD using modified zirconia MF membrane.

discharge rates with temperature was then determined. The feed side temperature varied from 75 to 95°C, while keeping the cooling system temperature.

4.6 Effects of temperature

The effects of temperature on permeate flow and release rates in the distillation of the air spacer membrane for aqueous solutions of NaCl were determined using the modified zirconium membrane microfiltration membrane at a feed rate of 2.6 m.s^{-1}. The temperature on the feed side varied from 75 to 95°C; keeping the cooling system temperature constant at 5°C. **Figure 6** shows the flux variations of a 1 mol.L^{-1} permeate of NaCl solution examined at different temperatures. Increasing the temperature of the source solution from 75 to 95°C led to an increase of permeate flux from 76 to

155 L.day^{-1}.m^{-2} for modified zirconia membrane, in what the vapor pressure differences increase with the increase of the temperature. The effect of variable temperature on the permeates of aqueous solutions with different concentrations of NaCl was investigated using the modified zirconia membrane. The results of all these experiments are presented in **Figure 7**. As observed, for each temperature, the permeate flow increases with the decrease in the concentration of the NaCl solution.

4.7 Effect of concentration

In order to study the effects of feed concentration on permeate flow, several series of experiments were carried out. The experimental conditions were cooling system was maintained at 5°C, a feed velocity of 2.6 m.s^{-1}, and a feed temperature of 95°C. We can observe from **Figure 8** that the permeate flux decreases when the feed NaCl

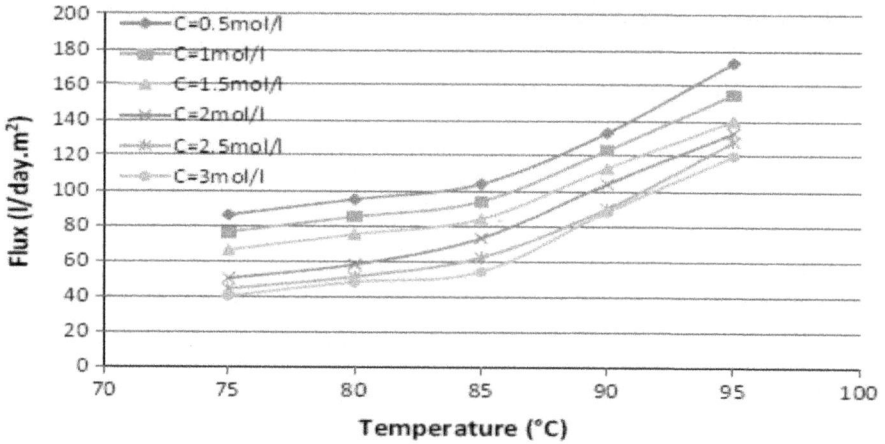

Figure 7.
Evolution of permeate flux as a function of the feed solution temperature for modified zirconia MF membrane at different NaCl concentrations.

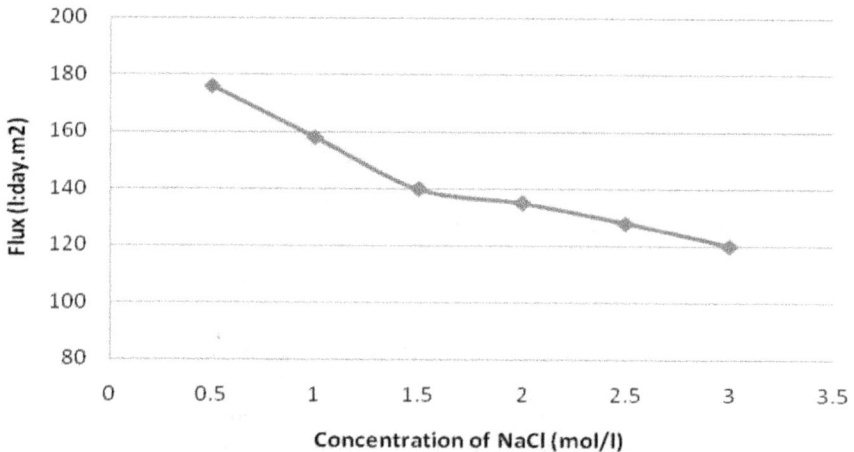

Figure 8.
Variation of the flux, with the NaCl concentration in desalination by AGMD using modified zirconia MF membrane.

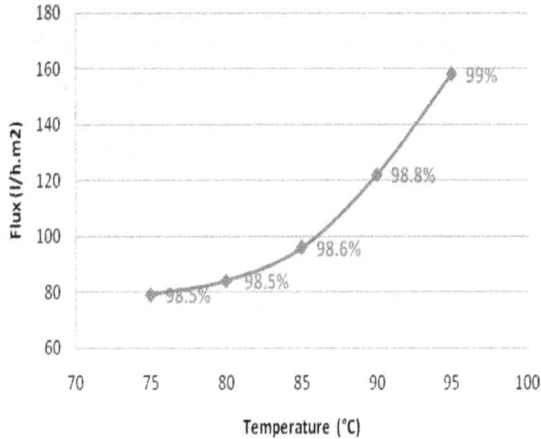

Figure 9.
Variation of the permeate flux as a function of the temperature. The values reported on the graph correspond to the rejection rates calculated for modified zirconia MF membrane after seawater filtration.

concentration increased from 0.5 to 3 mol.L^{-1}. Thus, increasing the salt concentration from 0.5 to 3 mol.L^{-1} led to a decrease of permeate flux from 173 to 120 L.day^{-1}.m^{-2} for modified MF membrane. Raoult's Law can be used to explain these results. From **Figure 8**, we observed that the air-gap membrane distillation efficiency decreased over 30% when the NaCl concentration increased from 0.5 to 3 mol.L^{-1}. According to Raoult's Law, the water vapor pressure over salt solutions is $P = P0^{*}(1 - Xsalt)$. It is concluded that the reduction of permeate flow to a percentage of 5 to 7% can be interpreted by this law. It should be noted that in all the results, the salt rejection was always greater than 99%. The rejection rate for MF-modified membrane is not modified when NaCl concentration varies.

4.8 Desalination of seawater

Seawater desalination aims to obtain fresh water for drinking. In this work, the treated seawater is taken from the sea of the region of SIDI MANSOUR Sfax (Tunisia). The measurements of the rejection rates and permeate flow were carried out by the filtered pilot used later. The feed side temperature was thus varied from 75 to 95°C, while keeping the cooling system temperature constant at 5°C. As it is shown in **Figure 9**, the rejection rate of NaCl is about 100% for microfiltration modified sand/zirconia membrane. These results proved that in the air-gap membrane distillation with aqueous solutions containing nonvolatile compounds such as NaCl, only water vapor is transported through the membrane.

5. Conclusion

Membrane distillation is a new technology used for desalination. This technique differs from other membrane technologies in that the driving force responsible for desalination is the difference in water vapor pressure across the membrane. In this research work, very encouraging results have been found for distillation experiments with a microfiltration modified sand/zirconia membrane. An important influence of the feed temperature and NaCl concentration on the permeate flux was observed. At the same time, very high salt rejection rates have been found in this research with grafted sand/zirconia ceramic membranes, the rejection rate of NaCl is about 100%. The membranes for membrane distillation are hydrophobic, which allows water vapor to pass. The vapor pressure gradient is created by heating the source water. It is expected that the total costs for drinking water with membrane distillation depend on the source of the thermal energy required for the evaporation of water through the membrane. Solar energy could very much help this process in our countries, which are very sunny resulting in a reduction of energy costs. Thus, membrane distillation could become competitive relative to other processes.

Author details

Sabeur Khemakhem
Faculté des Sciences de Sfax, Laboratoire des Sciences de Matériaux et Environnement, Université de Sfax, Sfax, Tunisie

*Address all correspondence to: khemakhem_sabeur@yahoo.fr

IntechOpen

References

[1] Larbot AB. Fundamentals on inorganic membranes: Present and new developments. Polish Journal of Chemical Technology. 2003;**6**:8-13

[2] Xiaowei W, Zhicheng Y, Li B, Tian G, Yuqin L, Meihua Z, et al. Fabrication of low cost and high performance NaA zeolite membranes on 100-cm-long coarse macroporous supports for pervaporation dehydration of dimethoxymethane. Separation and Purification Technology. 2022;**281**:119877

[3] Zhang Z, Haoyang Y, Mengxue X, Cui X. Preparation, characterization and application of geopolymer-based tubular inorganic membrane. Applied Clay Science. 2021;**203**:106001

[4] Khemakhem M, Khemakhem S, Ayedi S, BenAmar R. Study of ceramic ultrafiltration membrane support based on phosphate industry subproduct: Application for the cuttlefish conditioning effluents treatment. Ceramics International. 2011;**37**(8):3617-3625

[5] Picard C, Larbot A, Guida-Pietrasanta F, Boutevin B, Ratsimihety A. Grafting of ceramic membranes by fluorinated silanes: Hydrophobic features. Separation and Purification Technology. 2001;**25**:65-69

[6] Ye Z, Oriol R, Yang C, Sirés I, Li X-Y. A novel NH_2-MIL-88B(Fe)-modified ceramic membrane for the integration of electro-Fenton and filtration processes: A case study on naproxen degradation. Chemical Engineering Journal. 2022;**433**:133547

[7] Khemakhem S, Ben Amar R. Modification of Tunisian clay membrane surface by silane grafting: Application for desalination with Air process. Colloids and Surfaces, A: Physiochemical and Engineering Aspects. 2011;**387**:79-85

[8] Krajewski SR, Kujawski W, Dijoux F, Picard C, Larbot A. Grafting of $ZrO2$ powder and $ZrO2$ membrane by fluoroalkylsilanes. Colloids and Surfaces, A: Physiochemical and Engineering Aspects. 2004;**243**:43-47

[9] Schondelmaier D, Cramm S, Klingeler R, Morenzin J, Zilkens C, Eberhardt W. Orientation and self-assembly of hydrophobic fluoroalkylsilanes. Langmuir. 2002;**18**:6242-6245

[10] Yoshida W, Cohen Y. Topological AFM characterization of graft polymerized silica membranes. Journal of Membrane Science. 2003;**215**:249-264

[11] Fadeev AY, Yaroshenko VA. Wettability of porous silicas chemically modified with fluoroalkylsilanes according to data on water porosimetry. Colloid Journal. 1996;**58**:654-657

[12] Akamatsu Y, Makita K, Inaba H, Minami T. Water-repellent coating films on glass prepared from hydrolysis and polycondensation reactions of fluoroalkyltrialkoxysilane. Thin Solid Films. 2001;**289**:138-145

[13] Faibish RS, Cohen Y. Fouling-resistant ceramic-supported polymer membranes for ultrafiltration of oil-in-water microemulsions. Journal of Membrane Sciences. 2001;**185**:129-143

[14] Larbot A, Gazagnes L, Krajewski S, Bukowska M, Kujawski W. Water desalination using

ceramic membrane distillation. Desalination;**168**(2004):367-372

[15] Picard C, Larbot A, Tronel-Peyroz E, Berjoan R. Characterisation of hydrophilic ceramic membranes modified by fluoroalkylsilanes into hydrophobic membranes. Solid State Sciences. 2004;**6**:605-612

[16] Janknecht P, Widerer PA, Picard C, Larbot A. Ozone–water contacting by ceramic membranes. Separation and Purification Technology. 2001;**25**:341-346

[17] Filho CMT. Regilene de Sousa Silva, Carolina D' Ávila Kramer Cavalcanti, Miguel Angelo Granato, Ricardo Antonio Francisco Machado, Cintia Marangoni, Membrane distillation for the recovery textile wastewater: Influence of dye concentration. Journal of Water Process Engineering. 2022;**46**:102611

[18] Alklaibi AM, Lior N. Membrane-distillation desalination: Status and potential. Desalination. 2005;**171**:111

[19] Cabassud C, Wirth D. Membrane distillation for water desalination: How to choose an appropriate membrane. Desalination. 2003;**157**:307

[20] Chouikh R, Bouguecha S, Dhahbi M. Modeling of a modified air gap distillation membrane for the desalination of seawater. Desalination. 2005;**181**:257

[21] Cath TY, Dean Adams V, Childress AE. Experimental study of desalination using direct contact membrane distillation: A new approach to flux enhancement. Journal of Membrane Science. 2004;**228**:5

[22] Calabro V, Jiao BL, Drioli E. Theoretical and experimental study on membrane distillation in the

concentration of orange juice. Industrial Engineering Chemical Research. 1994;**33**:1803

[23] El-Bourawi MS, Ding Z, Khayet M. A framework for better understanding membrane distillation separation process. Journal of Membrane Science. 2006;**285**:4

[24] Daiminger U, Nitsch W, Plucinski P, Hoffmann S. Novel techniques for oil/water separation. Journal of Membrane Science. 1995;**99**:197-203

[25] Honga A, Fane AG, Burford R. Factors affecting membrane coalescence of stable oil-in-water emulsions. Journal of Membrane Science. 2003;**222**:19-39

[26] Van Hecke W, Joossen-Meyvis E, Beckers H, Wever HD. Prospects & potential of biobutanol production integrated with organophilic pervaporation – A techno-economic assessment. Applied Energy. 2018;**228**:437-449

[27] Mohammadi T, Kazemimoghadam M, Saadabadi M. Modeling of membrane fouling and flux decline in reverse osmosis during separation of oil in water emulsions. Desalination. 2003;**157**:369-375

[28] Zhong J, Sun X, Wang C. Treatment of oily wastewater produced from refinery processes using flocculation and ceramic membrane filtration. Separation and Purification Technology. 2003;**32**:93-98

[29] Ohya H, Kim JJ, Chinen A, Aihara M, Semenova SI, Negishi Y, et al. Effects of pore size on separation mechanisms of microfiltration of oily water using porous glass tubular membrane. Journal of Membrane Science. 1998;**145**:1-14

[30] Gryta M, Karakulski K, Morawski AW. Purification of oily

wastewater by hybrid UF/MD. Water Research. 2001;**35**:3665-3669

[31] Tang YP, Paul DR, Chung TS. Free-standing graphene oxide thin films assembled by a pressurized ultrafiltration method for dehydration of ethanol. Journal of Membrane Science. 2014;**458**:199-208

[32] Siefan A, Rachid E, Elashwah N. Faisal Al Marzooqi, Fawzi Banat, Riaanvan der Merwe, Desalination via solar membrane distillation and conventional membrane distillation: Life cycle assessment case study in Jordan. Desalination. 2022;**522**(15):115383

[33] Lawson KW, Lloyd DR. Membrane distillation. Journal of Membrane Science. 1997;**124**:1-25

[34] Gryta M. Osmotic MD and other membrane distillation variants. Journal of Membrane Science. 2005;**246**(2):145-156

[35] Khemakhem M, Khemakhem S, Ben Amar R. Surface modification of microfiltration ceramic membrane by fluoroalkylsilane. Desalination Water Treatment. 2014;**52**:1786-1791

[36] Anderson MA, Gieselmann MJ, Xu QY. Journal of Membrane Science. 1988;**39**:243

[37] Gregg SJ, Sing KSW. Adsorption, Surface Area and Porosity. London: Academic; 1982

Section 5

Challenges for Wastewater Treatment

Chapter 11

Fundamentals and Practical Aspects of Acid Mine Drainage Treatment: An Overview from Mine Closure Perspective

Gonzalo Montes-Atenas

Abstract

Acid mine drainage (AMD) is perhaps one of the most relevant challenges the mining industry has faced during the last few decades. This issue is particularly important in the scenario of mine closure where mining processes cease to be active, and the sustainability of the sites needs to be re-established. This chapter reviews the fundamentals behind the generation of AMD as well as a set of physicochemical phenomena (chemisorption, precipitation, neutralisation, etc.) usually considered by researchers to mitigate it. Mine closure conditions where human presence is seldom or frankly rare turn the wastewater treatment even more challenging as it cannot be intensive in the utilization of reagents, energy, or human resources. Therefore, from a practical standpoint, passive-like wastewater treatment strategies mimicking nature are preferred. Finally, insights with regards to the complexities behind the implementation of pilot plant and industrial wastewater treatment systems conformed by long-term reactive barriers and constructed wetlands are also revised.

Keywords: acid mine drainage, mine closure, heterogeneous reactions, reactive barriers, wetlands

1. Introduction

During last few decades, there has been an increasing awareness among the scientific community about the impact of carrying out mining activities [1, 2]. Before implementing standard ore exploitation activities, potential contaminant species remain restrained inside the original rock, however, such situation changes once mining activities kickoff and valuable material along with other toxic species are mobilised throughout the atmosphere or other media such as surface- or groundwaters. Among the latter, Acid Mine Drainage (AMD) has arisen as one of the most relevant multidisciplinary challenges in the mining industry [3]. The AMD corresponds to an aqueous stream which appears spontaneously from the natural contact, and therefore the natural interaction, between the surface of the rocks (or mineral particles) exhibiting at their surface primarily metal sulphide structures, and water

IntechOpen

either in the form of vapour or liquid in conjunction with other atmospheric gases (**Figure 1**) [4, 5].

Perhaps the major difference between AMD and other sorts of pollution is that the former is not directly produced by mining activities. Mining activities would inevitably produce, to some extent, solid wastes and then, the environment in contact with them would eventually trigger the generation of AMD. In other words, the misplacing of solid wastes coming from anthropogenic mining activities in nature itself spontaneously transforms it into a different system with increased toxicity. In this context, the appearance of AMD depends largely on the local atmospheric conditions. For instance, higher humidity or rainy weather will favour the generation of AMD compared to dry conditions [4].

From a historical standpoint, one of the first reports indicating the generation of AMD was published in 1895 by F.G. Holman who glanced at the presence of a waterflow coming out from a small mine site in Forbestown, Sierra Nevada, California [6]. During AMD formation several physical, chemical, and biological phenomena are triggered while solid wastes and environment interact and are commonly summarised by many authors as simple as "weathering" [7, 8]. Weathering, though, is a wide concept that encompasses all the characteristics related to the environment including climate and biosphere. This makes it a bit too general to fully predict the specifics of AMD (timespan to appear, chemical composition, etc.) and its instantaneous or long-term impact on the mine site surrounding areas. The locations where mine sites are placed present a variety of different climates like desertic, Mediterranean, or other. Therefore, when carrying out any study on AMD prediction, prevention, treatment or other, the ambient conditions used will be crucial to get proper results [9].

Before examining the fundamentals behind the generation of AMD, a brief analysis of where AMD might be generated from a mineral processing perspective will be presented. There are many situations where AMD may appear across the mineral processing line, especially when solid wastes appear. For instance, it is well known that base metals occurrence covers a wide range of mineral structures such as oxides, sulphides, and intermediate phases [10]. Although oxide minerals bearing ores may also produce AMD due to their susceptibility to undertake leaching and metal hydrolysis steps in aqueous aerated conditions, sulphide-bearing minerals are considered the major ones responsible for it. In that context, AMD is mainly associated with base

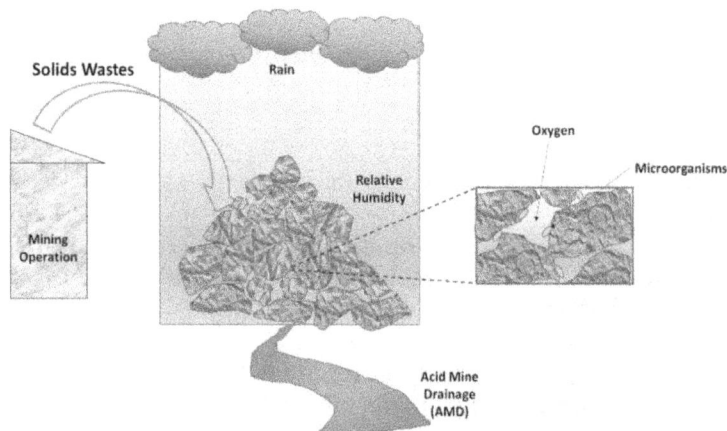

Figure 1.
Scheme of acid mine drainage (AMD) production.

metals and coal beneficiation plants where sulphide minerals occur as valuable or gangue material [11]. Only for exemplification purposes, and given its worldwide relevancy, the copper sulphide pyrometallurgical beneficiation path will be discussed. The traditional copper sulphide line of process usually includes blasting, rock size reduction (crushing and milling) and mineral selective separation commonly froth flotation [12]. **Figure 2** presents a block diagram of that line of ore processing, identifying the most relevant scenarios where AMD may be anticipated to take place.

From a practical perspective, it is all about sulphide-bearing material in the form of particles with different diameters being piled up which generate a certain natural porosity that will determine the access of atmospheric gases (or material weathering) towards the interior of the porous material. Different particle size distributions can be observed across the process line characterised in the picture by the maximum particle size only. Another common and better way to estimate a mean diameter biased to larger diameter values of a set of rocks or particles is through the Sauter mean diameter (d_{32}) [14]. As expected, the Sauter mean diameters follow a similar trend to the maximum particle diameter [Eq. (1)].

$$d_{32, \ ROM} \geq d_{32,pebbles} \geq d_{32,tailings} \tag{1}$$

where,

$$d_{32} = \frac{\sum_i d_i^3}{\sum_i d_i^2} \tag{2}$$

Eq. (2) is normally preferred as it considers the whole population of rocks or particles which brings up the significance of a correct sampling procedure, another crucial aspect of AMD. The mean diameter of a population of rocks or particles is

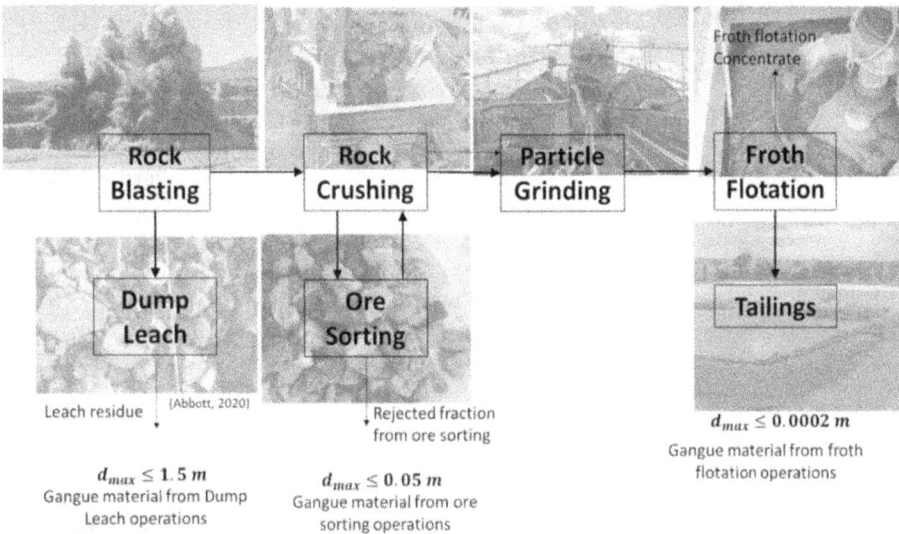

Figure 2.
Possible ore processing situations where AMD may appear, and the maximum diameter of rocks/particles involved in each scenario. Particle diameter is referred from [13].

studied with different techniques depending on the relevant particle sizes. For instance, after blasting operation optical methods are widely used while for tailing materials sieving processes are preferred [15]. Nevertheless, particle size is not the only key variable to look at. The content of the valuable and gangue species present in the solid waste is also a variable to take into consideration. For instance, copper grades of sulphide minerals bearing ores fed to a concentrator usually contain around 0.8% copper leaving final tailings with about 0.1% of the metal [16]. Low-grade copper sulphide ores fed to dump leach operations contain copper grades between 0.1 and 0.3% and the extraction may reach values of around 50% [17]. The requirements for ore sorting vary from one case to another but it is common to impose a cut-off grade of around 0.2% [15].

2. Fundamentals: physical, chemical, and biological aspects involved in AMD generation

2.1 Some physical aspects

From a mineral processing standpoint, the bigger the particles or rocks the less the chances of sulphide minerals from getting exposed to the environment. Such exposition is commonly associated with the concept of liberation. The liberation of a specific mineral was originally defined as the particle size threshold allowing the generation of particles composed of only one mineral [18]. From this definition, the liberation of each mineral present in an ore should be different as the presence of each mineral (or occurrence) changes from one to another. Such definition has been modified in time to better explain the efficiency of processes involved in metallurgy or mineral separation stages where the composition of the surfaces of particles is critical for their success [19]. For any geological occurrence of sulphide minerals (or any other mineral) the smaller the particle size the higher the liberation expected. The higher the liberation, the larger the exposure to the environment of surfaces containing such minerals and therefore the higher the chances of producing AMD. Nevertheless, there is in a way a trade-off between AMD flow rates and the potential of producing AMD.

Figure 3 shows two sets of particles having significant differences in particle diameter. The group of bigger particles will exhibit larger pore sizes although the air hold up may be similar for both case scenarios [20]. The set of particles exhibiting coarser sizes presents higher permeability than that of smaller sizes. Permeability depends on both static and dynamic properties of the porous medium and fluid characteristics. In the case of finer particles, capillary forces are more relevant which allow retaining more volume of the aqueous phase inside the porous material leaving the fluid phase and dissolved species to be transported slowly across through diffusive mechanisms favouring the acidification of the aqueous phase [21]. Coarser particles can produce higher flow rates reducing the residence time of fluid in contact with the surface of the particles. Then, there might be situations where the porous structure is fulfilled or saturated with water and the permeability can be modelled using the Darcy equation, however, for most case scenarios unsaturation would be frequently observed. So, permeability would be better modelled by the Soil-Water Critical Curve (SWCC) curve (**Figure 4**) which can be evaluated using three ranges of pressure values [Eqs. (3)–(5)] [22].

The parameters w_u and w_{aev} represent the water content for 1 kPa suction and for air-entry value, respectively. The first section [Eq. (3)] close to the y-axis represents

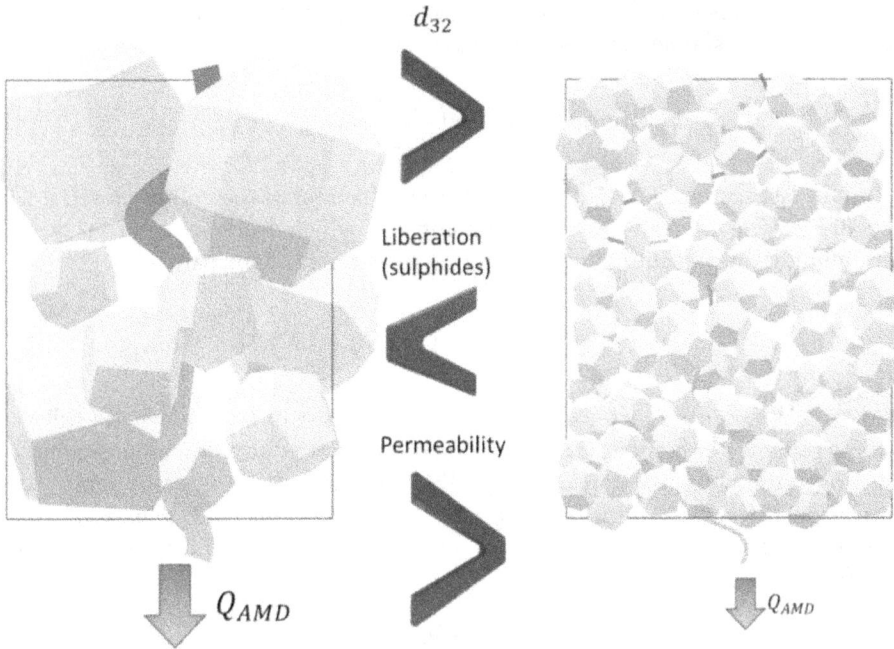

Figure 3.
The impact of particle size distribution on mineral liberation and on permeability and its association with AMD generation.

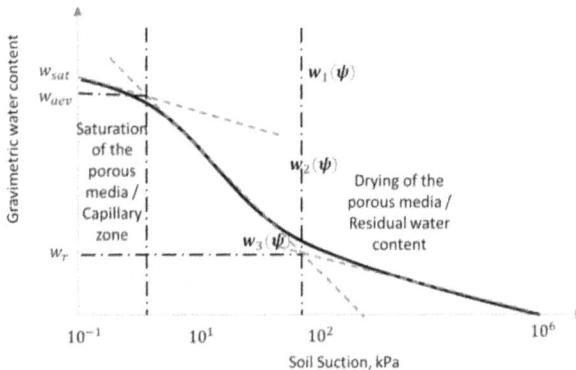

Figure 4.
SWCC Curve used to model the water content in an unsaturated porous media.

the zone where the porous media is fulfilled with the aqueous phase which could represent the case of tailings coming from flotation operations when they have been freshly disposed of.

$$w_1(\psi) = w_u - S_1 \log(\psi) \quad 1 \ \psi < \psi_{aev} \tag{3}$$

The second section [Eq. (4)] located immediately to the right-hand side represents the behaviour of the porous media when the air gets inside displacing the water present in the pores which could correspond to an intermediate case observed in tailings where, by simple syneresis, the water drains across the material.

$$w_2(\psi) = w_{aev} - S_2 \log \left(\frac{\psi}{\psi_{aev}} \right) \quad \psi_{aev} \; \psi < \psi_r \tag{4}$$

And the last section [Eq. (5)] represents the behaviour of the porous material when it gets dry leaving a certain residual water content, which describes a porous media where humidity is present mainly by wetting the surface of the particles.

$$w_3(\psi) = S_3 \log \left(\frac{10^6}{\psi} \right) \quad \psi_r \; \psi < 10^6 kPa \tag{5}$$

Eq. (5) corresponds to later stages of tailings, the case of larger particles still retaining some water content (i.e., material coming from dump leaching or ore sorting operations) or abandoned tailings. From a thermodynamic standpoint, the suction is a function of the partial pressure of the pore-water vapour and the density of the vapour which depends on the temperature and can be computed as in Eq. (6).

$$\psi = (pore - air\ pressure) - (pore - water\ pressure) + osmotic\ suction \tag{6}$$

And the permeability at residual water conditions can be computed from the matric (soil) suction as Eq. (7) [22].

$$k_w = \frac{k_s}{\left[\frac{u_a - u_b}{(u_a - u_b)_b} \right]^{n'} + 1} \tag{7}$$

One of the strong capabilities of Eq. (7) is that it correctly describes porous media with small particle sizes exhibiting lower permeability and higher water retention capacity [22, 23]. The latter would reduce the chances of producing large flow rates of AMD under these conditions [20]. Large flowrates would only then be possible in these conditions when water flows over the external surface of the piled-up porous material which reduces radically the exposed surface of the particles to the aqueous phase flowing around. Certainly, the magnitude of the drying conditions will depend on the water table present in each system. As it can be observed such description is based on semi-empirical mathematical models which is an indication that this is still a quite fruitful field of research.

Finally, the transport of liquid in porous materials built up of smaller particle sizes will expose the sulphide minerals in greater extension but they usually exhibit a higher hydrophobicity due to the presence of sulphur produced by oxidation reactions. This is known as natural hydrophobicity which occurs with much lower significance in the case of mineral oxides with stronger wettability properties. The liquid phase transferred through the porous medium needs to fill the voids displacing the air. Such subprocess is commonly referred to as imbibition and will be inhibited by the presence of sulphide hydrophobic surfaces which provide a first glance of how physics and chemistry are linked in these systems, but it is usually not considered. Indeed, physics and chemistry are frequently addressed by researchers separately. The relevancy of the chemistry and biology behind this process will be examined in the next subchapter.

2.2 Some chemical and biological aspects

There are several documents describing how sulphide minerals produce the so-called AMD, and the reader could refer to them for more information [4, 24]. Gas-solid and liquid-solid interactions are the major ones responsible for the significant differences between the chemical composition and structure of the bulk of the solid phase and the outmost surface layer arising from such interaction [25]. There are specific minerals that due to their instability under aerated conditions, notably metal sulphide minerals, are likely to produce enough acidity to stabilise several metals in dissolved state. Firstly, metal sulphide minerals would directly produce hydronium ions from their oxidation produced by the oxygen present in the atmosphere [Eq. (8)].

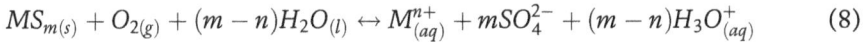

$$MS_{m(s)} + O_{2(g)} + (m-n)H_2O_{(l)} \leftrightarrow M^{n+}_{(aq)} + mSO_4^{2-} + (m-n)H_3O^+_{(aq)} \qquad (8)$$

Such acidity is then enhanced by metal hydrolysis reactions occurring at the bulk of the aqueous phase. Hydrolysis can be represented by Eq. (9).

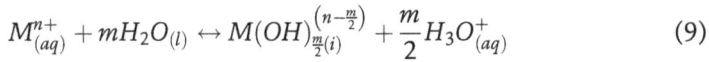

$$M^{n+}_{(aq)} + mH_2O_{(l)} \leftrightarrow M(OH)^{\left(n-\frac{m}{2}\right)}_{\frac{m}{2}(i)} + \frac{m}{2}H_3O^+_{(aq)} \qquad (9)$$

where the variable $m = 0, 2, 4, 6, 8, etc.$ Depending on the concentration, i may refer to a solid phase for $n = m$ leading to the precipitation of the metal hydroxide. If the dissolved metal is polyvalent, it may undergo subsequent oxidation stages due to the presence of dissolved oxygen in the system. The latter may raise other stronger mechanisms of oxidation of the sulphide minerals. That is the case of dissolved iron which can go from Fe(II) to Fe(III) in acidic aqueous solutions due to dissolved oxygen reduction. The latter increases the rate at which the metal sulphide dissolves producing more acidity simultaneously rising the concentration of sulphate ions in solution.

2.3 Application to pyrite and marcasite (FeS_2)

Probably the most reported case study that exemplifies AMD generation is that of pyrite and marcasite (FeS_2) which encompasses a series of processes that up to now are not fully understood, especially with regards to the state of surface or surface mediator being formed [26].

In any case, the oxidation reaction is usually described as Eq. (10).

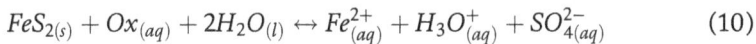

$$FeS_{2(s)} + Ox_{(aq)} + 2H_2O_{(l)} \leftrightarrow Fe^{2+}_{(aq)} + H_3O^+_{(aq)} + SO_4^{2-}{}_{(aq)} \qquad (10)$$

In Eq. (10) the oxidant, represented by the symbol Ox, can be oxygen or ferric ions if the thermodynamic potentials are suitable. The $FeS_{2(s)}$ in the reactants is only an over-simplification of what is really happening. Indeed, Holmes and Crundwell (2000) succeed in describing the oxidation of pyrite using the classic mixed potential theory but claimed that the state of surface also plays a relevant role in the process and is still not well understood [27]. **Figure 5** attempts to summarise the kinetics of the major electrochemical reactions taking place in the system. The oxidation reactions of pyrite in presence of sulphate ions start at potentials about 0.54 V vs SHE, which as mentioned later in this text becomes wider in presence of chloride ions. The oxygen reduction reaction in acidic conditions exhibits a Nernst potential of about 1.23 V vs SHE. Such reduction reaction is known to be from an electrochemical point of view

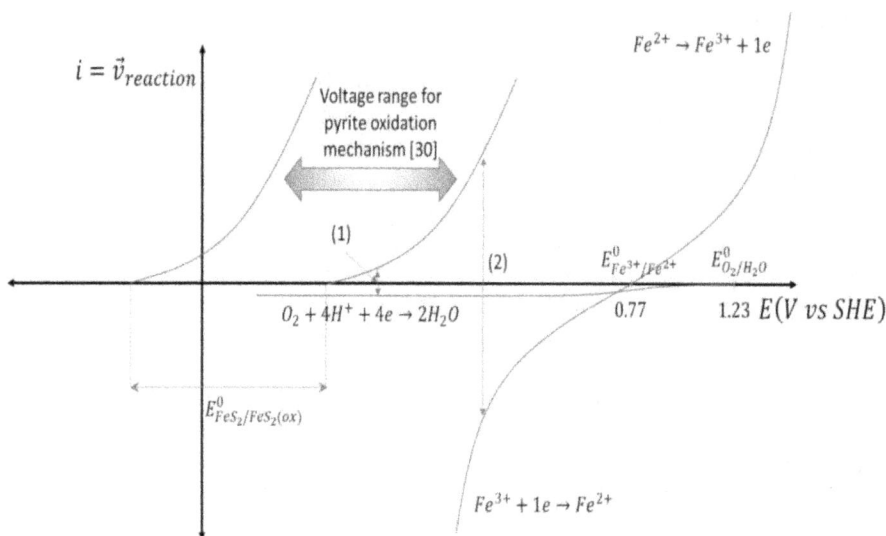

Figure 5.
Electrochemical scheme of the I vs E curve presenting the most relevant electrochemical reactions as well as the mixed potentials.

rather irreversible, so currents are only observed below 0.61 V vs SHE which is explained by the authors in terms of the low conductivity n-type pyrite semiconductor properties [28]. The latter is not observed in **Figure 5**. Instead, due to the significant reduction in overvoltage of the oxygen reduction, the current density would reach a maximum value of about 38.5 $mAcm^2$ considering a layer thickness of 0.05 cm, a concentration of 8 mgL^{-1}, and a diffusion coefficient of the unstirred aqueous phase of 2 $10^{-5}cm^2s^{-1}$. The current density observed by the authors is about 2% or that maximum value which is an indication that the diffusion layer thickness is much higher than that assumed. This only allows an exiguous net current that can be barely detected (in fact, in **Figure 5** it was enhanced for the reader to be able to see it!) at a mixed potential labelled as (1), which is controlled by the cathodic reaction [27]. The low current is also obtained in the situation presented by the authors where both cathodic and anodic reactions exhibit a thermodynamic potential close to each other. In fact, under these conditions, the reversibility of both reactions needs to be high to attain any relevant reaction rates. Other authors have indicated that the oxidation potential in presence of chloride ions, especially relevant in mineral processes implemented using seawater, ranges between −70 V and 530 V vs SHE [29]. Strikingly, under such a range of potentials, a series of reversible electrochemical adsorption/desorption steps take place whenever sweep rates of30 mVs^{-1} or higher is used. At sweep rates below 10 mVs^{-1} the authors found that oxidation steps are triggered indicating that precursors need time to be formed to undertake oxidation reactions. Under these experimental conditions, the pyrite oxidation takes place as a sequence of electron losses promoting the formation of several sulphite-like precursors occurring at both solid surface and bulk of aqueous phase.

Once ferric ions are formed in the aqueous phase, the oxidation rate of pyrite increases significantly not only because of the increase of the anodic overvoltage but also because of the high reversibility of the reduction reaction of ferric ions exhibits. Indeed, the mixed potential moves to higher voltages represented as (2) in **Figure 5**. Then, an accumulation of ferrous ions in solution may arise. The chances of regenerating the oxidant only by introducing oxygen would not be enough since the latter reaction still is mass transfer controlled. This is one of the most crucial issues the leaching of copper sulphide minerals presents which has been partially solved by microorganisms. In effect, it has been proved that bacteria, specifically, *thiobacillus ferroxidans*, would catalyse this reaction [30]. For many years it was not clear whether the catalysis is achieved by direct bacteria adsorption and oxidation of the mineral sulphide surface or indirect reaction through oxidation of ferrous ions happening at the bulk of the aqueous phase. Nowadays, such matter has been sorted out and it is known that it is the indirect oxidation mechanism that governs the oxidation of ferrous ions to ferric ions, and it constitutes the foundations of bioleaching of sulphide minerals [31]. During such studies on bioleaching, it has been learnt that the *thiobacillus ferroxidans* have a relatively narrow pH range in which they may adapt at their best and the presence of chloride ions at 2M or higher inhibits its growth [32].

Figure 6 summarises the main role of the two major types of bacteria, *thiobacillus ferrooxidans* and *thiobacillus thiooxidans*. Eqs. (10) and (11) are commonly coupled and a global reaction is obtained indicating how oxygen can oxidise elemental sulphur. These two reactions may also be studied separately. Dissolved oxygen oxidises a number of reduced chemical species such as Fe(II) or any other sulphide more susceptible to be oxidised than pyrite while sulphur can be at least oxidised by Fe(III). A simple stoichiometric analysis of both routes of reaction, without considering the cycle Fe(II)/Fe(III), indicates the simultaneous oxidation of 1 mol of elemental sulphur and the reduction of 2 mols of molecular oxygen would neutralise the local pH. In acidic aqueous solutions (as it may happen with AMD) it would be desirable to have more than 2 mols of oxygen reacting per mol of sulphur being oxidise. This rather simplistic analysis attempts to prove that the generation of AMD is not the inevitable outcome coming from these systems. Understanding and tuning the relevancy of the reactions at a fundamental level may also lead to inhibiting its formation.

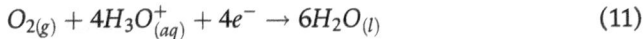

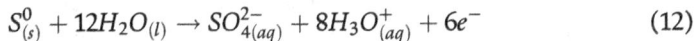

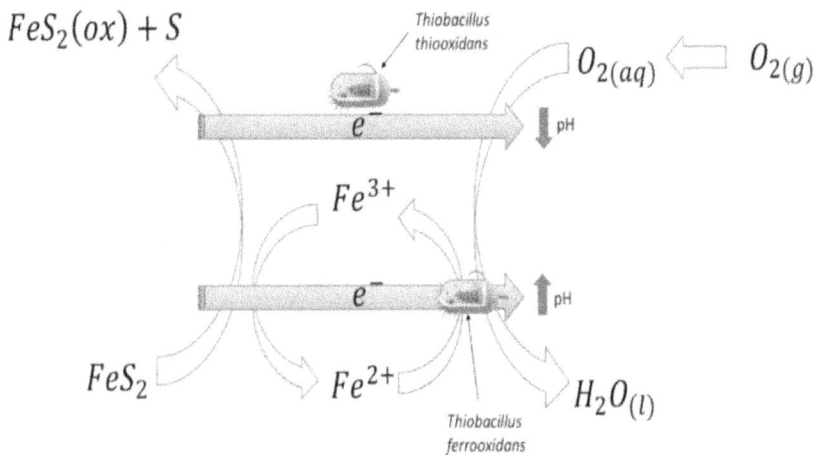

$$O_{2(g)} + 4H_3O^+_{(aq)} + 4e^- \rightarrow 6H_2O_{(l)} \tag{11}$$

$$S^0_{(s)} + 12H_2O_{(l)} \rightarrow SO^{2-}_{4(aq)} + 8H_3O^+_{(aq)} + 6e^- \tag{12}$$

Figure 6.
Conceptual simplified model for generation of AMD.

Another more realistic analysis would involve coupling these two reactions. In this case, 2 mols of elemental sulphur would react with 3 mols of molecular oxygen producing 4 mols of hydronium ions and 2 mols of sulphate which is also troublesome for AMD (to be discussed in Section 5). Simultaneous oxidation of 12 mols of ferrous ions using 3 mols of molecular oxygen would then neutralise the acidity provided by the overall sulphur oxidation reaction by oxygen.

Supposedly, in real systems oxygen is slowly transferred to surface sites inside the porous media where the interaction with sulphide minerals exposed would produce elemental sulphur, one of the most relevant products generated at the surface of the particles, which then is oxidized due to the presence of microorganisms. Additionally, bacteria require oxygen and eventually carbon dioxide for growth, which is also responsible for producing both sulphur and acid [33]. It looks like the key would lie in inhibiting the formation of elemental sulphur which is quite insoluble (about $0.6\ ngL^{-1}$) [34]. Nevertheless, the role of sulphur though is much more complicated than just the formation of sulphate ions [29]. Research studies have proved that thiosulphate would be an intermediate species that in acid media would be disproportionate to sulphite or sulphur dioxide and elemental sulphur, going back to the initial state. Then, handling better the presence of sulphite or sulphur dioxide then might be crucial. Thermodynamically, elemental sulphur could be avoided but to do that an increase in temperature or the reduction of total amount of sulphur in the system is needed which is something not easy to accomplish considering the scale at these systems are commonly implemented [35].

From all the above, it may infer that the chances of producing AMD cannot only be observed from a fluid dynamic or physical perspective. The gathering of key reactants needs to occur to produce it. Delays in the interaction between reactants given by the transport of oxygen, or carbon dioxide in less importance, will slow down the generation of acidity and therefore that of AMD.

It is precise because of this that many of the strategies to prevent AMD obey to block the reagents from coming into the porous material (**Figure 7**). For instance, with regards to the sulphide minerals present in the porous material authors have suggested removing it before piling up the solid wastes using froth flotation or any other selective separation

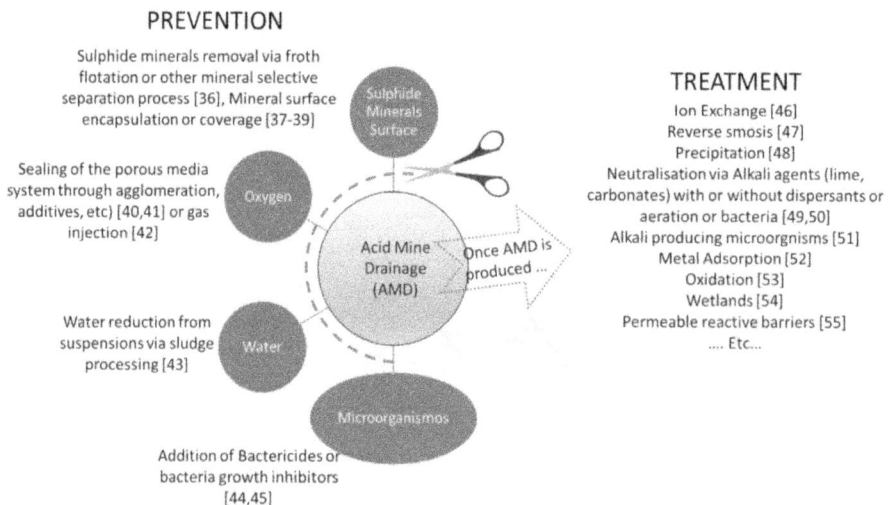

Figure 7.
Scheme of various strategies to prevent or treat AMD [36–55].

method [36]. Bear in mind that this requires preparing the material for the separation process such as milling to certain particle diameter and the use of appropriate reagents at certain dosages to run froth flotation operations adequately. Another option would be to cover up only the surface of the particles exposing the sulphide mineral to the gaseous phase to avoid any contact with oxygen inhibiting the AMD formation [40]. A similar but more extensive blockage would entail forming a cap enclosing the whole porous material preventing oxygen from entering the system [41]. Even more, some researchers have suggested using materials known as not acid producers (or NAP) and directly carrying out some neutralisation of the AMD. It also has been suggested to introduce some positive pressure on inert gases to keep the oxygen from entering the system [42]. From the use of water, perhaps the most studied strategy for suspensions of mineral particles involves reducing the water content of slurries to dose lower amounts of neutralisation reagents [43]. And finally, with respect to the microorganisms, it has been recommended the use of some bactericides to prevent the acid-forming bacteria to appear in the system [44] while other authors have focused their attention in using some bacteria growth inhibitors [45].

However, having implemented any of these paths to prevent AMD from appearing does not secure that it will not take place and if it occurs, several actions have been studied to deal with it. Furthermore, in many cases, these solid wastes are not appropriately disposed of over geomembranes or other impervious materials which forces continuously monitor surface- and ground-waters at the mining location outskirts. Whenever these waters acquire any properties resembling AMD, the wastewater treatment must act as a barrier to bringing the parameters of water quality back to their usual values. Even using such geomembranes does not ensure that the eventual AMD produced will be appropriately contained as the properties of these covers may also detrimentally evolve in time [56].

3. AMD in mine closure conditions

Mine closure is one of the subjects in the field of mine management that has gained notoriety over the last few decades [57]. The people's perspective of mining activities is negative when abandoned mine sites impact adversely the environment or the nearby communities [58]. **Figure 8A** presents the iconic case of the Grand Canyon, in the United States, where mining activities were developed between 1957 to early 1960s. It can be observed that a section of the original plant is still in place and the exploitation has increased the surface area exposed to the atmosphere. In the same picture the before and after of a gold exploitation mine site located in the northern part of Chile shows a similar situation. In this case, the open-pit mine, unfortunately, made the entire town of *Churrumata* move without procuring better conditions for the villagers (**Figure 8B**).

All mine sites have a definite lifetime. At the end of the mining exploitation, the site needs to be rehabilitated, ideally eluding any threat to the environment, or living organisms, vegetation, and nearby communities.

Behind this topic there are many concepts to address and it is difficult to summarise them in just a few lines. For instance, researchers have differentiated the terms mine closure and mine completion [59]. On one hand, mine closure is a procedure over a timespan where plant operation stops, and decommissioning is undertaken. On the other hand, mine completion refers to an aim of mine closure where the ownership is renounced by the mining lease and accepted by the next user of the land for a different purpose. These and other perspectives are still an ongoing theme for the whole society.

Figure 8.
Pictures of two abandoned mine sites. (A.1 and A.2) Iconic case of Grand Canyon, United States, Guano Point, Bat Cave Guano Mine active between 1957 and early 1960, nowadays it is a renown National Park, (B.1) Before and (B.2) after gold mining exploitation of an open pit mine built at the heart of the small town Churrumata *in the northern part of Chile, the mine site was active between 1984 and 2010 approximately.*

Despite the latter, for this quest to be successful, every government has stated a plan which considers not only technical requirements but also regulatory and legislation guidelines to prevent the sites to become hazardous reducing eventual further contamination in many years to come.

It was not necessary to go by a great deal of time before governments, mining companies and the whole society realised that the major threat behind the cease of a mining operation is the lack of planning. There are many reasons why mine sites close such as economics, geological, technical, regulatory, policy changes, social pressure, end of markets, etc. Therefore, it is not surprising that this stage of mine development requires a multidisciplinary set of actions. For instance, the actions relevant to the present subject might include:

     i. Focused brainstorming aims at identifying environmental values, gains or losses inherent to AMD.

ii. Preliminary evaluation of testing sulphide ore bodies for acid-based accounting and metals.

iii. Potential use of overburden to cap potential generation of AMD.

As the Mine Closure conditions refer to the situation where the mining activities cease to take place, it is expected that energy, material, or personnel are not going to be available to deal with AMD. Therefore, unassisted wastewater treatment needs to be implemented.

4. Major physicochemical phenomena used to mitigate the generation of AMD

The AMDs characteristics vary from site to site, however, for simplicity and without loss of generality the pH usually ranges between acid (close to pH 1 to 3) and slightly acid (around pH 5) which after wastewater treatment needs to reach pH values between 6.5 and 9 to be adequately discarded [60]. The acidity stabilises a concentration of certain heavy metals in aqueous solutions such as copper, zinc, chromium among others and metalloids like arsenic. Several physicochemical phenomena used to remove ions from an aqueous phase are succinctly reviewed in this section from both thermodynamic and kinetic classic modelling standpoints.

4.1 Neutralisation: coupled or not with chemical precipitation

Perhaps this is the most straightforward strategy to treat AMD wastewaters coming from mine sites. The idea is to remove from the aqueous phase dissolved species that might be toxic to human beings and the environment by adding salts with high saturation or solubility product constant (*Kps*) values but very small ones for the ions to be withdrawn. Considering that the AMD is acid, one of the priorities is to bring the pH of the water close to neutral and that is usually accomplished by dosing with an alkali reagent. The release of hydroxyl ions ($OH^-_{(aq)}$) to the aqueous solution enhances the hydrolysis of the metal ions promoting the generation of neutral species which will eventually precipitate. The most used reagent at an industrial scale is the quicklime which in presence of water is transformed into slaked lime [61]. In **Figure 9** a usual AMD containing a high concentration of acid ($H_3O^+_{(aq)}$), heavy metals ($M^{n+}_{(aq)}$) and sulphate ($SO^{2-}_{4(aq)}$) is presented. When adding quicklime, ultimately dissolved calcium ions would react with sulphate forming anhydrite-like structures while hydroxyl ions would form metal hydroxides and simultaneous adjust the pH to neutral values. The precipitation of sparingly soluble species would follow two sequential steps namely nucleation and crystal growth being the former, in many cases, more energy consuming than the latter. That is the reason why seeds and/or surface roughness of different materials would be desirable to improve the precipitation thermodynamic spontaneity and kinetics [62].

4.1.1 Thermodynamics

An initial concentration of metal is $C_{Me,0}$ and of a counter ion (anion, which can be sulphate) is $C_{ci,0}$. The flow rate of liquid enriched in these two ions is Q. For

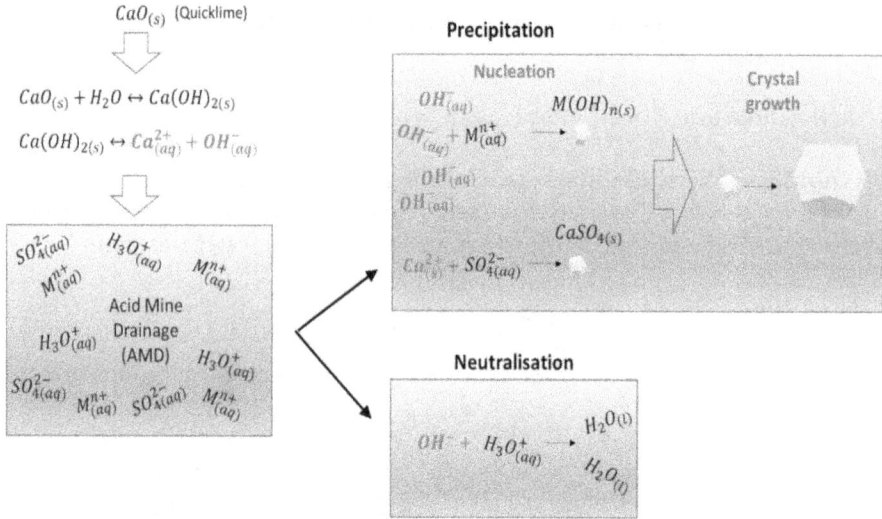

Figure 9.
Sketch of the role of different ions formed during the dosing of quicklime to acidic aqueous solutions.

computation purposes, the dose of leachate of a metal hydroxide (for simplicity, calcium hydroxide or slaked lime) of concentration C_x will be considered at a flow rate of q. Usually, as C_x is relatively high, the following flowrate relationship holds $q \ll Q$. The reagent with higher solubility compared to that of the salt to be precipitated, undertake an ionisation reaction such as that presented in Eq. (13).

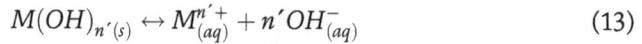

$$M(OH)_{n'(s)} \leftrightarrow M^{n'+}_{(aq)} + n'OH^{-}_{(aq)} \tag{13}$$

The mathematical condition required for the precipitation to occur is given by Eq. (13).

$$\frac{C_x Q_r}{Q} C_{ci,0} > Kps^{(1)} \tag{14}$$

Then, the efficiency of the treatment can be computed using Eq. (14).

$$\varphi = 100 \left\{ \frac{C_x Q_r}{2C_{ci,0}Q} + \frac{1}{2} \left[\sqrt{\left(\frac{C_x Q_r}{C_{ci,0}Q} - 1 \right)^2 - \frac{Kps^{(1)}}{C_{ci,0}^2}} - 1 \right] \right\} \tag{15}$$

The physical units of the different parameters in these equations need to be consistent. Plus, bear in mind that the efficiency of Eq. (15) is overestimated as only the major species were considered in this computation. Simultaneously, the precipitation of metals can be computed from Eq. (16). For the metal precipitation, Eq. (16) needs to be solved.

$$Kps^{(2)} = (C_{Me,0} - y)(n'C_x - ny)^n \tag{16}$$

Eq. (16) does admit an analytical solution only in very specific conditions, so it is better to solve it numerically. Although it looks like this strategy is quite promising as it would be able to remove sulphate, heavy metals and even neutralise the acidic conditions of the AMD, it is somehow misleading for at least four reasons [63, 64]:

i. The precipitates being formed are not necessarily stable.

ii. The dose of quicklime does not produce a perfect solution. It is most of the time leachate indicating that several particles are suspended and not instantaneously dissolved. This is to some point troublesome as the final pH will evolve in time reaching eventually much higher values.

iii. Even more, such loss of control of pH may lead to an over increase of pH generating anionic species of the metals which will be redissolved increasing again their concentration in the aqueous solution. Even more, in many cases, the minimum concentration of heavy metals reached by using quicklime does not satisfy the maximum concentration permitted by current regulations.

iv. Finally, there are many chances to get different quicklime dosing optimum points for metals and pH adjustment. Indeed, there are small chances that one single dosage of the reagent will exactly satisfy the minimum concentration of multiple heavy metals and the right pH at once. That is why this mechanism can only be used to approach best conditions for the correct discharge of the treated AMD.

During mine closure conditions this strategy is not directly recommended since the reagent dosing control is difficult to implement without personnel in place. However, the fundamentals behind this mechanism still hold. In the scenario of mine closure, the main idea would be to incorporate some sparingly soluble minerals with alkaline behaviour such as silicates, or others.

4.1.2 Kinetics

There are several studies focused on determining the dissolution rate of solids, especially that of quicklime in water. The first stage of dissolution is usually modelled by Eq. (17) [65].

$$\frac{da_i}{dt} = k(a_{i,sat} - a_i) \tag{17}$$

where a_i is the activity of the ion "i" as a function of time, $a_{i,sat}$ is the activity of the ion "i" in saturation conditions, t represents the time, and k is a specific rate constant obtained per unit of volume that has been defined as a function of the diffusion coefficient of the ion being transferred from the solid surface to the aqueous solution bulk (D), the thickness of the diffusion layer (δ), and the specific surface ($\frac{S}{V}$) as presented in Eq. (18) [66].

$$k = \left(\frac{S}{V}\right)\frac{D}{\delta} \tag{18}$$

The classic shrinking core model with reaction control can also be used (Eq. 18) [67–69].

$$1 - (1 - \alpha)^{\frac{1}{3}} = k_{cc}t \qquad (19)$$

where α represents the conversion of the reaction which corresponds to the volume fraction of solid that has been dissolved, and k_{cc} is the kinetic rate constant when the process is governed by chemical control.

4.2 Adsorption: Chemisorption

This mechanism aims at removing pollutants from an aqueous phase by fixing them onto a surface of a solid which is stable when immersed in the wastewater. The adsorption mechanism is one of the preferred reactions for wastewater treatment not only because low-cost adsorbents consisting of by-products or wastes from other industries may be used, but also because it may reach high removing efficiencies of dissolved molecules with final concentrations of a few parts per billion [69]. **Figure 10** presents several aspects to consider when picking up this mechanism. Different reactions between the adsorbent and the aqueous solution lead to the partial dissolution of the adsorbent affecting the local pH near its surface and its stability of the adsorbent suspension whenever forming small particles may occur [70].

There are several drawbacks behind the implementation of adsorption-based technologies. This technology requires optimising the contact between the solid phase and the aqueous phase containing the species to be adsorbed. Usually, piling up of adsorbent material in a column disposed of vertically or horizontally is preferred [71] but maintaining the permeability of the porous medium with time could become a challenge. It is also desirable to implement technologies using chemisorption rather than physisorption. Chemisorption has many advantages such as its specificity exemplified in **Figure 10** by the single and double binding shown for the metal and sulphate. That is, different adsorption sites would be used by different types of adsorbates reducing the competition for adsorption sites. Plus, the relatively high binding energy

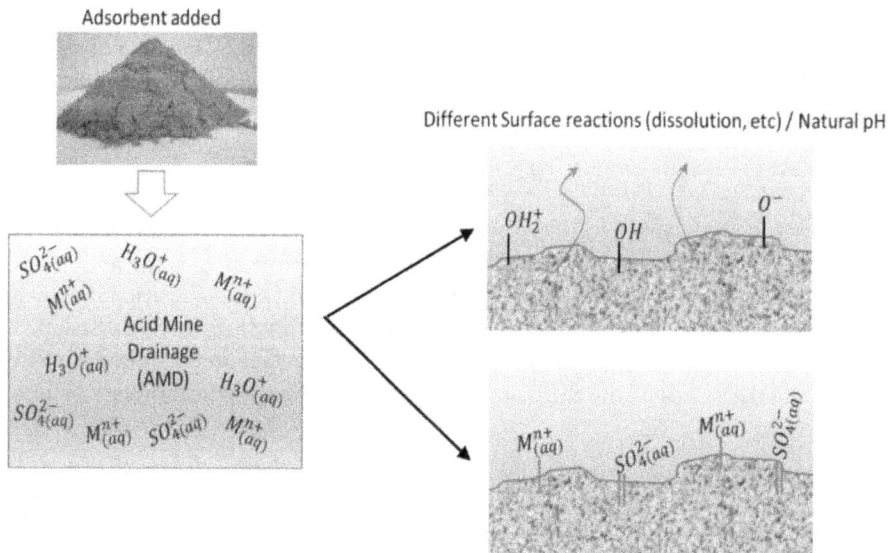

Figure 10.
Diagram of adsorption processes used in wastewater treatment.

associated with the adsorption process turns it quite irreversible from a kinetic stand-point which reduces the chances of pollutants desorption. Nevertheless, the main disadvantage would be that the eventual saturation of the adsorbent may be reached needing to move forward to a desorption stage to regenerate the adsorbent [72].

4.2.1 Thermodynamics

The thermodynamics of the adsorption process is explained in terms of the adsorption isotherm [73]. The adsorption isotherm is usually plotted in a graph where the y-axis represents the maximum quantity of adsorbed species per unit of the dry mass of adsorbent (also known as specific adsorption) while the x-axis presents the concentration of the species in equilibrium with the specific adsorption measured. All the data is obtained experimentally at a constant temperature and solids percent. There are many mathematical models that can be used to describe the process having each of the conditions and assumptions that as much as possible must represent the specifics of the process under study. The most used adsorption isotherms cited by researchers are the Langmuir and Freundlich isotherms as Eqs. (20) and (21) [74].

$$q_{i,eq,L} = \frac{S_m K_L a_{i,eq}}{1 + K_L a_{i,eq}} \tag{20}$$

where $q_{i,eq,L}$ corresponds to the volume (or mol) of adsorbate i at the surface of the adsorbent in equilibrium conditions, $a_{i,eq}$ is the equilibrium real concentration of the adsorbate in $molL^{-1}$, K_L is the Langmuir isotherm constant commonly associated with the binding energy, S_m es the amount of adsorbate required to form a monolayer.

$$q_{i,eq,F} = K_F a_{i,eq}^{1/n} \tag{21}$$

where K_F is the Freundlich empirical constant usually associated with the sorption capacity, and n is the sorption intensity.

4.2.2 Kinetics

One general case to model adsorption kinetics is presented in Eq. (23) [75]

$$r = \frac{(kinetic\ factors)(process\ potential)}{(adsorption\ factor)} \tag{22}$$

wherein the numerator there is a description of the classic law of mass action and in the denominator, the inhibition of the adsorption rate procured by the blockage of surface sites of other species in the system is incorporated. For example, the mathematical model for adsorption kinetics of one species labelled with the underscore "i" can be described as in Eq. (24) in the case of chemical reaction control.

$$r = k\theta_i = k\frac{K[a_i]}{1 + K[a_i] + \sum_{j \neq i} K_j[a_j]} \tag{23}$$

where k is the specific kinetic constant, θ_i is the fractional occupancy of adsorption sites by the main species "i", K and a_i are the Langmuir adsorption constant referred

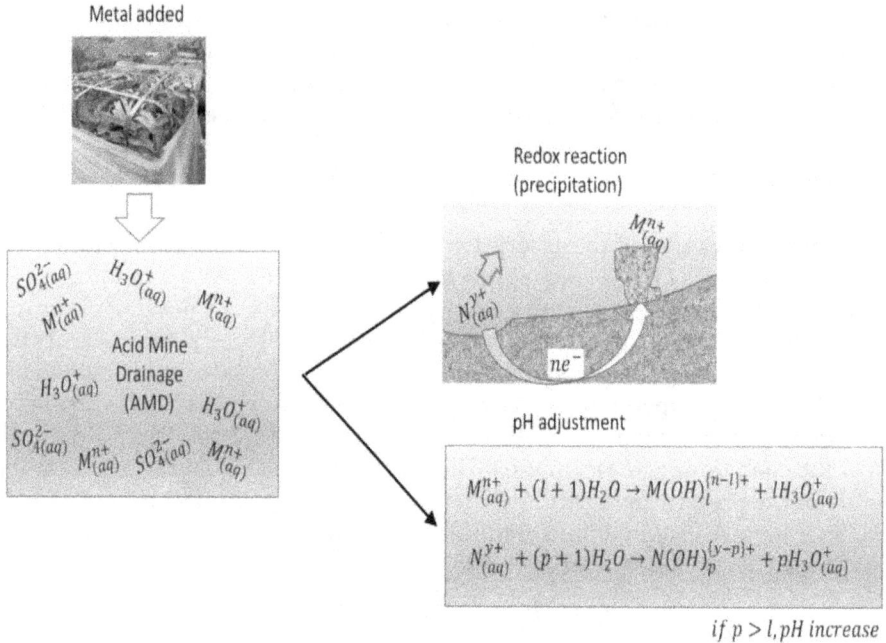

Figure 11.
Redox reactions in wastewater treatment.

to the main ion and its activity in the aqueous phase while K_j and a_j are their equivalent but for other ions competing for adsorption sites.

4.3 Redox reactions

This mechanism is highly valuable for certain ions which as product of the electron transfer type of reaction may directly precipitate or produce precursors for precipitation (**Figure 11**). The reduction of metals is somehow difficult to implement unless certain scrap of metallic wastes contain metals with low standard reduction potentials, also called less noble elements. If the metal is removed from the aqueous solution exhibits higher hydrolysis constants than the metal being released into the aqueous medium, the resulting pH of the solution should increase. One example of this would be the removal of polynuclear lead (II) ions by iron (II) ions [76–78].

4.3.1 Thermodynamics

The idea is to couple two electrochemical half-reactions, one half-reduction reaction and one half-oxidation reaction, having the former a higher standard reduction potential than the latter. As a requirement, since this wastewater treatment must occur without any energy input, the reaction must evolve spontaneously. Such a condition is presented in Eq. (25) [79].

$$E_{RED} - E_{OX} > 0 \tag{24}$$

Using Nernst equation, Eq. (25) would be of use to assess a first analysis of the impact of varying activities of different ions, partial pressures, or temperature on the spontaneity of the process can be assessed.

4.3.2 Kinetics

Electrochemical kinetics of spontaneous redox reactions are commonly studied in terms of the mixed potential theory and the corresponding current density. The mixed potential which is not a thermodynamic parameter is obtained from equalising the anodic and cathodic reactions [Eq. (26)]

$$I_{anodic} = |I_{cathodic}| \tag{25}$$

In a complex system, there could be several reduction reactions and oxidation reactions occurring simultaneously in different locations within the system, therefore several mixed potentials may be installed. The open rest potentials, in this case, will attempt to follow such mixed potentials and depending on the conductivity of the species formed at the solid interface and at aqueous solution bulk such tracking down will be faster or slower [80].

5. Practical aspects of semi-passive wastewater treatment: coupling long-term permeable reactive barriers with horizontally constructed sub-surface wetlands

This section is devoted to the design of AMD treatment systems which can operate in mine closure situations, close to stand-alone and unassisted systems to treat wastewaters. These treatment paths attempt to gather many of the mechanisms previously revised acting simultaneously to clean up AMD streams. Since most of these mechanisms need to take place spontaneously, many of these wastewater treatment systems attempt to mimic nature. Particularly, this subchapter addresses some aspects of two of these systems: Long-Term Permeable Reactive Barriers (LTPRB) and Horizontally Constructed Sub-Surface Wetlands (HCSSW).

On one hand, wetlands are one of the preferred passive wastewater treatment strategies to be implemented as a tertiary wastewater treatment [81]. Although it is considered a mature technology by many authors, it is one of the most difficult systems to model and understand. The latter is not only because of the many physicochemical interactions simultaneously taking place between all the species belonging to the system but also because of the multiple roles the local biota may play. For instance, sulphate ions are difficult to reduce using inorganic species only [82]. Indeed, looking at the Pourbaix diagrams, sulphate ions are stable over the whole pH range either in their acid form or not. It has been pointed out, though, that such reduction can be accomplished at the surface of organic material where carbon has a pivotal role. Indeed, it is well known that anaerobic systems, as well as aqueous media set in contact with solid metals, promote the growth of sulphate-reducing bacteria [83]. Carbon, among all its functions such as respiration, fermentation, methanogenesis, denitrification, and iron reduction would have a key role in sulphate reduction [84]. Sulphate reduction reactions are summarised by Eqs. (27), (28).

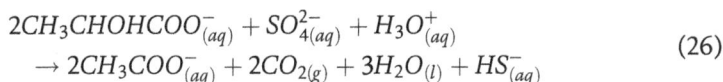

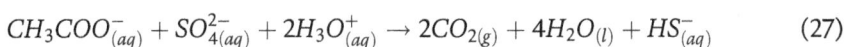

$$2CH_3CHOHCOO^-_{(aq)} + SO^{2-}_{4(aq)} + H_3O^+_{(aq)}$$
$$\rightarrow 2CH_3COO^-_{(aq)} + 2CO_{2(g)} + 3H_2O_{(l)} + HS^-_{(aq)} \tag{26}$$

$$CH_3COO^-_{(aq)} + SO^{2-}_{4(aq)} + 2H_3O^+_{(aq)} \rightarrow 2CO_{2(g)} + 4H_2O_{(l)} + HS^-_{(aq)} \tag{27}$$

These reactions, though, are not in total agreement with classic electrochemical fundamentals. The standard electrode potential of the sulphate reduction is -220 mV vs SHE which is not fully consistent with the stability region of sulphate ions declared in Pourbaix diagrams [35]. Authors have indicated that such reduction is complex and involves metastable products [35, 85]. The reduction reaction would then consist of at least two reactions in series which are triggered by the sulphate activation by ATP sulphurylase increasing the potential to about -60 mV where the reduction from sulphate to sulphite is achieved. However, the reduction of sulphite to sulphide is yet not fully understood [86]. In addition, another disadvantage of these two reactions is, in principle, the production of carbon dioxide identified as a greenhouse gas. Additionally, authors have pointed out that the low performance in eliminating phosphorous may also be observed for other contaminants which usually increase the requirements in terms of residence time and/or surface lands available to implement these systems [87]. Whenever these systems are not available naturally, constructed wetlands are engineering-designed which can be implemented vertically or horizontally [84]. The latter corresponds to the case study to be described in the next section.

On the other hand, long-term permeable reactive barriers have captured interest from the scientific community since it houses several materials to treat wastewaters securing the correct quality of groundwater resources. The phenomena embedded in this type of strategy are mainly chemical or biological degradation, precipitation, and adsorption to immobilise contaminants [88]. Due to the similarities in dealing with organic matter between reactive barriers and wetlands, sulphate reduction bacteria can also be promoted in these systems. Additionally, the permeability of these systems needs to be secured. Unreactive or low-reaction alkali materials are used as a fixed bed introducing more reactive materials inside the pores that can range from specifically designed materials to wastes from other industries such as ferrihydrite-bearing soils or nanostructured calcium silicate adsorbent, among others [89, 90]. Since the growth of vegetation is not present in these systems, the permeability may be designed to avoid dead volumes or volumes with low mixing capabilities. Long term reaction kinetics is still a matter to do research on. Considering that a few reactions are associated to oxidation mechanisms by oxygen, and given the relatively low concentration of the gas, particularly in low permeability media, atmospheric corrosion perspective could improve the knowledge on these matters [91].

6. Physicochemical complexities behind coupling long-term reactive barriers with horizontally constructed sub-surface wetlands

In a mine site located in the northern part of Chile, a pilot plant of AMD treatment designed by the company ISMP SpA consisting of LTPRB-HCSSW combined has been installed ideally to secure the water quality of surface waters for a timespan of a few years. It has been widely proved that different types of vegetation have absorption capabilities for different metals such as that shown in **Figure 12** which corresponds to Phragmites Australis.

The inlet pH was 5.0 and the aqueous solution flowrate is $1\ Ls^{-1}$. The pilot plant consisted of 30 m^3 effective volumes of LTPRB-HCSSW combined. The system has been designed in a way to promote and enhance the rate at which sulphate reduction reaction [Eq. (29)] is produced triggering as a secondary mechanism the heavy metal precipitation as metal sulphides. To accomplish this, the acidity is provided by introducing metallic scrap into the LTPRB. The dissolution of the metallic scrap by oxygen

Figure 12.
Picture of the AMD treatment pilot plant system implemented before being covered (a), and close-up to one of the Phragmites Australis used in the HCSSW (b).

reduction is inhibited and it is expected to especially be driven by complexation reactions.

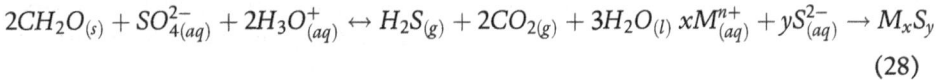

$$2CH_2O_{(s)} + SO_{4(aq)}^{2-} + 2H_3O_{(aq)}^+ \leftrightarrow H_2S_{(g)} + 2CO_{2(g)} + 3H_2O_{(l)} \; xM_{(aq)}^{n+} + yS_{(aq)}^{2-} \rightarrow M_xS_y$$
$$(28)$$

with

$$mx - 2y = 0$$

Metal sulphide precipitation, though, is required to be formed as much as possible at pH values where hydrolysis of sulphide ions is low which could be accomplished by evaluating the competitiveness for sulphide complexation within the system. Otherwise, the acidification of the aqueous phase could again take place following Eq. (30).

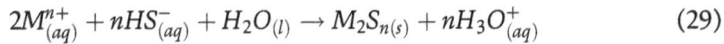

$$2M_{(aq)}^{n+} + nHS_{(aq)}^- + H_2O_{(l)} \rightarrow M_2S_{n(s)} + nH_3O_{(aq)}^+ \qquad (29)$$

Preliminary results indicate that LTPRB removes sulphate at between 10 and 30 times the rate reported for sulphate removal observed using wetlands only [92, 93]. The HCSSW allowed stabilising of the pH between 6 and 8. Preliminary computations indicate that the volume of control used is about one or two orders of magnitude lower than classic wastewater treatments. All these systems are complex by nature, but they could be engineering-designed from the beginning to enhance/inhibit reactions to avoid AMD. Now, considering the residence time of the AMD flowing through the system, is there any chance to adjust all these mechanisms to act standing alone at the appropriate rates enabling a wastewater treatment to last for a few years by itself keeping as much as possible the permeability of the porous media? This is certainly an opportunity still to be accomplished.

7. Conclusions

Acid Mine Drainage (AMD) formation is yet a process ill-understood. The AMD occurs spontaneously, and it is highly dependent on the local atmospheric conditions which makes it difficult to predict any of its characteristics. Preliminary strategies aiming at forming caps around these solid wastes could be considered a good first step towards preventing the formation of AMD. Nevertheless, in cases where AMD is already formed new strategies for isolating the wastes need to be considered.

Although the precursors of AMD such as sulphide minerals, and notably pyrite, water and oxygen are known to be involved, the physical chemistry and the biology linked to its production need to be studied in more detail and integrated, particularly in long-term reaction kinetics of the different mechanisms taking place.

Several strategies have been suggested to treat AMD. The condition of mine closure takes this challenge to the next level requiring a solution that cannot be intensive in the use of personnel, energy, or reagents.

Strategies involving passive wastewater treatment technologies which attempt to somehow mimic natural systems look promising.

Nowadays, the difference between passive and active wastewater treatment has become a thin line. On one hand, even passive wastewater treatment strategies require to some point the involvement of human resources. On other hand, new long-term permeable reactive barriers have been pointed out as wastewater treatment strategies than can gather several aspects of passive treatment systems such as low maintenance requirements to work and, simultaneously, exhibit fast wastewater treatment kinetics. Some practical aspects associated with implementing long-term permeable barriers coupled with constructed wetlands were presented but improvements with regard to the efficiency of these strategies to remove sulphate, heavy metals and other contaminants are still a matter of study.

Finally, perhaps the most relevant conclusion that can be drawn from this chapter is that addressing AMD generation, prevention or treatment is in fact a multidisciplinary topic where the conjunction of many specialities occurs such as chemistry, physics, hydrology, microbiology, electrochemistry, among others.

Acknowledgements

The author would like to dedicate and acknowledge the contribution to this study of Professor Fernando Valenzuela Lozano. For his invaluable friendship, his continued mentorship and all the technical and inspiring discussions across many years already, regarding hydrometallurgy and especially wastewater treatment of AMD. I would like to specially acknowledge the helpful input of Dr. Marcelo Sepulveda and Mr. Cesar Arredondo to this work. In addition, many thanks to Mr. Mario Solari and Mr. Thomas Ph. Chirino for providing pictures from industrial case scenarios and real-life portraits which had significantly increased the value of this manuscript. And, last but not least many thanks to Ms. Ana Maria Rojo for her assistance and good ideas to put together this chapter.

Conflict of interest

The author declares no conflict of interest.

Notes/thanks/other declarations

Not Applicable.

Author details

Gonzalo Montes-Atenas
University of Chile, Santiago, Chile

*Address all correspondence to: gmontes@ing.uchile.cl

IntechOpen

References

[1] Allan RJ. Impact of mining activities on the terrestrial and aquatic environment with emphasis on mitigation and remedial measures. In: Förstner U, Salomons W, Mader P, editors. Heavy Metals: Environmental Science. Berlin: Springer; 1995

[2] Palmer MA, Bernhardt ES, Schlesinger WH, Eshleman KN, Foufoula-Georgiou E, Hendryx MS, et al. Mountaintop mining consequences. Science. 2010;**327**:148-149

[3] Evangelou VP. Pyrite Oxidation and Its Control, Solution Chemistry, Surface Chemistry, Acid Mine Drainage (AMD), Molecular Oxidation Mechanisms, Microbial Role, Kinetics, Control, Ameliorates and Limitations, Microencapsulation. Boca Raton, Florida, USA: CRC Press, Taylor and Francis Group; 1995. pp. 1-293

[4] Geller W, Klapper H, Salomons W. Acidic Mining Lakes, Acid Mine Drainage, Limnology and Reclamation. Berlin: Springer-Verlag; 1998. pp. 1-435

[5] Weiss FT, Leuzinger M, Zurbrugg C, Eggen RIL. Chemical pollution in low- and middle-income countries. Swiss Federal Institute of Aquatic Science and Technology. 2016;**4**:67-101

[6] Holman FG. Notes on certain water-worn vein-specimens. Am Inst. Mg. Engn. Trans. Atlanta Meeting. 1985;**25**: 514-518

[7] Raymond PA, Oh N-H. Long term changes of chemical weathering products in rivers heavily impacted from acid mine drainage: Insights on the impact of coal mining on regional and global carbon and sulfur budgets. Earth and Planetary Science Letters. 2009;**284**: 50-56

[8] Matsumoto S, Shimada H, Sasaoka T. Interaction between physical and chemical weathering of argillaceous rocks and the effects on the occurrence of acid mine drainage (AMD). Geoscience Journals. 2017;**21**:397-406

[9] Dold B. Acid rock drainage prediction: A critical review. Journal of Geochemical Exploration. 2017;**172**: 120-132

[10] Tarbuck EJ, Lutgens FK, Tasa D. Earth: An Introduction to Physical Geology. 12th ed. New Jersey: Pearson; 2014. p. 875

[11] Wang Z, Xu Y, Zhang Z, Zhang Y. Review: Acid Mine Drainage (AMD) in Abandoned Coal Mines of Shanxi, China. Water. 2021;**13**:8

[12] Dunne RC, Kawatra SK, Young CA, editors. SME Mineral Processing and Extractive Metallurgy Handbook. Colorado, USA: Society for Mining, Metallurgy, and Exploration, Inc.; 2019. p. 2312

[13] Mineral Comminution Circuits. Their operation and optimization, Julius Kruttschnitt Mineral Research Centre Monographs, p. 250

[14] Sauter J. Determining the efficiency of atomization by its fineness and uniformity. In: Forsschingsarbeiten auf dem Gebiete des Ingenieurwesens No. 279, 1926. Vol. 396. Washington: Technical Memorandums National Advisory Committee for Aeronautics; 1927. p. 24

[15] Gupta A, Yan D. Editors, Mineral Processing Design and Operations. 2nd ed. Cambridge, Massachusetts, USA: Elsevier; 2016. p. 850

[16] Zanin M, Grano S. Benchmarking the flotation performance of ores. Minerals Engineering. 2012;**26**(1):70-79

[17] Rosenbaum JB, McKinney WA. In situ recovery of copper from sulfide ore bodies following nuclear fracturing. In: Symposium on Engineering with Nuclear Explosives, Las Vegas, Nevada. 1970. p. 877

[18] Barbery G. Mineral Liberation Measurement, Simulation ad Practical Use in Mineral Processing. Editions GB: Quebec; 1991. p. 351

[19] Wills B, Napier-Munn T. Mineral Processing Technology: An Introduction to the Practical Aspects of Ore Treatment and Mineral Recovery. Maryland Heights, MO: Elsevier Science & Technology Books; 2006. p. 444

[20] Are KS. Biorchar and soil physical health. In: Abrol V, Sharma P, editors. Biochar: An Imperative Amendment for Soil and the Environment. Rijeka, Croatia: Intech Pub.; pp. 21-34

[21] Lee T-K, Ro H-M. Estimating soil water retention function from its particle-size distribution. Geosciences Journal. 2014;**18**:219-230

[22] Fredlung DG, Rahardjo H, Fredlund MD. Unsaturated Soil Mechanics in Engineering Practice. New Jersey, USA: Wiley; 2012. p. 926

[23] Lu N, Likos WJ. Unsaturated Soil Mechanics. 1st ed. New Jersey, USA: Wiley; 2004. p. 556

[24] Evangelou VP. Pyrite Oxidation and Its Control. Boca Raton, London, New York: CRC Press, Taylor and Francis Group; 1995. p. 293

[25] Montes-Atenas G, Mielckzarski E, Mielczarski JA. Composition and structure of iron oxidation surface layers

produced in weak acidic conditions. Journal of Colloid and Interface Science. 2005;**289**:157-170

[26] Singer PC, Stumm W. Acid mine drainage: Rate-determining step. Science. 1970;**167**:1121-1123

[27] Holmes PR, Crundwell FK. The kinetics of the oxidation of pyrite by ferric ions and dissolved oxygen: An electrochemical study. Geochimica et Cosmochimica Acta. 2000;**64**(2): 263-274

[28] Charlot G, Badoz-Lambling J, Tremillon B. Les reactions electrochimiques, Les methodes electrochimiques d´analyse. Paris: Masson et Cie Editeurs; 1958. p. 395

[29] Kelsall GH, Yin Q, Vaughan DJ, England KER, Brandon NP. Electrochemical oxidation of pyrite (FeS2) in aqueous electrolytes. Journal of Electroanalytical Chemistry. 1999;**471**: 116-125

[30] Nyavor K, Egiebor NO, Fedorak PM. Bacteria oxidation of sulfides during acid mine drainage formation: A mechanistic study. In: EPD Congress 1996, Warren GW, The Minerals, Metals and Materials Society. 1995. pp. 269-287

[31] Rawlings DE. Biomining: Theory, Microbes and Industrial Processes. New York: Springer-Verlag Berlin Heildelberg, GmbH; 1997. p. 302

[32] Vorreiter A, Madgwick JC. The effect of sodium chloride on bacterial leaching of low-grade copper ore. Proceedings of the Australian Institute of Mining and Metallurgy. 1982

[33] Adams DJ, Pennnington P, McLemoe VT, Wilson GW, Tachie-Menson S, Gutierrez LAF, et al. The role

of microorganisms in acid rock drainage. SME Annual Meeting. 2005:1-8

[34] Boulegue J. Solubility of elemental sulfur in water at 298 K. Short Communication, Phosphorus and Sulfur. 1978;5:127-128

[35] Bailey LK. Electrochemistry of Pyrite and Other Sulfides in Acid Oxygen Pressure Leaching, PhD thesis. The University of British Columbia; 1977. p. 138

[36] Bois D, Bussiere B, Kongolo M, Poirier P. A feasibility study on the use of desulphurized tailing to control acid mine drainage. CIM Bulletin. 2005; 98(20):1-8

[37] Nyavor K, Egiebor NO. Suppression of pyrite oxidation by fatty acid mine treatment. In: Hager J, Hansen B, Imrie W, Pusatori J, Ramachandran V, editors. Extraction and Processing for the Treatment and Minimization of Wastes. San Francisco, California, USA: The Minerals, Metals & Materials Society; 1993. pp. 773-790

[38] Warren LA, Haack EA. Microbial geoengineering in acid rock drainage (ARD), Waste Processing and Recycling in Mineral and Metallurgical Industries V. In: Fifth International Symposium, 43rd Annual Conference of Metallurgists of CIM. Hamilton, Ontario, Canada. pp. 501-507

[39] Tabelin CB, Corpuz RD, Veerawattananum S, Ito M, Hiroyoshi N, Igarashi T. Formation of Schwertmannite. like and scorodite like coatings on pyrite and its implications in acid mine drainage control. In: IMPC 2016, XXVIII International Mineral Processing Congress Proceedings. 2016. pp. 1-12

[40] Misra M, Kumar S, Neve C. Mitigation of acid mine drainage by

agglomeration of reactive tailings. In: EPD Congress. The Minerals, Metals and Materials Society; 1992. pp. 137-155

[41] Lamontagne A, Fortin S, Poulin R, Tasse N, Lefebvre R. Layered co-mingling for the construction of waste rock piles as a method to mitigate acid mine drainage—Laboratory Investigations. In: Fifth International Conference on Acid Rock Drainage. Denver, CO; 2000. pp. 1087-1094

[42] Ameglio L, Barrie H. Acid rock drainage prevention using inert gas mixture technology. In: Tailings and Mine Waste Management for the 21st Century. Sydney, NSW; 2015. pp. 85-95

[43] Dube C, Banerjee K. Sludge conditioning technology to reduce sludge and the cost of acid mine drainage treatment. In: Tailings and Mine Waste Management for the 21st Century. Sydney, NSW; 2015. pp. 105-109

[44] Kleinmann RLP, Erickson PM. Control of Acid Drainage from coal refuse using anionic surfactants, RI 8847. In: Bureau of Mines Report of Investigations. United States Department of the Interior; 1983. p. 16

[45] Olson GJ, Clark TR, Mudder TI, Logsdon M. A novel approach for control and prevention of acid rock drainage. In: Sixth ICARD. Cairns, QLD, Australia; 2003. pp. 789-799

[46] Jay WH. Application of ion exchange polymers in copper cyanide and acid mine drainage, Hydrometallurgy 2003. In: Young CA, Alfantazi AM, Anderson CG, Dreisinger DB, Harris B, James A, editors. Fifth International Conference in Honor of Professor Ian Ritchie, Vol. 1: Leaching and Solution Purification. TMS (The Minerals, Metals and Materials Society). pp. 717-728

[47] Wilmoth RC, Hill RD. Mine drainage pollution control by reverse osmosis. In: SME Fall Meeting and Exhibit. Birmigham, Alabama; 1972. p. 28

[48] Laubscher C, Petersen FW, Smit JP. Treatment of acid mine drainage through chemical precipitation. In: Lorenzen L, Bradshaw DJ, editors. XXII International Mineral Processing Congress. Cape Town, South Africa; 2003. pp. 1814-1820

[49] Shimada H, Sasaoka T, Matsui K, Kusuma GJ, Oya J, Takamoto H, et al. Fundamental study of acid mine drainage control using flyash. In: Mine Planning and Equipment Selection (MPES) Conference. Freemantle, WA, Australia; 2010. pp. 247-254

[50] Taylor RM, Restarick C, Ennis I, Robins RG. The green precipitate process for remediation of acid mine drainage. In: SME Annual Meeting. Salt Lake City, Utah, USA; 2000. p. 12

[51] Kuyucak N. Microorganisms, biotechnology and acid rock drainage—Emphasis on passive-biological control and treatment methods. Minerals and Metallurgical Processing. 2000;**17**(2): 85-95

[52] El-Ammouri E, Distin PA, Rao SR, Finch JA, Ngoviky K. Treatment of acid mine drainage sludge by leaching and metal recovery using activated silica. In: Fifth International Conference on Acid Rock Drainage. Denver, CO; 2000. pp. 1087-1094

[53] Sato M, Robbins EI. Recovery/removal of metallic elements from acid mine drainage using ozone. In: 5th International Conference on Acid Rock Drainage. Denver, Colorado; 2000. pp. 1095-1100

[54] Gusek JJ. Reality check: Passive treatment of mine drainage an

emerging technology of proven methodology? In: SME Annual Meeting. Salt Lake City, Utah, USA; 2000. pp. 1-10

[55] Fytas K, Lapointe F, McConchie D. The use of permeable reactive barriers for treating acid mine effluents. In: 2nd International Conference on Advances in Mineral Resources Management and Environmental Geotechnology. Hania, Greece; 2006. pp. 573-578

[56] Gulec SB, Edil TB, Benson CH. Effect of acidic mine drainage on the polymer properties of an HDPE geomembrane. Geosynthetics International. 2004;**11**(2):60-72

[57] Mroueh U-M, Vahanne P, Wahlstrom MM, Kaartinen T, Juvankoski M, Vestola E, et al. Environmental techniques for extractive industries. In: Mine Closure Handbook. Espoo; 2008. p. 169

[58] Sukarman RAG. Ex-coal mine lands and their land suitability for agricultural commodities in South Kalimantan. Journal of Degraded and Mining Lands Management. 2020;7(3):2171-2183

[59] Bell C, Lawrence K, Biggs B, Bingham E, Bouwhuis E, Currey N, et al. Mine closure and completion. In: Leading practice sustanaible development program for the mining industry. Commonwealth of Australia; 2006. p. 63

[60] United States Environmental Protection Agency. Available from: https://www.epa.gov/caddis-vol2/cadd is-volume-2-sources-stressors-response s-ph#tab-4

[61] Othman A, Sulaiman A, Sulaiman SK. The use of quicklime in acid mine drainage treatment. Chemical Engineering Transactions. 2017;**56**: 1585-1590

[62] McGinty J, Yazdanpanah N, Price C, ter Horst JH, Sefcik J. Nucleation and crystal growth in continuous crystallization. In: The Handbook of Continuous Crystallization. Chapter 1, London, UK: Royal Society of Chemistry; 2020. pp. 1-50

[63] Skousen J. Overview of acid mine drainage treatment with chemicals. In: Jacobs JA, Lehr JH, Testa SM, editors. Acid Mine Drainage, Rock Drainage, and Acid Sulphate Soils (Causes, Assessment, Prediction, Prevention, and Remediation). Chapter 29, New Jersey, USA: John Wiley & Sons; 2014. pp. 327-337

[64] Qasem NAA, Mohammed RH, Lawal DU. Removal of heavy metal ions from wastewater: A comprehensive and critical review. Clean Water. 2021;**4**:36

[65] Noyes AA, Whitney WR. The rate of solution of solid substances in their own solutions. Journal of American Chemical Society. 1897;**19**:930-934

[66] Dokoumetzidis A, Macheras P. A century of dissolution research: From Noyes and Whitney to the biopharmaceutics classification system. International Journal of Pharmaceutics. 2006;**321**:1-11

[67] Sohn HY, Wadsworth ME. Rate Processes of Extractive Metallurgy. New York and London: Plenum Press; 1979. p. 472

[68] Giles DE, Ritchie IM, Xu B-A. The kinetics of dissolution of slaked lime. Hydrometallurgy. 1993;**32**: 119-128

[69] Montes-Atenas G, Valenzuela F. Wastewater treatment through low-cost adsorption technologies. In: Farooq R, Ahmad Z, editors. Physico-chemical Wastewater Treatment and Resource Recovery. Rijeka, Croatia, London: Intech; 2017. pp. 213-238

[70] Elimelech M, Gregory J, Jia X, Williams RA. Particle Deposition and Aggregation: Measurements, Modelling and Simulation. Oxford, UK: Butterworth Heinemann, Ltd; 1995. p. 441

[71] Cooney DO. Adsorption Design for Wastewater Treatment. Boca Raton, Boston, London, New York, Washington DC: Lewis Publisher, CRC Press LLC; 1999. p. 189

[72] Montes-Sotomayor S, Montes-Atenas G, Garcia-Garcia F, Valenzuela M, Valero E, Diaz O. Evaluation of an Adsorption–Desorption Process for Concentrating Heavy Metal Ions from Acidic Wastewaters. Adsorption Science & Technology. 2009;**27**:513-521

[73] Defay R, Prigogine I. Surface Tension and Adsorption. New York: John Wiley and Sons, Inc; 1966. p. 432

[74] Montes-Atenas G, Valenzuela F, Montes S. The application of diffusion–reaction mixed model to assess the best experimental conditions for bark chemical activation to improve copper (II) ions adsorption. Environmental Earth Sciences. 2014;**72**(5):1625-1631. DOI: 10.1007/s12665-014-3066-3

[75] Walas SM. Reaction Kinetics for Chemical Engineers. 1st ed. Massachusetts, USA: McGraw-Hill; 1989. p. 338

[76] Wulfsberg G. Principles of Descriptive Inorganic Chemistry. California, USA: Brooks/Cole Pub; 1987. p. 461

[77] Baes CF, Mesmer RE. The Hydrolysis of Cations. New York, London, Sydney, Toronto: Wiley Interscience Pub., John Wiley and Sons; 1976. p. 491

[78] Cruywagen JJ, van de Water RF. The hydrolysis of lead(II). A potentiometric and enthalpimetric study. Talanta;**40**(7): 1091-1095

[79] Vetter KJ. Electrochemical Kinetics: Theoretical and Experimental Aspects. New York, San Francisco, London: Academic Press; 1967. p. 808

[80] Brenet J. Introduction a l'electrochimie de l'equilibre et du non equilibre. Paris: Masson; 1980. p. 155

[81] Vymazal J. Constructed wetlands for wastewater treatment: Five decades of experience. Environmental Science and Technology. 2011;**45**:61-69

[82] Ma Q, Ellis GS, Amrani A, Zhang T. Theoretical study on the reactivity of sulfate species with hydrocarbons. Geochimica et Cosmochimica Acta. 2008;**72**(18):4565-4576

[83] Barton LL. Sulfate-Reducing Bacteria: Biotechnology Handbooks. Vol. 8. New York, and London: Plenum Press; 1995. p. 336

[84] Kadlec RH, Wallace S. Treatment Wetlands. 2nd ed. Boca Raton, Florida: CRC Press; 2008. p. 1016

[85] Brock TD, Madigan MT, Martinko JM, Parker J. Biology of Microorganisms. 7th ed. Englewood Cliffs, NJ: Prentice-Hall; 1994. p. 986

[86] Muyzer G, Stams AJM. The ecology and biotechnology of sulphate-reducing bacteria. Natural Review of Microbiology. 2008;**6**:441-454

[87] Rodriguez-Dominguez MA, Konnerup D, Brix H, Arias CA. Constructed Wetlands in Latin America and the Caribbean: A review of experiences during the last decade. Water. 2020;**12**:1744

[88] Roehl KE, Meggyes T, Simon F-G, Stewart DI. Long-term Performance of Permeable Reactive Barriers. Oxford, UK: Elsevier; 2005. p. 326

[89] Valenzuela F, Basualto C, Sapag J, Ide V, Luis N, Narvaez N, et al. Adsorption of pollutant ions from residual aqueous solutions onto nano-structured calcium silicate. Journal of the Chilean Chemical Society. 2013;**58**: 1744-1749

[90] Karapinar N. Magnetic separation of ferrihydrite from wastewater by magnetic seeding and high-gradient magnetic separation. International Journal of Mineral Processing. 2003;**71**: 45-54

[91] Leygraf C, Wallinder IO, Tidblad J, Graedel T. Atmospheric Corrosion. 2nd ed. New Jersey, Canada: John Wiley and Sons Inc; 2016. p. 374

[92] Eger P. Designing wetland and treatments systems for long term treatment of mine drainage—An impossible dream? In: SME Annual Meeting. Salt Lake City, UT, US; 2005. p. 9

[93] Gammons CH, Zhang J, Wang P. Attenuation of Heavy Metals in Constructed Wetlands, Butte, Montana. In: Proceedings, 1998 Pacific Northwest Regional Meeting of the American Society of Agricultural Engineers: Engineering Biological Processes for Environmental Enhancement. 1998. pp. 1159-1168

Section 6

Recent Advances in Wastewater Treatment Applications

Chapter 12

Application of Vermifiltration for Domestic Sewage Treatment

Lubelihle Gwebu and Canisius Mpala

Abstract

Climate change has led to water shortages in semi-arid regions. SDG 13 was advocates for wastewater reuse. Zimbabwe uses centralised conventional sewage treatment systems. Vermifiltration combines filtration process and earthworms in sewage water treatment. Vermifiltration is efficient, viable, requires less expertise and can be decentralised. Vermifiltration technique was used in treating domestic septic tank sewage water. Design parameters and efficiency were determined and characterised Vermifiltered water parameters were compared against the Environmental Management Agency Statutory Instrument 6 irrigation water standards. Vermifilter media contained gravel and composted soil with 20g Eseinia fetida earthworms per litre of soil. Treatments were septic tank raw water, vermifilter and control biofilter. A duplicate analysis was conducted. Hydraulic retention time was 1 hour 40 minutes and hydraulic loading rate 163l/m2/hour. Disposed wastewater did not meet required EMA standards. Both filters were effective in treating domestic sewage. There was a significant difference between untreated and treated wastewater. Vermifilter and the control, significantly (p < 0.01) treated pH, turbidity, total dissolved solids total suspended solids, biological oxygen demand, nitrates, phosphates and total coliforms properties. Vermifiltered water met EMA standards for irrigation and non-potable water uses. Phytoremediation can be incorporated in the designs to increase efficiency.

Keywords: biofilters, *Eseinia fetida*, vermifiltration, wastewater treatment, physico-chemical parameters

1. Introduction

Climate change has led to water shortages in arid and semi-arid regions. Sustainable development goal (SDG) 13 on climate change has been set to combat climate change by advocating for wastewater reuse. Zimbabwe depends on conventional sewage treatment systems, which are expensive, centralised, inefficient due to increasing population dynamics, high maintenance and need of high expertise. Countries such as Australia and China have adopted the vermifiltration technique, which combines the filtration process and earthworms in the treatment of sewage water. Vermifiltration is efficient, cost-effective, requires less expertise and can be decentralised. This study applied the vermifiltration technique on the treatment of domestic sewage water from a septic tank. The aim was to determine the design parameters of a vermifilter, characterising its efficiency in the removal of selected physico-chemical, microbiological parameters and comparing vermifiltered water against the Environmental Management Agency (EMA) irrigation water standards.

1.1 Wastewater management

Reclaimed water is a product of treated wastewater that includes industrial and domestic effluent [1]. Wastewater treatment is a technique for sewage water management. It involves the removal of pollutants in wastewater that are a threat to the environment. It involves chemical, biological and physical processes for the removal of the pollutants in water using infiltration systems, trickling filters, wastewater stabilisation ponds and septic tanks. One of the main products is reclaimed water and sludge that is deposited into the environment [2].

The use of conventional systems for the treatment of wastewater is costly, requires high maintenance, is centralised and has no resource recovery of reclaimed water. However, some countries have adopted vermifiltration as an alternative method for the treatment of wastewater [3].

1.2 Vermifiltration

Vermifiltration is a wastewater treatment method that incorporates the use of earthworms and the infiltration system [4]. The earthworms act as bio-filters and are capable of degrading, digesting and decomposing organic waste [5]. This is achieved through promoting growth of beneficial decomposing bacteria, biological stimulation, aeration and chemical degradation. Studies have shown the efficiency of vermifiltration in controlling pH, chemical oxygen demand (COD), total dissolved solids (TDSs), biological oxygen demand (BOD5), turbidity and chlorides (Cl) [4, 6]. It was first established at a University in Chile by Professor Jose Toha in 1992 [7]. Other countries are now using the treatment system, and it has been recommended for developing countries [4].

It is an efficient, inexpensive technique, non-labour intensive, requires low expertise, can be decentralised and is environmentally friendly. One of the products is reclaimed water that can be used for irrigation, landscaping, fire protection, flushing and vermicompost [1, 6]. Therefore, this study intended to recover the wastewater for non-potable water uses through the application of vermifiltration for domestic sewage treatment from septic tanks at Lupane State University (LSU).

LSU and Lupane Town do not have any wastewater treatment plant that enables the access of reclaimed water. Septic tanks are used for wastewater treatment, and thus, there is no use of either recycled or reclaimed water. The university uses approximately 30,000 litres of water per day from the hostels and the dining hall that goes to the septic tanks, and it is not reused. The use of septic tanks could also contribute to the pollution of underground water and contamination of the soil. The wastewater treatment plants currently used in Zimbabwe need high maintenance. With the growing population at the LSU campus, there will be decrease in the treatment efficiency of the septic tanks due to overload [8, 9].

1.3 Justification

Vermifiltration is a low-cost method that is eco-friendly, non-labour intensive, and it enables the treatment and reuse of wastewater [10]. LSU wastewater is disposed into the environment using septic tanks, thus increasing chances of pollution of the soil and underground water. Vermifiltration induces a decrease in COD, chlorides, TDS, TSS and pH in an efficient manner. The water is disinfected and clean enough to be used for farm irrigation and other non-potable uses. Some developed countries are using the system, and it has been highly recommended for developing countries since it is economically feasible and does not require a lot of maintenance and expertise. This study would therefore enable LSU to be able to reuse its wastewater for multiple tasks such as landscaping, construction, fire protection, irrigation of sport grounds and crops. This technique could be further utilised at household level, thus promoting decentralisation of wastewater treatment. The application of this low-cost sewage treatment system would go a long way in availing wastewater for irrigation and other non-potable water uses especially in the semi-arid areas.

1.4 Broad objective

The main objective was to apply and characterise a vermifiltration system for domestic sewage treatment at LSU.

1.4.1 Specific objectives

The specific objectives were to determine the design parameters of a vermifilter (hydraulic loading rate, hydraulic retention time, soil type and earthworm species); to characterise the efficiency of the vermifilter for the removal of the physicochemical and microbiological parameters (pH, BOD, turbidity, nitrates, phosphates, TDS, TSS and total coliforms) from the wastewater in the septic tank and to compare the physicochemical and microbiological parameters of the vermifiltered water against EMA standards.

2. Review of literature

2.1 Wastewater management

Wastewater management is the process of managing effluent through treatment, safe disposal in the environment and for reuse [11]. Wastewater consists of industrial, domestic and agricultural effluent. Wastewater contaminants include plant nutrients, pathogenic micro-organisms, heavy metals, organic pollutants, biodegradable organics and micro pollutants [6]. Wastewater management facilitates complete removal of pollutants for environmental protection, human and animal health [12]. The World Health Organisation (WHO) Guidelines [13] and SDGs advocate for efficient wastewater treatment and management to cater for future projected water scarcity, safe drinking water, sanitation and sustainable environmental management [14].

Wastewater mismanagement has detrimental effects on the environment, people, animals and aquatic life. Water pollutants disposed into the environment promote water pollution [15]. Wastewater is a drought-resistant resource in households, industry and agriculture if the sludge and water can be treated and reused [16]. Wastewater can be reused as a nutrient source in agriculture, irrigation, energy (biogas) and soil conditioning [4]. Efficient wastewater treatment promotes prominent levels of sanitation and sustainable development within nations.

2.2 Sewage treatment

Biological, physical and chemical techniques are used in wastewater treatment [11]. These include natural water purification processes that occur in oceans, lakes and streams. Treatment systems can be centralised or decentralised depending on the economy, technological advancement and population [17]. Zimbabwe's treatment systems include infiltration, waste stabilisation ponds, trickling filters, activated sludge systems, septic tanks and pit latrines [8].

2.3 Wastewater management in Zimbabwe

Wastewater management in Zimbabwe is governed through legislation and policed by the Environmental Management Agency. There are regulations and policies that stress on the management of wastewater. These include the Environmental Management Act CAP 20:27) and other regulations. The regulations are the Statutory Instrument 6 of 2007 on Effluent and Solid Waste Disposal, Public Health Act (CAP 15:09), Urban Councils Act (CAP 29:150, Municipal bye-laws and Rural District Councils Act (CAP 29:1) [1]. These acts and bye-laws help in the control and monitoring of efficient treatment and disposal of wastewater. In most urban areas, the treatment of wastewater is overseen by the town municipality. The aim is meeting the standards for disposal and reuse of the reclaimed water for other non-potable uses.

2.3.1 Sewage treatment plants used

Wastewater management consists of various means for the treatment of the wastewater. The treatment methods include biological, physical and chemical techniques [12]. These systems incorporate the natural processes that occur naturally in oceans, lakes and streams in water purification. The treatment systems can either be centralised or decentralised depending on the economy of the country, technological advancement and its population [17]. The treatment plants vary in preference in terms of land requirements, sludge production, efficiency, reliability, affordability and energy consumption.

The treatment systems for wastewater used in Zimbabwe include the infiltration systems, waste stabilisation ponds, trickling filters, activated sludge systems, septic tanks and pit latrines [8, 18]. These systems treat both industrial and domestic effluents; however, some are limited to domestic effluent such as the waste stabilisation ponds and pit latrines. These conventional systems for wastewater treatment are mostly centralised where the municipality controls and monitors the treatment processes. In urban areas, there are sewerage pipelines from households and industries that are directed to the designated treatment plants. In industries, it is a must for the effluent to be firstly pre-treated before being directed to the municipality pipeline. This is to protect the sewerage pipelines and the treatment plants from being corroded and clogged by the toxic effluent from the industries.

The conventional system plants used require heavy maintenance and are faced with tremendous pressure due to high populations [19]. The treatment plants are therefore viewed inefficient in the treatment. This is supported by the pollution of water courses such as Lake Chivero in Harare and Umguza River in Bulawayo. These have been invaded by *Eichhornia crassipes* (water hyacinth) due to pollution caused by the inadequately treated effluent. Another recent cause was of the outbreak of cholera which was due to the mixing of drinking water with sewage effluent. This shows the inefficiency of the wastewater management in Zimbabwe [20].

2.3.2 Parameters monitored

The treatment plants are set to meet certain standards in the treatment process for the disposal and reuse of the waste water [1]. These standards are set against EMA standards such as SI6 of 2007, SAZ and WHO standards. The Statutory Instrument 6 of 2007 on Effluent and Solid Waste disposal is the one mainly used in Zimbabwe. **Table 1** shows the categorised limits using colours for disposal. The parameters monitored include physical, chemical and microbiological pollutants. These include pH, turbidity, suspended solids, dissolved solids, dissolved oxygen, nitrogen, phosphates, ammonia, alkalinity, BOD, COD, coliforms and metals. The treated effluent should meet the set safe standards for disposal and reuse. The treatment plants are designed to reduce and control the pollutants found within the effluent [19].

Table 1 shows Statutory Instrument (SI) 6 of 2007 on Effluent and Solid Waste Disposal in Zimbabwe.

Generally, the treatment plants in the country are expensive to maintain. Most of them in urban areas are centralised, hence the loading density becomes very high to maintain optimum treatment. This leads to the recurring sewer bursts perpetuating water pollution [18]. The lack of the proper management of sludge is another problem which finds its way in the landfills contributing to the production of methane [21]. There is no resource recovery in some treatment plants such as the septic tanks.

Parameter		Blue	Green	Yellow	Red
pH		6–9	5–6	4–5	0–4
		9–10	10–12	12–14	
Turbidity	NTU	≤5	—	—	—
Chloride	Cl	≤250	≤300	≤400	≤500
Phosphate	P	≤0.5	≤1.5	≤3.0	≤5.0
Nitrate	N	≤10	≤20	≤30	≤50
Soluble Solids		≤25	≤50	≤100	≤150
Dissolved oxygen % sat		≥60	≥50	≥30	≥15
Total Dissolved Solids		≤500	≤1500	≤2000	≤3000
Faecal Coliforms/100 ml		≤1000	>1000	>1500	≥2000

Key: Blue = Good-quality effluent suitable for disposal to the environment.
Green = Satisfactory quality effluent.
Yellow = Poor quality effluent not suitable for disposal to the environment.
Red = Very-poor-quality effluent attracting heavy disposal fees.

Table 1.
Environmental management agency (effluent and solid waste disposal) regulations SI 6 of 2007).

Decentralisation and resource recovery of the wastewater from the treatment plants can be the solution to the current situation of Zimbabwe. With the growing populations and water scarcity due to climate change, alternatives to cater for the challenges have to be put into place. Decentralisation and resource recovery can be utilised by everyone and capitalised at household level [22]. Other countries are using the vermifiltration treatment system which enables resource recovery and decentralisation of wastewater treatment, and it has been recommended for adoption by developing countries.

2.4 Vermifiltration

Vermifiltration is an infiltration method combined with the biological means of treating wastewater. The main actors in the vermifiltration process are earthworms. The earthworms act as the biofilters that destroy the waste. Studies on vermifiltration for domestic and industrial effluent have been carried out in other countries, and its efficiency has been proven [5, 10]. It was first established at a University in Chile by Professor Jose Toha in 1992 [7]. Some studies on vermifiltration that have been that done showed efficient removal of BOD, COD, suspended solids, total dissolved solids in wastewater of 80–90% removal [4, 6, 10]. A few studies on vermifiltration for the treatment of domestic effluent have been done in Zimbabwe. The studies showed a percentage reduction of above 70% of pH, BOD, COD, TDSS and turbidity [4, 9]. The use of vermifiltration proved to be more efficient and environmentally friendly as compared with the conventional systems used. However, less has been done on implementing it on treatment plants such as septic tanks that have no recovery of the wastewater.

2.4.1 Treatment process in vermifiltration

The treatment process in vermifiltration involves the vermicomposting process and the microbial processes in the removal waste loadings in the effluent. The vermifiltration media is usually characterised of granular materials which could be gravel, quartz of layers of various sizes, ceramsite and soil which is inoculated with earthworms [23]. The granular materials help in the removal of pollutants through filtration. The earthworms play the role of degrading the waste and activating the growth of microbiological organisms that decompose the waste [24, 25].

2.4.2 Earth worm action

There are various earthworm species that can be used for vermifiltration, since they are well known as the waste and environment managers [26, 27] The earthworms are long, narrow, bilateral, cylindrical segmented species with no bone formation. They have millions of nitrogen fixing and decomposing microbes in their guts. They have chemoreceptors that assist them for the searching of food. The types of earthworms that are effective in wastewater treatment include the African Night Crawler *(Eudrillas eugenor)*, red tiger *(E. andre)*, Indiana blue worm *(Feihoye excavatoxa)* and tiger worm *(Eisenia fetida)*. These earthworms have proven to be efficient; however, most studies have shown that E. *fetida* works more efficiently compared with the other earthworm species.

E. fetida worms are epigeic versatile waste eating worms with digestive enzymes such as protease, alkaline, phosphates and cellulose and use their bodies as bio filters. They feed on organic waste in wastewater; promote growth of decomposing bacteria and increase aeration and biological stimulation in the bioreactor. They have a very wide temperature tolerance of 20–25°C and can live in organic waste with a range of 60–75% moisture content. They prefer dark moist soil and can tolerate pH as low as 4. In their action, the earthworms excrete microbial organisms which consist of nutrients such as phosphates and nitrates that are further utilised by the micro-organisms for the decomposition of the waste [4, 6]. The earthworms also feed on the sludge from the effluent and degrade suspended solids that are trapped on top of the filter and fed on the soil microbes. Adsorption and stabilisation of inorganic solids, dissolved and suspended solids occur through biodegradation within the media inhabited by earthworms.

The burrowing earthworms increase aeration processes; thus, it enhances the filtration and soil stabilisation to be more efficient. Choking and production of foul smells are prevented [6, 28]. The earthworms generally act as the biofilters in the system by adequately treating the effluent [29].

2.4.3 Hydraulic retention time (HRT), hydraulic loading rate (HLR) and flowrate

The hydraulic loading rate (HLR) is the volume of wastewater applied per unit area of the bed per unit time. It is influenced by the volumetric flow rate, number of live adult earthworms functioning per unit area and their health. On the other hand, the hydraulic retention time (HRT) is the time taken by the wastewater in interaction with the worms. It depends on the flow rate of the wastewater, porosity of the soil and the volume of the soil profile. The higher the retention time, the more efficient the treatment system becomes. The HRT is also influenced by HLR; the higher the HLR, the lower the retention time, thus a decrease in treatment efficiency occurs.

2.5 Resource recovery from vermifiltered wastewater

Vermifiltration leads to treatment of wastewater that can be reused and vermicompost is formed as a by-product. The treatment of wastewater by vermifiltration has proven to be more effective as in some studies, it has achieved approximately more than 90% removal of some pollutants. This makes the water reusable for non-portable uses [7]. With the increasing demand and shortage for water, the reuse of wastewater is an alternative for combating water scarcity and the protection of the available water sources.

The vermifiltered water is clean enough and contains some nutritive nutrients that could be of use in agriculture. Irrigation can be carried out using the reclaimed water. In drought seasons, irrigation can also be practised continuously with the use of reclaimed water [22]. The water will be containing nutrients such as nitrates which increase the fertility of the soil. The plants will not be attacked by water nor nutritive stress. Where there are treatment plants with no reuse of wastewater, vermifiltration can be implemented and developmental projects of irrigation will emerge. The reclaimed water can also be for fire protection as compared with the use of fire hydrants and fire extinguishers. This again aids in water preservation and the use of environmentally friendly techniques. This can be done at household level through the decentralisation of wastewater treatment by the introduction of vermifiltration. Fire awareness and preparedness will therefore be improved and implemented at large scale.

The reclaimed water can be used for domestic use for the flushing system. This reduces the pressure on the freshwater resources used. The loading in the treatment plants and sewage pipes will be reduced, thus preventing the occurrence of sewage bursts. This will be due to the reuse of the wastewater for flushing system. This will also control the occurrence of diseases since during load shedding, people tend to use undesignated areas for relief, thus perpetuating disease emanation. The use of reclaimed water thus can act as a substitute and alternative for the use of freshwater for the flushing system [30].

The reclaimed water and the vermicasts from vermifiltration act as fertiliser in agriculture [31]. This water can be used for landscaping in areas which are dry. Some areas which have water shortages and infertile soils can have the opportunity of practising landscaping. The wastewater can be reused for the watering of ornamental flowers and introduce loans. The infertile soils will be made arable by the treated effluent. The maintenance of the landscape will be much easier as the production of the reclaimed water is throughout the year [32, 33].

3. Methodology

3.1 Site description

The study was carried out at Lupane State University (18.9300^{0}S, 27.7593°E) in Zimbabwe. This is a semi-arid area with annual rainfall of 550 mm, high summer and low winter temperatures. The university depends on septic tanks for sewage treatment. The design experiment followed a duplicate analysis setup with three treatments: septic tank effluent raw water (RW), vermifilter (VF) and a control (CF). Biofilters were designed and set up (two vermifilters and the control filter). Twenty-litre plastic buckets measuring 27 cm x 29 cm x 32 cm were used as biofilter containers. Media consisted of four layers of 10–14 mm gravel, 4-8 mm gravel, river sand, black soil with *Eisenia festida* and 3 cm top layer. Layers were 4 cm, 4 cm, 10 cm, 20 cm and 3 cm thick, respectively. Twenty grams of earthworms per litre of black soil were used. The control biofilters did not have earthworms.

3.2 Measured parameters and method of measurement

Total dissolved solids, pH, TSS, turbidity, BOD after 5 days, nitrates and phosphates and total coliforms (TCs) were measured using by the oven drying method, an electronic pH meter, Lovibond portable turbidity meter, BOD5 dilution method; Beckman Conlter UV/VIS Spectrum and Multiple Tube Fermentation Technique.

3.3 Data analysis

Water quality parameter analysis was done before and after treatment. Genstat 14.0 was used for statistical analysis. The Shapiro–Wilk test ($p < 0.05$) was used to check for normality of data. One-way analysis of variance (ANOVA) at 5% level of significance was used to test for significant differences in related parameters per treatment. Results were compared against the EMA standards (SI6 of 2007 on Effluent and Solid Waste Disposal).

4. Findings and conclusions

Design hydraulic retention time was 1 hour 40 minutes, with a hydraulic loading rate of 163 l/m²/hour. The design parameters of the vermifilter are shown in **Table 2**.

The designed vermifilter and control biofilters were efficient in treating domestic sewage for the selected chemical parameters. Analysis of variance revealed that the vermifilter and control filter significantly (p < 0.001) treated the chemical properties of the domestic sewage water (**Table 3**).

All chemical parameters levels (pH, BOD, NO₃, PO₄ and chlorides) of treated sewage using the two filters were significantly different from the untreated sewage. However, no significant differences were observed on the chemical properties of

Parameter	Property
Capacity of vermifilter	27 cm x 29 cm x 32 cm
First gravel layer	10–14 mm (4 cm thick)
Second gravel layer	4–8 mm (4 cm thick)
Third layer (sand)	2 mm (4 cm thick)
Forth layer (Black soil and *E fetida* earthworms)	20 cm
Top layer (free space)	3 cm
Hydraulic loading rate	$163 \text{ L m}^2 \text{ h}^{-1}$
Hydraulic retention time	1.4 L h^{-1}
Water discharge	2.1 ml s^{-1}

Table 2.
Parameters of the designed vermifilter.

Treatment	pH	BOD	NO₃	PO₄
Untreated sewage (RW)	7.71a	49.9a	9a	6.9a
Treated sewage using vermifilter (VF)	6.79b	6.8b	376b	0.38b
Treated sewage using control filter (CF)	6.52b	7.5b	256c	0.23b
Overall mean	7	21.4	214	2.47
F probability	<0.001	<0.001	<0.001	<0.001
LSD	0.44	8.47	105.4	0.72
% CV	5.1	32.2	40.1	23.7

Table 3.
Analysis of variance of the efficiency of the vermifilter and control biofilter in treating domestic sewage on the selected chemical parameters.

sewage water treated using the two filters. The percentage range reduction in BOD on the VF was 68–97% whilst in the CF it was 76–98%, PO_4 in VF it was 90–99%, in the CF 95–98%. An increase was observed in nitrates in the VF of 96–98% and CF 62–82%.

4.1 Effect of vermifilter and control biofilter on the turbidity of the sewage water

The turbidity level of sewage water treated using the two filters was significantly different from untreated sewage. However, no significant differences were observed in turbidity of water treated using the two filters. The percentage reduction was higher in the VF at 83–99% and at the CF it was 77–97% (**Figure 1**).

Bars shown using different letters were significantly different at 5% level of significance. The vertical bars shown at the top of each bar show the standard error of difference of means.

4.2 Effect of vermifilter and control biofilter on total dissolved solids in sewage water

The study reveals that the vermifilter and control biofilter significantly treated the TDS from the domestic sewage (**Figure 2**). The TDS level of treated sewage showed no significance difference in the VF and CF. The VF shows an efficiency of 44–90% and in the CF it was 70–93% (**Figure 2**).

Bars shown using different letters were significantly different at 5% level of significance. The vertical bars shown at the top of each bar show the standard error of difference of means.

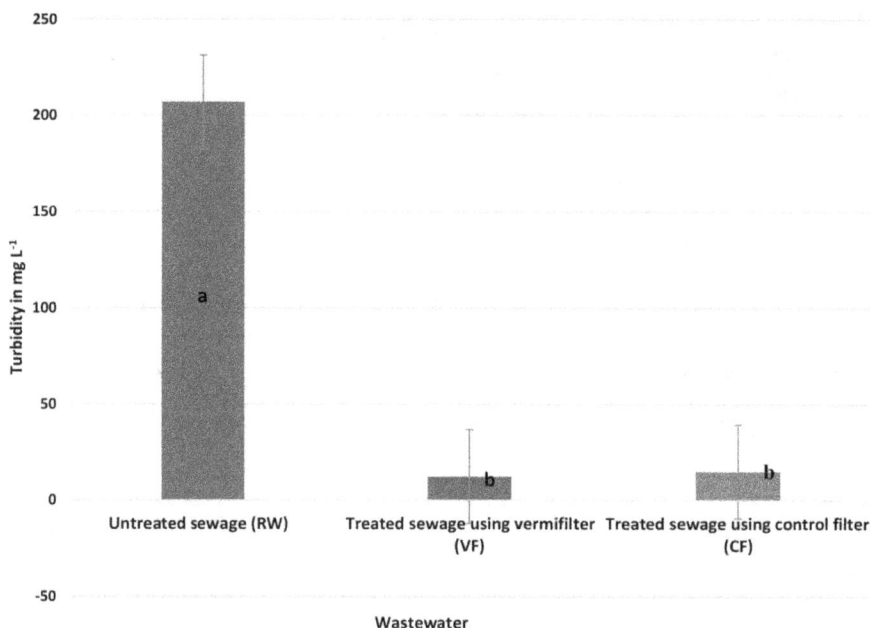

Figure 1.
Turbidity of influent and effluent.

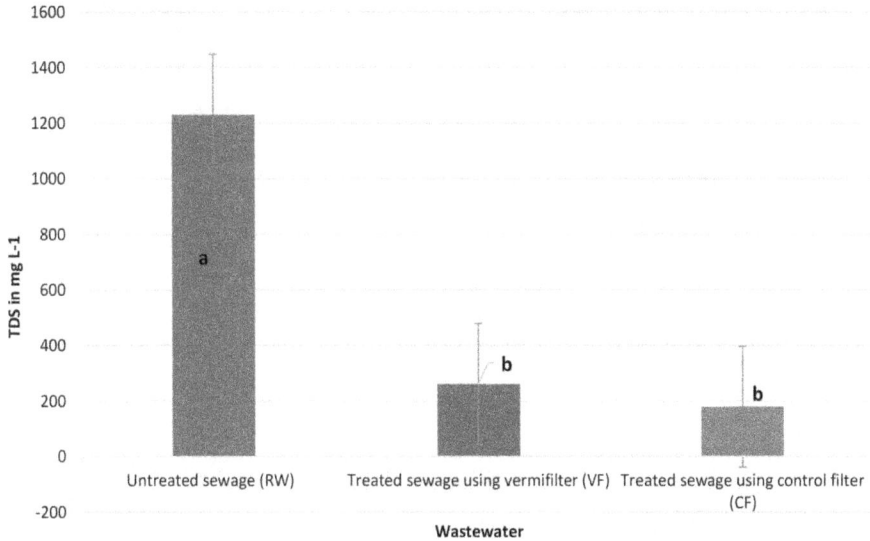

Figure 2.
TDS of influent and effluent.

4.3 Effect of vermifilter and control biofilter on total suspended solids of sewage water

There was a significant difference in the level of TSS in the untreated sewage and the treated sewage from the two biofilters. The two biofilters had no significant difference in their operation of removing total suspended solids. The percentage reduction in TSS was higher in the VF at 48–97% and 24–97% in the CF (**Figure 3**).

Bars shown using different letters were significantly different at 5% level of significance.

The vertical bars shown at the top of each bar show the standard error of difference of means.

4.4 Efficiency of vermifilter and control filter in treating total coliforms in domestic sewage

The vermifilter and control efficiently treat the available total coliforms in untreated sewage. The total coliforms in untreated sewage were significantly different from total coliforms from the two filters. The VF and CF, however, did not have any significant difference in their treatment porosity. The percentage reduction in the VF was 78–98% whilst in the CF it was 97–98% (**Table 4**).

Septic tank effluent did not meet the required EMA SI6 Standards for disposal into the environment. Both filters were effective in treating domestic sewage. There was significant difference between untreated and treated wastewater on selected physico-chemical and microbiological parameters. Vermifilters and control significantly (p < 0.01) treated the physico-chemical (pH, turbidity, TDS, TSS, BOD, nitrates, phosphates) and microbiological (total coliforms) properties of domestic sewage water.

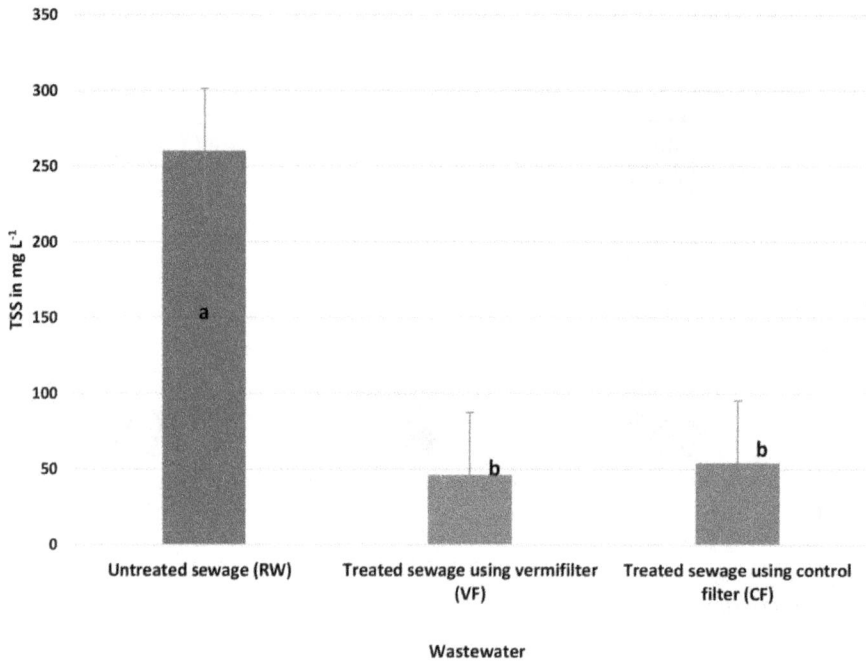

Figure 3.
TSS in influent and effluents.

Treatment	TC
Untreated sewage (RW)	180a
Treated sewage using vermifilter (VF)	14b
Treated sewage using control filter (CF)	4b
Overall mean	66.1
F probability	<0.001
LSD	9.46
%CV	11.6

Table 4.
Analysis of variance on the efficiency in total coliform removal of a vermifilter and control filter in treating domestic sewage.

4.5 Comparison of effluent with EMA guidelines (SI 6 of 2007 on effluent and solid waste disposal)

Table 5 shows that the domestic sewage from the septic tank was not within the EMA standards for effluent disposal except for pH and nitrates.

4.6 Discussion

4.6.1 Change in chemical parameters of the wastewater

There was a decrease in the pH, phosphates and BOD after treatment in both the VF and CF. An increase occurred in the level of nitrates after treatment.

Parameter	Raw Water	Vermifilter	Control Filter	EMA
pH	7.71	6.79	6.52	6–9
TDS	1230	261.3	179.5	≤500
TSS	259.7	46.9		≤25
Turbidity	207	12.79	14.99	≤5
BOD	49.87	6.79	7.47	≤30
Nitrates	9.01	275.8	250.8	≤10
Phosphates	6.80	0.38	0.23	≤0.5
Total Coliforms	180+	14	4	≤1000

The EMA Statutory Instrument 6 on Irrigation water standards uses the colour codes blue, green, yellow and red to indicate the level of pollution. Hence the colour shows whether the parameter is with the permissible limits or not.

Table 5.
Sewage quality parameters for disposal of the effluent relative to EMA standard guidelines.

4.6.2 pH

The pH of the untreated domestic sewage was within the EMA standards. It is due to the decomposition of organic matter that occurs in the septic, thus reducing acidity and alkalinity of the domestic sewage. The treated sewage both from water from the CF and VF had a pH range that decreased though it was within the EMA standards. These results in the CF were due to the biological reactions that occurred through the infiltration process. The pH of the vermifilteted water drew closer to neutrality due to the vermicasts produced by the earthworms which are more neutral. Similar results were obtained by [4, 6].

4.6.3 Phosphates

Analysis of variance revealed that the biofilters were significantly efficient in treating the domestic sewage. The decrease in phosphates in the VF is attributed to the availability of anoxic conditions, biological metabolism and iron corrosion. The earthworms accumulate phosphate in the gut system, hence reduce the level of phosphates in the water.

4.6.4 Nitrates

The analysis of variance results showed that there was a significant difference in the untreated sewage and the treated sewage using the VF and CF. However, a significance difference also existed within the two filters. Generally, the nitrates in raw sewage are supposed to decrease during the treatment as they are supposed to be converted from nitrate and further decomposed to nitrogen gas and water through the nitrification and denitrification processes. The CF and VF had an increase in nitrates due the removal of ammonia through adsorption which leads to the formation of nitrates through biological nitrification. The earthworms also have a lot of nitrifying and denitrifying bacteria in the gut of worms.

Therefore, the presence of oxygen, aerobic *Nitrobacter* and the worm cast oxygenates the influent, thus leading to the increase of nitrates [34]. The nitrates formed are, however, made available to the plants. This is due to the secretion of polysaccharides, proteins and other nitrogenous compounds; these mineralise the nitrogen and make it available to the plants [35]. Similar results were obtained by [6, 9].

4.6.5 Bod

Vermifilter was significantly effective in reducing BOD which is attributed to the absorption of the organic waste by the worms through their body wall and the uptake through their gut. There is high effective symbiotic relationship between the earthworms and the soil microbes. They accelerate and increase the rate of decomposition of the organic matter [36, 37]. The muscular action of the foregut of the earthworms plays a crucial role. The organic matter is degraded, homogenised, conditioned, and biological activity in filter is improved. A decrease in BOD in the control filter is due to the filtration process and the action of bacteria that grow in the filter as well as the presence of the biofilm that is formed.

4.6.6 Decrease in physical parameters

There was a significant difference in all the selected physical parameters on the untreated and treated sewage after experimentation.

4.6.7 Turbidity

The analysis of variance results revealed that the vermifilter and the control filter were effective in treating domestic sewage. The decrease in turbidity is due to the adsorption of both the macroscopic and microscopic suspended solids in the filters. The adsorption occurs in the soil and gravel. The VF was more efficient due to the role of earthworms. The earthworms feed on the organic matter, thus reducing organic waste and proportionally controlling turbidity.

4.6.8 Total dissolved solids

The vermifilter and the control filter had a significant effect on the treatment of TDS in the domestic sewage water. The VF and CF water were within the EMA standards. The decrease of the TDS in the VF is attributed to the ingestion of the organic and inorganic solid particles in the RW by the earthworms. The earthworms excrete finer particles. The TDSs both in the CF and VF are trapped through adsorption and stabilised during the infiltration process [38].

4.6.9 Total suspended solids

There was a significance difference in the total suspended solids in untreated and treated sewage. A decrease in the CF is a result of the TSS sticking on the surfaces of individual filter media such as gravel and sand through adsorption and trapping of the suspended solids [39]. The removal of TSS in the VF was more effective as the TSS trapped on the filter reprocessed by the earthworms and fed to the microbes in the soil for further decomposition. Earthworms burrow within the soil and this increase aeration, thus enhancing efficient filtration and soil stabilisation [29].

4.6.10 Change in the total coliforms

The coliforms decreased in the VF due to the ability of earthworms to release coelomic fluids. These fluids have antibacterial properties and are capable of destroying all the pathogens in the biomass. The earthworms ingest the organic matter which they process through culling up the harmful microorganisms. The end products are deposits which are mixed with minerals and beneficial microbes free from pathogenic particles in the soil [39]. The decrease is also attributed to gizzard and intestinal enzymes of the earthworms. They are secreted through their body wall [40]. The decrease in the CF of total coliforms was due to the action of the biofilm that destroys the bacteria in the wastewater.

4.7 Conclusion

The water from the VF and CF was both within allowable limits of the SI 6 standards by EMA, except for the nitrates and turbidity. Total suspended solids were in the blue allowable range in the VF, whilst in the CF it was on the yellow range which is a threat to the environment. The vermifilter met the blue and yellow allowable limits that are even permissible for irrigation. The vermifiltered water can therefore be used for irrigation of parks and landscaping.

The vermifilter and the control filter were both effective in treating the domestic sewage from the septic tank. All the physicochemical and microbiological parameters were within the EMA regulation standards for disposal. They ranged between the blue-yellow range of acceptance. However, the most effective filter was the vermifilter.

The control filter is not sustainable since with time it clogs, discharge of effluent rate lowers and there was a foul smell. It is rich in nutrients that are nitrates and phosphates that are in an available form to the plants. There is a potential for the recycling of the sludge in the septic for use as manure and production of biogas.

The study shows that there is a potential for the recycling of the sludge in the septic for use as manure and production of biogas. It is recommended to carry out water quality assessments on borehole water and soil to see if there is any contamination as a result of the use of septic tanks.

Design improvement of the vermifilter can be done by incorporating phytoremediation as to increase treatment efficiency as to control the increase in nitrate concentration.

Acknowledgements

The authors appreciate the assistance given by Mr. R. Mudziwapasi for guidance in the experimental setup; Mr. E. Ndlovu for data analysis; L. Musaradenga, Mr. T. Kachote and Miss G. Dube, who assisted in laboratory analysis. Thanks to Mr. D. Sibanda and Mr. C. Moyo from the Works and Physical Planning Department for the provision of designing tools and their assistance.

Conflict of interest

The authors declare no conflict of interest.

Author details

Lubelihle Gwebu and Canisius Mpala*
Department of Crop and Soil Sciences, Faculty of Agricultural Sciences, Lupane State University, Lupane, Zimbabwe

*Address all correspondence to: mpalacanisius@yahoo.com; cmpala@lsu.ac.zw

IntechOpen

References

[1] Nhapi I. A framework for the decentralized management of wastewater in Zimbabwe. Physics and Chemistry of the Earth. 2004;**29**:1265-1273

[2] Nhapi I, Hoko Z, Siebel M, Gijzen HJ. Assessment of major water and nutrient flows in the Chivero catchment area, Zimbabwe. Physics and Chemistry of the Earth. 2002;**27**:783-792

[3] Sinha RK, Sunil H, Bharambe G, Brahambhatt G. Vermistabilization of sewage sludge (biosolids) by earthworms: Converting a potential biohazard destined for landfill disposal into a pathogen-free, nutritive and safe biofertilizer for farmsISSN 0734–242X. Waste Management & Research. 2010; **28**:872-881. DOI: 10.1177/0734242X09342147

[4] Manyuchi MM, Kadzungura L, Boka S. Pilot studies for Vermifiltration of 1000m³/day of sewage wastewater. Asian Journal of Engineering and Technology. 2013;**01**(01):10-16

[5] Lakshmi P, Anupama S, Kumar JMA, Vidyashree KG. Treatment of sugar industry wastewater in anaerobic downflow stationary fixed film (DSFF) reactor. International Journal of Applied Biology. 2014;**16**(1):9-14

[6] Sinha RK, Bharambe G, Chowdhary U. Sewage Treatment by Vermifiltration with Synchronous Treatment of Sludge by Earthworms: a Low-Cost Sustainable Technology over Conventional Systems with Potential for Decentralization; the Environmentalist; UK. Vol. 28. USA: Springer; 2008. pp. 409-420; Published Online 8 April

[7] Li X, Xing M, Yang J, Huang Z. Compositional and functional features of humic acid-like fractions from vermicomposting of sewage sludge and cow dung. Journal of Hazardous Materials. 2011;**185**:740-748

[8] Nhapi I, Siebel M, Gijzen HJ. The impact of urbanisation on the water quality of Lake Chivero, Zimbabwe, journal of the chartered institute of. Water and Environmental Management. 2004;**18**(1):44-49

[9] Mudziwapasi R, Mlambo SS, Kuipa PK, Chigu NL, Utete B. Potential synchronous detoxification and biological treatment of raw sewage in small scale Vermifiltration using an Epigeic earthworm, *Eisenia fetidas*. Resources and Environment. 2016;**6**(1): 16-21

[10] Xing M, Li X, Yang J. Treatment perfomance of small-scale vermifilter for domestic wastewater and its relationship to earthworm growth, reproduction and enzyme activity. African Journal of Biotechnology. 2010;**9**(44):7513-7520

[11] Bani R, Amoatet P. Waste Water - Evaluation and Management. In: Einschlag FSG, editors. Wastewater Management. London: InTech; 2011. pp. 379-398, 470. ISBN 978-953-307-233-3

[12] Metcalf and Eddy, Inc. Wastewater Engineering. New York: McGraw Hill; 1972

[13] WHO. Global Water Supply and Sanitation Assessment Report. Geneva, Switzerland: World Health Organisation; 2000

[14] OECD Report. Brazil's journey from the Earth's Summit to Rio. Development Co-operation Report 2012 Lessons in Linking sustainability and Development

Contents Brazil's journey from the Earth Summit to Rio+20 Izabella Teixeira Chapter 2.

[15] Amoatey P, Bani R. Wastewater Management Department of Agricultural Engineering, Faculty of Engineering Sciences, University of Ghana, Ghana Australia. The Environmentalist. 2011;**22**(2):261-268

[16] Drechsel P, Scott CA, Raschid-Sally L, Redwood M, Bahri A, editors. Wastewater Irrigation and Health: Assessing and Mitigation Risks in Low-Income Countries. UK: Earthscan-IDRC-IWMI; 2010 www.idrc.ca/en/ev-149129-201-1-DO_TOPIC.html

[17] United Nations Environment Programme ANNUAL REPORT. www.unep.org/annualreport Published: February 2011 © 2011 United Nations Environment Programme

[18] Thebe TA, Mangore EN. Wastewater production, treatment and use in Zimbabwe. 2012. Department of Civil and Water Engineering, National University of Science and Technology, Bulawayo

[19] Utete B, Kunhe RM. Ecological integrity of a peri-urban river system, Chiraura River in Zimbabwe. Journal of Water Resources and Ocean Science. 2013;**2**(5):56-61

[20] Chirisa I, Bandauko E, Matamanda A, Mandisvika G. Decentralized domestic wastewater systems in developing countries: The case study of Harare (Zimbabwe). Applied Water Science. 2017;**2017**(7):1069-1078. DOI: 10.1007/s13201-016-0377-4

[21] Makoni S, Hoko Z, Mutengu S. An assessment of the public health hazard potential of wastewater reuse for crop production. A case of Bulawayo city, Zimbabwe. Physics and Chemistry of the Earth Parts A/B/C. 2007;**32**(15): 1195-1203

[22] Kathryn CH, Gregory KE, Brian B, Mike G. Water reuse: Using reclaimed water for irrigation. Communications and Marketing, College of Agriculture and Life Sciences, Virginia Polytechnic Institute and State University, Virginia Cooperative Extension. 2009

[23] Rajiv S, Sunita KA, Krunal C, Vinod C. Vermiculture Technology is Emerging as an Environmentally Sustainable, Economically Viable and Socially Acceptable Technology all Over the World. 2010;**1**:155-172. DOI: 10.4236/ti.2010.13019

[24] Binet F, Fayolle L, Pussard M. Significance of earthworms in stimulating Soilmicrobial activity. Biology and Fertility of Soils. 1998;**27**(1): 79-84

[25] Singleton PW, Bohlool BB. Effect of salinity on nodule formation by soybean. Plant Physiology. 1984;**74**:72-76

[26] Darwin C, Darwin F, Seward AC. More letters of Charles Darwin. In: Darwin F, Seward AC, editors. A record of his work in a series of hitherto unpublished letters. Vol. 11. New york: D. Appleton and Company;

[27] Martin GN. Effect of Wastewater Disposal by Land - Irrigation and Ground Water Quality in the Burnham Templeton Area New Zealand. New Zealand Journal of Agricultural Research. 1976;**21**(3)

[28] Zhao JY, Yan C, Li LY, Li JH, Yang M, Nie E, et al. Effect of C/N ratios on the performance of earthworm eco-filter for treatment of synthetic domestic ewage. Environmental Science and

Pollution Research International. 2012;
19(9):4049-4059

[29] Sinha RK, Agarwal S, Asadi R, Carretero E. Vermiculture technology for environmental management: Study of action of earthworms *Elsinia foetida, Eudrilus euginae* and *Perionyx excavatus* on biodegradation of some community wastes in India and Australia. The Environmentalist, U.K. 2002;**22**(2): 261-268

[30] U.S. EPA. Wastewater Technology Fact Sheet: Anaerobic Lagoons. Washington, D.C., USA: U.S. Environmental Protection Agency, EPA 832-F-02-009; 2002

[31] FAO. FAO. http://www.fao.org/fa ostat/en/#home

[32] Lin SD. *Water and Wastewater Calculations Manual*. 2nd ed. New York, USA: McGraw-Hill Companies, Inc., ISBN 0-07-154266-3; 2007

[33] Nhapi I, Gijzen H. Wastewater Management in Zimbabwe. Proceedings of the 28th WEDC Conference on Sustainable Environmental Sanitation and Water Services. Kolkata, India, 2002. 28th WEDC Conference Kolkata (Calcutta), India

[34] Wang F, Wang C, Li M, Gui L, Zhang J, Chang W. Purification, characterization and crystallization of a group of earthworms fibrinolytic enzymes from *Eisenia fetida*. Biotechnology. 2003;**25**(13):1105-1109

[35] Baisa O, Nair J, Mathew K, Ho GE. Vermiculture as a tool for domestic wastewater management. Water Science and Technology. 2003;**489**(11–12): 125-132

[36] Sinha RK, Bharambe G, Bapat PD. Removal of high bod & cod loadings of primary liquid waste products from dairy industry by vermifiltration technology using earthworms. Indian Journal of Environmental Protection (IJEP). 2007;**27**(6):486-501; ISSN 0253-7141

[37] Kumar T, Ankur RR, Bhargava KS, Prasad H. Performance evaluation of vermifilter at different hydraulic loading rate using river bed material. Ecological Engineering. 2014;**62**:77-82

[38] Pali S, Swapnali R, Mane S. Treatment of grey and small scale industry waste water with the help of vermifilter. Civil Engineering and Urban Planning: An International Journal (CiVEJ). 2015;**2**(1)

[39] Bobade AP, Ansari KS. The use of Vermifiltration in wastewater treatment: A review. Journal of Civil Engineering and Environmental Technology. 2016; **3**(2):164-169

[40] Khwairakpam M, Bhargava R. Vermitechnology for sewage sludge recycling. Journal of Hazardous Materials. 2009;**161**(2–3):948-954

Chapter 13

Emerging Human Coronaviruses (SARS-CoV-2) in the Environment Associated with Outbreaks Viral Pandemics

Chourouk Ibrahim, Salah Hammami, Eya Ghanmi and Abdennaceur Hassen

Abstract

In December 2019, there was a cluster of pneumonia cases in Wuhan, a city of about 11 million people in Hubei Province. The World Health Organization (WHO), qualified CoVid-19 as an emerging infectious disease on March 11, 2020, caused by severe acute respiratory syndrome coronavirus 2 (SARS-CoV-2) which spreads around the world. Coronaviruses are also included in the list of viruses likely to be found in raw sewage, as are other viruses belonging to the Picornaviridae family. SRAS-CoV-2 has been detected in wastewater worldwide such as the USA, France, Netherlands, Australia, and Italy according to the National Research Institute for Public Health and the Environment. In addition, the SARS-CoV-2 could infect many animals since it has been noticed in pigs, domestic and wild birds, bats, rodents, dogs, cats, tigers, cattle. Therefore, the SARS-CoV-2 molecular characterization in the environment, particularly in wastewater and animals, appeared to be a novel approach to monitor the outbreaks of viral pandemics. This review will be focused on the description of some virological characteristics of these emerging viruses, the different human and zoonotic coronaviruses, the sources of contamination of wastewater by coronaviruses and their potential procedures of disinfection from wastewater.

Keywords: SARS-CoV-2, human coronaviruses, zoonotic coronaviruses, disinfection procedures, wastewater

1. Introduction

The recent pandemic of the highly contagious coronavirus disease 2019 (COVID-19) caused by a novel severe acute respiratory syndrome coronavirus (SARS-CoV-2) has developed devastating consequences on human health, economy, and ecosystem services as an important public health concern [1]. Until 13 January 2022, over 307,373,791 cases have been reported, including over 5,492,154 deaths [2]. SARS-CoV-2 is a beta coronavirus that belongs to the family *Coronaviridae* and the order Nidovirales [3]. Coronaviruses

(CoVs), and the newly discovered SARS-CoV-2, are spherical or pleomorphic enveloped viruses with a diameter of 100–160 nm [4] characterized by spike proteins projecting to the virion surface [5]. The primary structure comprises four structural proteins: spike (S), envelope (E), membrane (M), and nucleocapsid (N) encoded at the 3' end of the viral genome [5, 6]. SARS-CoV-2 has a single-stranded genomic RNA (gRNA) positive-sense, approximately 30 kb [7]. This gRNA is among the largest RNA genomes known with a 5'-cap structure and a 3'-poly (A) tail acting like an mRNA for the immediate translation of viral polyproteins [8]. The 5' and 3' ends of the gRNA contain highly structured untranslated regions (UTR) that regulate RNA replication and transcription. There is one stem-loop and one pseudo-knot present in the 3'-UTR region mutually exclusive, since their sequences overlap, while seven stem-loop structures are included in the 5'-UTR region. The SARS-CoV-2 genome contains 14 open reading frames (ORFs), preceded by transcriptional regulatory sequences (TRS). The two main transcriptional units, ORF1a and ORF1ab, encode polyprotein replicase 1a (PP1a) and polyprotein 1ab (PP1ab), respectively [9]. These polyproteins are co- and post-translationally processed into 16 nonstructural proteins (NSPS), most of which drive viral genome replication and sub-genomic mRNA (sgmRNA) synthesis [8]. Usually, the transmission of SARS-CoV-2 was accounted occurs through inhalation of respiratory droplets diffused by coughing or sneezing from an infected patient, and through direct contact with contaminated surfaces or objects [10]. This respiratory syndrome leads to developing several chronic or acute disorders in patients such as fever, cough, fatigue, anosmia and ageusia, dyspnea, chest pain, muscle pain, chills, sore throat, rhinitis, headache [11], and gastrointestinal symptoms (nausea, vomiting, diarrhea) [11]. SARS-CoV-2 genomic RNA has been detected in patient stool and urine samples [12], suggesting the possibility that the virus may be transmitted via the fecal-oral route, besides droplet and fomite transmission. Therefore, the mode of transmission of SARS-CoV-2 becomes crucial paramount for human and environmental health. Recently, the world faces a large wave of COVID-19 infections caused by the highly contagious variant of SARS-CoV-2 Omicron can infect people even if they are vaccinated. The record number of people catching the COVID-19 from the beginning of the pandemic has left health systems under severe strain especially in developed countries. However, It has proven the particular importance of WBE (wastewater-based epidemiology) in monitoring the circulation and the transmission of the epidemic in the community, it could provide an early warning sign that reflects possible disease outbreaks in a community [13, 14] and an effective tool for epidemiological surveillance of SARS-CoV-2 viral diversity in samples and to anticipate the detection of certain mutations before they are detected in clinical samples. This communication will provide a basis for understanding SARS-CoV-2 and other viruses from the environmental perspective to design alternative strategies to counteract enteric virus transmission and to reduce the severity of the pandemic [15]. In this review, we present the different human and zoonotic coronaviruses, the sources of contamination of wastewater by coronaviruses, and their potential procedures of disinfection and eradication from wastewater.

2. Human coronaviruses

Coronaviruses (CoV) are enveloped viruses with a single positive-strand RNA genome (~26–32 kb). They belong to the subfamily *Ortho-Coronaviridae* of the family *Coronaviridae* and are classified into four genera: *Alpha coronavirus (α), Beta coronavirus (β), Gamma-coronavirus (γ)* and *Delta-coronavirus (δ)* [16].

Until the present time, seven human coronaviruses (HCoVs) can be transmitted between humans. Human alpha coronaviruses, 229E and NL63, and beta-coronaviruses, OC43, and HKU1 are common respiratory viruses usually causing mild upper respiratory illness. Unlike these, the three other human beta-coronaviruses, severe acute respiratory syndrome coronavirus (SARS), Middle East respiratory syndrome coronavirus (MERS), and SARS-CoV-2, are highly pathogenic in humans [17]. All seven HCoVs are the product of a Spillover, they have a zoonotic origin from bats, mice, or domestic animals. Multiple justifications support an evolutionary origin of all HCoVs from bats, where viruses are non-pathogenic and well adapted, so they show great genetic diversity. Genome analysis of the virus identified a high sequence similarity with Chinese bat coronaviruses (highest homology to bat coronavirus RaTG13). Beta coronavirus phylogenetic tree showing that SARS-CoV-2 is related to bat coronaviruses ZC45 and ZXC21. SARS-CoV-2 showed 99% sequence homology with pangolin CoV according to the findings of a research team from the South China University of Agriculture [18].

2.1 HCoV-229E and HCoV-OC43

HCoV-229E is the first strain isolated from patients with upper respiratory tract contamination in the year 1966 [19]. Patients infected with HCoV-229E showed cold symptoms, including headache, sneezing, malaise, and sore throat, with fever and cough in 10–20% of cases [20]. Later, in 1967, HCoV-OC43 was disengaged from organ culture, resulting in a sequential entry in the cerebrum of nursing mice. The clinical features of HCoV-OC43 infection give off an impression of resembling those caused by HCoV-229E, which are indistinguishable from diseases with other respiratory tract pathogens such as influenza A viruses and rhinoviruses [21]. Both HCoV-229E and HCoV-OC43 circulate globally, and they are prevalently diffused during the cold period in a moderate climate [22]. Developing these two viruses is less than one week, straggled by around a 2-week disease [21]. According to a human volunteer study, healthy individuals infested with HCoV-229E developed a slight common cold [23].

2.2 SARS-CoV

The first case of SARS-CoV-1 was discovered in late 2002 in Guangdong Province of China. SARS is an infectious disease caused by a virus belonging to the coronavirus family, SARS-CoV-1. The SARS epidemic has expanded across many countries and continents and caused about 8096 reported cases with 774 deaths. The incubation period of SARS-CoV-1 was 4 to 7 days and the peak of viral load was estimated on the 10th day of illness.

Patients infected with SARS-CoV-1 showed initial symptoms of myalgia, headache, fever, malaise, and chills, followed by dyspnea, cough, and respiratory distress as late symptoms of lymphopenia. However, deranged liver function tests and elevated creatine kinase are common laboratory abnormalities of SARS [24, 25]. The insectivorous bat has been identified as an animal reservoir of the SARS coronavirus. The intermediate host that allowed the virus transmission to humans is the masked palm civet, a wild animal sold in markets and eaten in southern China [26].

2.3 HCoV-NL63 et HCoV-HKU1

In late 2004, HCoV-NL63 was isolated from a 7-month-old child in the Netherlands. It was initially prevalent in young children, the elderly, and immunocompromised patients with respiratory illnesses [27]. The common symptoms of the disease caused by HCoV-NL63 are coryza, conjunctivitis, fever, and bronchiolitis [28]. It is distributed globally and it has been estimated that HCoV-NL63 accounts for nearly 4.7% of common respiratory diseases, and its peak incidence occurs during early summer, spring, and winter [22].

In the same year, HCoV-HKU1 was isolated in Hong Kong from a 71-year-old man hospitalized with pneumonia and bronchiolitis [29]. HCoV-HKU1 was reported to be associated with acute asthmatic exacerbation besides community-acquired pneumonia and bronchiolitis [30]. Alike to HCoV-NL63, HCoV-229E, and HCoV-OC43, HCoV-HKU1 was found worldwide, producing mild respiratory diseases [30].

These four community-acquired HCoVs have been well accustomed to humans and are less probable to mutate to produce exceptionally pathogenic diseases, however, accidents can occur for unclear details as in the uncommon case of subtype HCoV-NL63, which is more virulent and has recently been reported to cause severe lower respiratory tract infection in China [31].

As it has been shown for HCoV-NL63, HCoV-229E, and HCoV-OC43, HCoV-HKU1 has a worldwide distribution and causes mild respiratory diseases [30]. However, the subtype HCoV-NL63, which was found to be more virulent, caused recently severe respiratory tract infection in China [31].

2.4 MERS-CoV

In 2012, a new respiratory virus called MERS-CoV for Middle East Respiratory Syndrome coronavirus was detected in the lung of a 60-year-old patient who developed acute pneumonia and renal failure in Saudi Arabia [32, 33].

The virus was then reported in several countries in the Middle East. Since then, 1219 cases have been diagnosed, resulting in 449 deaths. Few cases have been detected in Europe, including 2 cases in France and 3 cases in Tunisia in 2013 [34]. Later in 2015, 186 confirmed cases were reported in South Korea. Compared to SARS, MERS is a similar disease with a progressive acute pneumonia. However, unlike SARS, many patients with MERS also developed acute renal failure [32, 33]. Over 30% of patients also showed gastrointestinal symptoms, such as diarrhea and vomiting [32, 33].

As of February 2020, over 2500 confirmed cases were accounted for with an intense case fatality of 34.4%, making MERS-CoV one of the most pathogenic viruses known to humans [35].

2.5 SARS-CoV-2

SARS-CoV-2 was first reported in a group of pneumonia patients of unknown etiology who witnessed their visit to Huanan Seafood Wholesale Market in December 2019 in Wuhan, Hubei Province, China [36, 37]. At the beginning of the 2020s, the 2019 coronavirus disease emerged around the world and become a pandemic, which disrupts human activity through general confinements and strict sanitary measures. The incubation of SARS-CoV-2 lasts from 2 to 14 days [38]. The most frequent signs in patients were fever, cough, fatigue, anosmia, ageusia, muscle pain, chills, sore throat, rhinitis, and headache head [11]. Additionally, gastrointestinal symptoms have been reported, diarrhea, stomach pain, vomiting, nausea, and poor appetite [11]. The nucleotide sequence of SARS-CoV-2 revealed about 51.8 and 79.0% of similarity with MERS-CoV and SARS-CoV-1, respectively, and is closely related to SARS-like coronavirus of bald origin—mouse (bat-SL-CoVZC45) with 87.6–89%

identity [16–25, 27–33, 35–39]. Recent studies strongly support the hypothesis that SARS-CoV-2 may have originated in bats and may have undergone host jumping to another intermediate mammal, including pangolins (Manis javanica) [10, 40]. Phylogenetic analyzes of the SARS-CoV-2 genome revealed this virus bind with the same human cellular receptor (ACE2: Angiotensin-Converting Enzyme 2) as same as SARS-CoV-1 to get into host cells with 10 times affinity higher than SARS-CoV-1,

WHO label	Pango lineage	Clade/lineage GISAID	Clade next strain	First samples	Listed designation date
Lambda	C.37	GR/452Q.V1	20D	Peru, Dec. 2020	14 June 2021
Mu	B.1.621	GH	21 h	Colombia, Jan. 2021	30 August 2021

Table 1.
Variants to follow VOI of SARS-CoV-2 [46].

Lines PANGO*	Clade GISAID	Clade next strain	First samples listed	Date of designation
B.1.427 B.1.429	GH/452R. V1	21C	U.S.A., Mar 2020	VOI: 5 Mar, 2 July 2021 VUM: 6 Apr 2021
R.1	GR	—	Many countries, Jan 2021	07 Apr 2021
B.1.466.2	GH	—	Indonesia, Nov 2020	28 Apr 2021
B.1.1.318	GR	20B	Many countries, Jan 2021	02 June 2021
B.1.1.519	GR	20. B/S.732 A	Many countries, Jan 2020	02 June 2021
C.36.3	GR	—	Many countries, Jan 2021	16 June 2021
B.1.214.2	G	—	Many countries, Nov 2020	30 June 2021
B.1.1.523	GR	—	Many countries, May 2020	14 July 2021
B.1.619	G	20 A/S.126 A	Many countries, May 2020	14 July 2021
B.1.620	G	—	Many countries, Nov 2020	14 July 2021
C.1.2	GR	—	South Africa, May 2021	01 Sept 2021
B.1.617.1§	G/452R. V3	21B	India, Oct 2020	VOI: 4 Apr 2021 VUM: 20 Sept 2021
B.1.526	GH/253G. V1	21F	U.S.A., Nov 2020	VOI: 24 Mar 2021 VUM: 20 Sept 2021
B.1.525	G/484 K. V3	21D	Many countries, Dec 2020	VOI: 17 Mar 2021 VUM: 20 Sept 2021

Table 2.
Variants under intensive care or VUM of SARS-CoV-2.

while MERS-CoV uses another (DPP4) [41, 42] ... Genetic variability of SARS-CoV-2 can be the consequence of nucleotide incorporation errors by viral RNA polymerase, genomic editing by cellular restriction factors, or even homologous recombination. The expansion of SARS-CoV-2 variants was observed in fall 2020 [43]. The evolutionary mutation rate of SARS-CoV-2 is estimated at 1.10^3 nucleotide substitutions per site per year [44] equivalent to approximately one substitution every two weeks in the genome [45]. Currently, the WHO considered five variants as "worrying," which were first detected in England, South Africa, and then later in Brazil (two variants were observed there, including P1 classified as worrying). In October 2020, a fourth variant (Delta) appeared in India received particular attention. This country of 1.3 billion people has seen an explosion of cases is resisted by other nations. At the end of November 2021, it was the Omicron variant, detected in South Africa, which caused the recent wave and concern all over the world.

Besides other variants called VOI ("variant under investigation" or "variant of interest" in English) has been detected in multiple countries and identified with mutations that lead to amino acid changes associated with phenotypic changes (confirmed or suspected) responsible for community transmission or multiple confirmed cases or clusters (**Table 1**) [46].

And diverse variants called under evaluation, or VUM ("variant under monitoring") exhibit genetic changes suspected of affecting the characteristics of the virus, indicating that it may pose a future risk without evidence of phenotypic or epidemiological repercussions being clear at this time, and which should be investigated repeated evaluation and enhanced surveillance pending confirmation of new evidence (**Table 2**) [46].

3. Animal coronavirus diseases

3.1 Pet coronavirus (dogs and cats)

Canine coronaviruses: Canine enteric coronavirus (CCoV) typically infects dogs, especially those housed in large groups such as kennels, shelters, and breeding facilities. CCoV belongs to the family *Coronaviridae*, order *Nidovirales*, and was first isolated in 1971 during an outbreak of gastroenteritis in military dogs and then perceived as a pathogen of dogs [47]. Dogs attracted with CCoV developed self-limiting enteritis with mild diarrheal disease. Two types of canine coronaviruses are known: CCoV is a member of the Alpha-coronavirus genus [48], and canine respiratory coronavirus (CRCoV) belonging to the beta coronavirus genus [49]. CCoV is closely associated with infectious gastroenteritis virus (TGEV) of pigs, ferret coronavirus, and feline coronavirus (FCoV) [48], while CRCoV is more related to bovine coronavirus [50]. All enteric CCoVs (along with the related viruses of cats, pigs, and ferrets) are given the similar strain designation (Alpha-coronavirus-1) from a taxonomic perspective. However, there are two distinct serotypes of CCoV: type I and type II [51, 52].

Cat coronaviruses: Two alpha coronaviruses are known for cats: feline enteric coronavirus or feline enteric peritonitis (FECV) associated with mild or asymptomatic diarrhea and can mutate to cause more serious feline infectious peritonitis (FIP) or feline infectious peritonitis (FIPV). Subclinical carriers of FECV handle the shedding and transmission of the virus to different felines through the fecal-oral route. Experiences realized on young, ancient, or immunocompromised subjects have revealed two clinical forms of FIP, a wet form, and a dry form. The wet form developed ascites which can be clinically apparent (abdominal distention) and confirmed by ultrasound. The dry form is associated with granulomatous lesions of various locations, basically affecting the eye, the central nervous system, the liver, and kidneys. In Addition, uveitis with keratin deposits on the level of the cornea was reported. The FIP is lethal and there is no particular treatment or vaccine marketed in France for this disease. Cats can also be infected with other coronaviruses such as SARS-CoV-2, transmissible porcine gastroenteritis, canine coronavirus, or human coronavirus 229E [53].

3.2 Coronaviruses of production animals

The bovine, porcine, and avian coronaviruses are mainly affecting production animals. These coronaviruses belong to alpha, beta, gamma, and/or delta-coronaviruses.

Bovine Coronavirus: Bovine coronaviruses (BCoVs) are pneumo-enteric viruses that infect the upper and lower respiratory tract of cattle and wild ruminants and it is rejected in feces and nasal secretions.. In cattle, BCoV causes 3 different clinical syndromes in cattle: calf diarrhea, winter dysentery with hemorrhagic diarrhea in adults and respiratory infections in cattle of different ages and the bovine respiratory disease complex or shipping fever of feedlot cattle. Distinction between these syndromes is not yet possible as antigenic or genetic specific markers have not been recognized. To our knowledge, no BCoV vaccines to prevent respiratory BCoV diseases in cattle [54].

Swine Coronavirus: Porcine coronaviruses are members of three genera: alpha-, beta- and delta- coronavirus. Since 1946, transmissible gastroenteritis virus or TGEV in alpha coronaviruses has been discovered. It developed severe, often fatal enteritis in piglets. Symptoms of vomiting and profuse diarrhea are registered for piglets less than a week old and the mortality rate was about 100%.

Later, in 1971, porcine epidemic diarrhea or DEP (Porcine enteritis disease virus or PEDV) was first described in England and caused watery diarrhea occasionally accompanied by vomiting [55]. Since 2013, a severe pathogenic variant of DEP has affected North America and then spread throughout the world, causing serious economic losses. This disease has been included in France to the list of first-category health hazards for emerging animal species. Several variants of the TGE virus are the source of several strains of porcine respiratory coronavirus (Porcine respiratory coronavirus or PRCV), responsible for discreet respiratory disorders [53]. Among these viruses, one of them infected almost most European pig herds in 1984. Two other coronaviruses were identified in 2016 also with digestive tropism [55]: the porcine enteritis coronavirus (swine enteritis coronavirus or SeCoV), a recombinant virus containing a TGEV genome in which the gene S is replaced by that of PEDV, and the porcine acute diarrhea virus (swine acute diarrhea syndrome or SADS) [56].

3.3 Avian coronaviruses

CoVs in the chicken: Among avian coronaviruses classified as gamma and delta-coronaviruses, the first avian coronavirus (infectious bronchitis virus or IBV virus) was described in 1931 and caused many economic losses in poultry farms (laying eggs and broilers). The disease is characterized by various lesions and damages in the genital tract with a drop in the rate of laying, malformation of the eggs, and low mortality. Besides other systems may be affected such as the respiratory tract, kidneys, etc. [53].

CoV of turkeys: Aside from IBV in chickens, the main avian species in which CoV has been definitively associated with the disease are the turkey, pheasant, and guinea fowl. Turkey coronavirus (TCoV) has been known, since the 1940s, to cause of enteric disease in turkeys in the USA. This disease is reported is found worldwide [57, 58]. Turkeys of all ages can be infected with high mortality in young poults. Most frequently reported clinical signs include decreased feed and water intake, wet droppings, diarrhea, and loss of body weight. TCoV is likewise associated with poultry enteritis and mortality syndrome (PEMS) which means high mortality, growth retardation, and immune dysfunction. Inbreeding turkeys, aberrant egg-laying performance is related to TCoV disease, similar to that seen in IBV infections of chickens [57, 58].

CoVs of pheasants (PhCoV): In pheasants, respiratory and renal problems have been associated with infections with CoV. PhCoV is closely related closely related to IBV and TCoV [59].

CoVs of guinea fowl (GfCoV): GfCoV causes acute enteritis, showing a high death rate, possibly pancreatic degeneration, and fulminating disease in guinea fowl [60]. Genetically, GfCoV shows similarity to both IBV and TCoV, however, differences were observed in the spike gene, and a common ancestor has been suggested for the three viruses [60].

3.4 SARS-CoV-2 animal infections

Under natural conditions, SARS-CoV-2 infection has been observed in owner-infected animals. This case is called Spillback when infections are gained by animals through contact with humans. Few confirmed cases of SARS-CoV-2 in pets were reported in diverse countries: France (2 cats), Spain (2 cats), Germany (1 cat), Russia (1 cat), China (2 dogs and a cat in Hong Kong), Belgium (4 cats), the United States (31 cats and 24 dogs), United Kingdom (one cat), Japan (4 dogs), Chile (one cat), Canada (one dog), Brazil (one cat), Denmark (a dog), Italy (a dog) [61]. A recent French study has shown for the first time a significant circulation of SARS-CoV-2 in a population of pets (34 cats and 13 dogs) whose owners were infected with COVID-19 [62].

Experimental conditions have revealed that pigs and poultry are resistant to every inoculation with SARS-CoV-2 [63] while rabbits (which are also pets or laboratory animals) [64], and other laboratory animals include the golden hamster (*Mesocricetus auratus*) and rhesus macaque (*Macaca mulatta*) have been susceptible to SARS-CoV-2 [65]. Contrarily, laboratory mice and rats were resistant to SARS-CoV-2 [66]. The SARS-CoV-2 could also threaten many species beyond the great apes. In January 2021, gorillas at the San Diego Zoo were tested positive for COVID-19. None of the animals died, luckily, but were suffered from high fevers, lethargy, and cough like humans [67].

On November 29, 2021, recent reports have proven that SARS-CoV-2 has been transmitted from humans to wild white-tailed deer in the United States, but conversely, no cases of transmission from deer to humans have been reported. All were "apparently in good health," and "showed no clinical signs of the disease" [67].

In the United States, 4 tigers and 3 lions were probably infected by humans in a zoo in the Bronx. They presented mild respiratory symptoms. Since then, a tiger and a puma have been also reported infected [68]. Recent studies from the Friedrich-Loeffler Institute in Germany reported raccoon dogs (canids bred in China for their fur) previously susceptible to SARS-CoV-1 were also susceptible to SARS-CoV-2 and could contaminate other raccoon dogs by direct contact with no clinical signs. These animals can be intermediate hosts potentially involved in the emergence of COVID-19 [69]. On the other side, infected mink farms by SARS-CoV-2 were detected (2 on April 26, 33 to August 14, then 52 to September 14). Two million mink were then culled by the Dutch authorities. As of September 1, the first human cases contaminated by mink were reported [70] 66 of the 97 employees of these farms tested positive for SARS-CoV-2, with whole-genome sequencing revealing mink-like variants in 47 cases [71].

4. Detection of SARS-CoV-2 in wastewater

Wastewater-Based epidemiology (WBE) has been successfully used to investigate polio circulation within the community. This novel biomonitoring tool has been successfully used to evaluate international poliovirus vaccine campaigns and to investigate the use of some illicit drugs. Additionally, this tool has been successfully used to detect the occurrence of hepatitis and norovirus outbreaks [72, 73].

The environmental circulation of viruses as human pathogens has been given more attention since the first occurrence an spread of Severe Acute Respiratory Syndrome Coronavirus 1 appeared (SARS-CoV-1) in 2003 and Middle East Respiratory Syndrome (MERS) in 2012. Even more focus on the development of surveillance systems of viruses in the environment has been reported since the first occurrence of COVID-19 in December 2019 in Wuhan, China [73, 74].

Since most patients infected with SARS-Cov-2 might be asymptomatic, rapid and accurate detection of potential virus carriers is a critical step to suppress the risk of disease transmission at an early stage of the disease [75]. SARS-CoV-2 has been shown to replicate actively in enterocytes of the human intestine, where there is the highest expression of ACE2 in the human body and the virus is excreted in the feces [76]. SARS-CoV-2 RNA has been detected worldwide in raw wastewater and sometimes in treated wastewater, which could imply potential environmental transmission via the water cycle [77–79]. SARS-CoV-2 RNA has been reported in wastewater treatment plants in various nations around the world such as Australia, Italy, Spain, the Netherlands, the United States, Japan, Germany, the Arab Emirates States, Istanbul, and Brazil [12]. The duration of the shedding through feces can be as long as 33 days, with a decreased shedding rate, ranging from 106 to 1012 gc/L, which is lower than some other infectious viruses, like MERS-CoV, and SARS-CoV-1 [80, 81].

Detection of SARS-CoV-2 RNA in wastewater was performed by PCR-based methods such as reverse transcription-polymerase chain reaction (RT-PCR) and digital PCR using the amplification of parts of the viral genome, such as the genes coding for the nucleocapsid [82] and the viral envelope [83]. To gain insights into the fate and transport of SARS- CoV-2 in WWTFs, the general workflow for SARS-CoV-2 testing in wastewater is conducted in the following order sample collection, sample concentration, RNA extraction and analysis, and data reporting [84, 85]. Molecular detection of viral RNA involves three major steps:

Viral concentration/enrichment: A viral enrichment step is recommended before RNA extraction because of the potential low concentration of viral titer in the wastewater. Viral particles are concentrated and recovered by polyethylene glycol (PEG) precipitation, or by filtration using 0.2 μm filters [7] ultrafilters [78], and ultracentrifugation [86]. Direct RNA extraction from electronegative membranes (0.45 μm) is another method that can be used [87]. For virus concentration, a variety of techniques have been explored, including polyethylene glycol (PEG)-NaCl precipitation, ultrafiltration, $AlCl_3$, flocculation, and others [88]. Because of its better selectivity and tolerance to PCR inhibitors in wastewater, PEG-NaCl precipitation is the most widely used technique [89].

RNA extraction: RNA extraction has typically been performed using commercial kits from a variety of supplies. The most commonly used RNA extraction kits are the RNeasy Power Microbiome kit [78, 87, 90], the BioMérieux Nuclisens kit [78], Power Fecal Pro-Kit [78], and RNeasy Power water Kit [87].

Amplification of viral RNA: Amplification of viral RNA extracted from wastewater was performed with a set of five primers/probes. These primers and probes target different regions of the viral particle in **Table 3**.

Gene	Probe	Sequence	References
The nucleocapsid (N)	2019-nCoV_N1-F 2019-nCoV_N1-R 2019-nCoV_N1-P	5'-GACCCCAAAATCAGCGAAAT-3' 5'-TCTGGTTACTGCCAGTTGAATCTG-3' 5'-FAM-ACCCCGCATTACGTTTGGTGGACC-ZEN/ Iowa Black-3'	[7, 79, 92] [7, 79]
The nucleocapsid (N)	2019-nCoV_N2-F	5'-TTACAAACATTGGCCGCAAA-3	
	2019-nCoV_N2-R	5'-GCGCGACATTCCGAAGAA-3'	
	2019-nCoV_N2-P	5'-FAM—ACAATTTGCCCCCAGCGCTTCAG—ZEN/ Iowa Black-3'	
The the nucleocapsid (N)	2019-nCoV_N3-F	5'-GGGAGCCTTGAATACACCAAAA-3'	[7, 79]
	2019-nCoV_N3-R	5'-TGTAGCACGATTGCAGCATTG-3	
	2019-nCoV_N3-P	5'-FAM-AYCACATTGGCACCCGCAATCCTG-ZEN/Iowa Black-3'	
Envelop (E)	E_Sarbeco_F	5'-ACAGGTACGTTAATAGTTAATAGCGT-3'	[79, 93]
	E_Sarbeco_R E_Sarbeco_P1 Cor-p-F2 (+) Cor-p-F3 (+)	5' — ATATTGCAGCAGTACGCACACA-3' 5'-FAM—ACACTAGCCATCCTTACTGCGCTTCG—ZEN/Iowa Black-3' 5'-CTAACATGCTTAGGATAATGG-3' 5'-GCCTCTCTTGTTCTTGCTCGC-3'	[94]
	Cor-p-R1 (−)	5'-CAGGTAAGCGTAAAACTCATC-3'	
ORF1ab		5' — CCCTGTGGGTTTTACACTTAA-3' 5'-ACGATTGTGCATCAGCTGA-3' 5'-FAM-CCGTCTGCGGTATGTGGAAAGGTTATGG -BHQ1–3'	[95–97]

Table 3.
Primers/probes used for amplification of SARS-CoV-2 RNA in wastewater [91].

Varying results have been reported using these primer/probe sets targeting different parts of the viral genome. For example, [79] found that primer N1 resulted in positive amplification of all study sites (6), but primers N3 and E resulted in positive amplification of 5 and 4 study sites, respectively. However, Rimoldi et al. [97] found a high frequency of positive amplification targeting the ORF1ab gene, compared to only three positive wastewater samples for the N and E genes. As a result, our findings are equivocal in terms of the optimum primer/probe combination for viral RNA amplification in wastewater. This could be attributed to the sensitivity of primers/probes, PCR inhibitors in wastewater samples from different regions/sites, and the potential stability of the virus and viral genome in these different areas [98]. Droplet digital PCR is another molecular technique used for the detection of coronaviruses in clinical and sewage samples. This was found to have an improved, more sensitive, and more accurate lower limit of detection than RT-PCR for environmental samples [99, 100].

Khan et al. [101] discovered that smaller sample volumes (50–100 ml), 30% (w/v) PEG-NaCl, a 12-hour incubation interval, and a 24-hour storage period resulted in improved RNA recoveries in terms of N1 and N2. RNA concentrations were always at least one order of magnitude greater in RT-qPCR than in RT-ddPCR. However, under all test conditions, both RT-qPCR and RT-ddPCR revealed that RNA is generally absent in the sludge samples, resulting in a false-negative result.

5. Risks of environmental transmission of SARS-CoV-2 in wastewater

Fecal-oral transmission of SARS-CoV-2 is yet to be approved, but additional research is essential to clarify the potential risks of the novel coronavirus in sanitation systems. The SARS-CoV-2 virus has been detected in fecal samples and effluents. Contaminated drinking water, contaminated raw, undercooked aquatic aquaculture, sewage-irrigated food, and vector-mediated transmission are all possible sub-pathways of the fecal-oral mode of transmission. Seepage from sanitation systems (pit latrines and septic tanks), landfill leachates without geomembrane protection toward shallow groundwater systems can pollute drinking water sources. In other types of coronaviruses, one study found 99.9% percent fatality after 10 days in tap water at 23°C and over 100 days at 4°C. This data also suggests that coronaviruses have a longer survival duration in tap water than in wastewater [102].

The exposure of humans to viruses, including SARS-CoV-2 through bioaerosol and wastewater aerosols has been highlighted. For example, a laboratory study investigating the persistence of SARS-CoV-2 in aerosols showed that the virus keeps its viability and infectivity in aerosols for up to 16 h [103].

Therefore, human and animal exposure to SARS-CoV-2 via wastewater aerosols could be significant in shared sanitation systems, especially in crowded informal settlements in developing countries [104]. Various studies have registered the prevalence of SARS-CoV-2 in urban and rural sewer systems. This wastewater might contaminate fresh water; it can pass through untreated effluent discharged to surface waters or leak and affect the supply of traditionally treated graywater. These recycled urban waters also represent possible modes of transmission [104].

In some regions with a high prevalence of COVID-19 disease, SARS-CoV-2 was prevalent in surface water, including both saltwater and freshwater. Coronaviruses from anthropogenic activities were confirmed in different water bodies [102, 105]. Marine and fresh aquatic foods such as fish and crustaceans may be contaminated by raw wastewater. Marine foods from coastal areas receiving untreated wastewater, aquatic food acquired from surface aquatic systems receiving raw or partially treated wastewater, and raw wastewater-irrigated salad crops are all possible sources of food transmission.

Raw wastewater-aquacultural systems and raw wastewater irrigation of crops consumed raw, such as salads, are two more techniques that promote food contamination. However, more research is needed to determine the prevalence and durability of SARS-CoV-2 in marine and surface aquatic systems, as well as food derived from these sources. Such research should also look into the effects of various food pre-treatments and culinary processes on SARV-CoV-2 persistence. Studies based on genomic and phylogenetic analyses are needed to evaluate whether SARS-CoV-2 may leap from aquatic environments to humans. This is important given the interactions between humans and wildlife, including the widespread consumption of aquatic and terrestrial animals [106].

6. Disinfection and eradication procedures of SARS-CoV-2 in wastewater

Information from the general suppression of viruses and surrogates of coronaviruses could be used, with caution, to give additional information on the possible suppression of these viruses. For example, [107] observed that activated sludge treatment (ASP) processes in subtropical conditions removed over 3 logs 10 of enteric viruses. ASP is a commonly used wastewater treatment process around the world [108–110]. This treatment process includes primary settling, biological degradation, and secondary clarification [107, 111]. Ye et al. [112] demonstrated that during ASP processes, the highest removal of coronaviruses can occur at the primary settling stage.

For example, a sewage pond system [113] reported an average reduction of 1 log10 of viruses for 14.5 to 20.9 days of retention. Besides adsorption on particles, a longer HRT (hydraulic retention time) may be required for coronavirus inactivation in wastewater. Because coronaviruses adsorb to solid surfaces, a large concentration can be expected in the sludge. Anaerobic digestion of sludge, which is a typical sludge treatment method, reduces pathogenic bacteria. The most commonly used membrane technologies in wastewater treatment are microfiltration (0.1–0.2 μm) and ultrafiltration (0.005 ≈ 10 μm). There are reports of microfiltration membranes with larger pore sizes (0.2 to 0.4) being used [114]. The best membrane technology for coronavirus removal is ultrafiltration with an average viral particle diameter of 120 nm (0.12 μm) and an envelope diameter of 80 nm (0.08 μm) [115]. Adsorption of coronaviruses on wastewater solids can enhance their removal. Tertiary wastewater treatment processes such as chlorination and UV treatment can also result in further removal of remaining coronaviruses in wastewater [98]. Chlorine has been reported to inactivate viruses through the cleavage of the virus capsid protein backbone, inhibiting the injection of the viral genome into host cells [116, 117].

The inactivation of coronaviruses by UV irradiation has also been reported in several studies [118–120]. Enveloped viruses, like coronaviruses, are more sensitive to UV than non-enveloped viruses. The mechanism by which UV inactivates coronaviruses is the generation of pyrimidine dimers which damage nucleic acid [94]. Methods of disinfection used in the drinking water treatment inactivate efficiently SARS-CoV-2 in water [121]. However, there is a need to investigate and ameliorate the performance of disinfection technologies to be adopted for the inactivation of SARS-CoV-2 in municipal and hospital wastewater to reduce the related risk of possible infections [121].

7. Conclusion

There has been a significant expansion that proved pathogenic viruses in the wastewater and/or treatment plants, including the novel coronavirus. Understanding the destiny of SARS-CoV-2 in wastewater treatment plants has arisen as an issue of extreme importance. The epidemiological surveillance of these viruses in wastewater would help to prevent the spread of the viral disease while producing safe treated water for reuse. Thus, the performance of various treatment procedures is now being explored to reduce viral disease outbreaks.

Potential dangers of SARS-CoV-2 transmission through water infrastructure are a major source of concern in the environmental setting, and detection and eradication will play a key role in limiting the virus's spread in the population. To comprehend the early warning of outbreaks and to effectively inactivate before emerging, the virus must be a regular criterion for routine monitoring with other quality metrics with environmental samples. A regulatory framework that incorporates environmental systems will help to protect the global community from future outbreaks and transmissions.

Author details

Chourouk Ibrahim[1,2]*, Salah Hammami[3], Eya Ghanmi[1,2] and Abdennaceur Hassen[2]

1 Faculty of Mathematical, Physical and Natural Sciences of Tunis, University of Tunis El Manar, Tunis, Tunisia

2 Centre of Research and Water Technologies (CERTE), Laboratory of Treatment and Wastewater Valorization, Techno Park of Borj-Cédria, Sulayman, Tunisia

3 National School of Veterinary Medicine at Sidi Thabet, University of Manouba, Tunis, Tunisia

*Address all correspondence to: ibrahimchourouk@yahoo.fr

IntechOpen

References

[1] Alafeef M, Dighe K, Moitra P, Pan D. Monitoring the viral transmission of SARS-CoV-2 in still waterbodies using a lanthanide-doped carbon nanoparticle-based sensor Array. ACS Sustainable Chemistry & Engineering. 2022;**10**(1):245-258. DOI: 10.1021/acssuschemeng.1c06066

[2] ECDC COVID-19 Situation Updates Worldwide, as of Week 1. 2022. Available from: https://www.ecdc.europa.eu/en/geographical-distribution-2019-ncov-cases [Accessed: January 2022]

[3] Xu X, Yu C, Qu J, Zhang L, Jiang S, Huang D, et al. Imaging and clinical features of patients with 2019 novel coronavirus SARS-CoV-2. European Journal of Nuclear Medicine and Molecular Imaging. 2020;**47**:1275-1280. DOI: 10.1007/s00259-020-04735-9

[4] Chen Y, Liu Q, Guo D. Emerging coronaviruses: Genome structure, replication, and pathogenesis. Journal of Medical Virology. 2020;**92**(4):418-423. DOI: 10.1002/jmv.25681

[5] Beniac DR, Andonov A, Grudeski E, Booth TF. Architecture of the SARS coronavirus profusion spike brief communications. Nature Structural & Molecular Biology. 2006;**13**(8):751-752. DOI: 10.1038/nsmb1123

[6] Delmas B, Laude H. Assembly of coronavirus spike protein into trimers and its role in epitope expression. Journal of Virology. 1990;**64**(11)

[7] Wu F, Zhang J, Xiao A, Gu X, Lin Lee W, Armas F, et al. SARS-CoV-2 titers in wastewater are higher than expected from clinically confirmed cases. Msystems. 2020. DOI: 10.1128/mSystems.00614-20

[8] Sola I, Almazán F, Zúñiga S, Enjuanes L. Continuous and discontinuous RNA synthesis in coronaviruses. Annual Review of Virology. 2015, 2015;**2**:265-288. DOI: 10.1146/annurev-virology-100114-055218

[9] Gorbalenya AE, Enjuanes L, Ziebuhr J, Snijder EJ. Nidovirales: Evolving the largest RNA virus genome. Virus Research. 2006;**117**(1):17-37. DOI: 10.1016/j.virusres.2006.01.017

[10] Hozhabri H, Sparascio FP, Sohrabi H, Mousavifar L, Roy R, Scribano D, et al. The global emergency of novel coronavirus (SARS-CoV-2): An update of the current status and forecasting. International Journal of Environmental Research and Public Health. 2020;**17**(16):1-35. DOI: 10.3390/ijerph17165648

[11] Ciechanowicz P, Lewandowski K, Szymańska E, Kaniewska M, Rydzewska GM, Walecka I. Skin and gastrointestinal symptoms in COVID-19. Przeglad Gastroenterologiczny. 2020;**15**(4):301-308. DOI: 10.5114/pg.2020.101558

[12] Tanhaei M, Mohebbi SR, Hosseini M, Rafieepoor M, Kazemian S, Ghaemi A, et al. The first detection of SARS-CoV-2 RNA in the wastewater of Tehran, Iran. Environmental Science and Pollution Research. 2021;**28**(29):38629-38636. DOI: 10.1007/s11356-021-13393-9

[13] Olive G, Lertxundi U, Barcelo D. Early SARS-CoV-2 outbreak detection by sewage-based epidemiology. Science of the Total Environment. 2020;**732**:139298. DOI: 10.1016/j.scitotenv.2020.139298

[14] O'Bannon D. Femmes dans la qualité de l'eau. 2020. Available from:

https://link.springer.com/content/
pdf/10.1007/978-3-030-17819-2.pdf

[15] Mohan SV, Hemalatha M, Koppert H,
Ranjith I, Kumar AK. SARS-CoV-2 in
environmental perspective: Occurrence,
persistence, surveillance, inactivation,
and challenges. Chemical Engineering
Journal. 2021;**405**:126893. DOI: 10.1016/j.
cej.2020.126893

[16] Ren L-L, Wang Y-M, Wu Z-Q, Xiang
Z-C, Guo L, Xu T, et al. Identification
of a novel coronavirus causing severe
pneumonia in human: A descriptive
study. Chinese Medical Journal. 2020;
133(9):1015-1024. DOI: 10.1097/CM9.
0000000000000722

[17] Chu DKW, Pan Y, Cheng SMS,
Hui KPY, Krishnan P, Liu Y, et al.
Molecular diagnosis of a novel
coronavirus (2019-nCoV) causing
an outbreak of pneumonia. Clinical
Chemistry. 2020;**66**(4):549-555.
DOI: 10.1093/clinchem/hvaa029

[18] Cyranoski D. Did pangolins spread
the China coronavirus to people? Nature.
2020. DOI: 10.1038/d41586-020-00364-2

[19] Hamre D, Procknow JJ. A new virus
isolated from the human respiratory
tract. Proceedings of the Society for
Experimental Biology and Medicine.
1966;**121**(1):190-193. DOI: 10.3181/
00379727-121-30734

[20] Tyrrell DA, Cohen S, Schlarb JE.
Signs and symptoms in common colds.
Epidemiology and Infection.
1993;**111**(1):143-156. DOI: 10.1017/
s0950268800056764

[21] McIntosh K, Dees JH, Becker WB,
Kapikian AZ, Chanock RM. Recovery
in tracheal organ cultures of novel
viruses from patients with respiratory
disease. Proceedings. National Academy
of Sciences. United States of America.

1967;**57**(4):933-940. DOI: 10.1073/
pnas.57.4.933

[22] Su S, Wong G, Shi W, Liu J, Lai ACK,
Zhou J, et al. Epidemiology, genetic
recombination, and pathogenesis of
coronaviruses. Trends in Microbiology.
2016;**24**(6):490-502. DOI: 10.1016/j.
him.2016.03.003

[23] Bradburne AF, Bynoe ML,
Tyrrell DA. Effects of a "new" human
respiratory virus in volunteers. British
Medical Journal. 1967;**3**(5568):767-769.
DOI: 10.1136/bmj.3.5568.767

[24] Cheng VC, Lau SK, Woo PC,
Yuen KY. Severe acute respiratory
syndrome coronavirus as an agent of
emerging and reemerging infection.
Clinical Microbiology Reviews.
2007;**20**(4):660-694. DOI: 10.1128/
CMR.00023-07

[25] Peiris JS, Lai ST, Poon LL, et al.
Coronavirus as a possible cause of the
severe acute respiratory syndrome.
Lancet. 2003;**361**(9366):1319-1325.
DOI: 10.1016/s0140-6736 (03)13077-2

[26] Available from: https://www.pasteur.
fr/fr/centre-medical/fiches-maladies/sras

[27] van der Hoek L, Pyrc K, Jebbink MF,
Vermeulen-Oost W, Berkhout RJ,
Wolthers KC, et al. Identification of
a new human coronavirus. Nature
Medicine. 2004;**10**(4):368-373.
DOI: 10.1038/nm 1024

[28] Abdul-Rasool S, Fielding BC.
Understanding human coronavirus
HCoV-NL63. Open Virology Journal.
2010;**4**:76-84. DOI: 10.2174/187435790
1004010076

[29] Woo PC, Lau SK, Chu CM,
Chan KH, Tsoi HW, Huang Y, et al.
Characterization and complete genome
sequence of a novel coronavirus,
coronavirus HKU1, from patients

with pneumonia. Journal of Virology. 2005;**79**(2):884-895. DOI: 10.1128/JVI.79.2.884-895.2005

[30] Lau SK, Woo PC, Yip CC, Tse H, Tsoi HW, Cheng VC, et al. Coronavirus HKU1 and other coronavirus infections in Hong Kong. Journal of Clinical Microbiology. 2006;**44**(6):2063-2071. DOI: 10.1128/JCM.02614-05

[31] Wang Y, Li X, Liu W, Gan M, Zhang L, Wang J, et al. Discovery of a subgenotype of human coronavirus NL63 associated with severe lower respiratory tract infection in China. Emerg Microbes Infect. 2018;**9**(1):246-255. DOI: 10.1080/22221751.2020.1717999

[32] Hilgenfeld R, Peiris M. From SARS to MERS: 10 years of research on highly pathogenic human coronaviruses. Antiviral Research. 2013;**100**(1):286-295. DOI: 10.1016/j.antiviral.2013.08.015

[33] Gao H, Yao H, Yang S, Li L. From SARS to MERS: Evidence and speculation. Frontiers in Medicine. 2016;**10**(4):377-382. DOI: 10.1007/s11684-016-0466-7

[34] Available from: https://www.pasteur.fr/fr/centre-medical/fiches-maladies/mers-cov

[35] Ye ZW, Yuan S, Yuen KS, Fung SY, Chan CP, Jin DY. Origines zoonotiques des coronavirus humains. International Journal of Biological Sciences. 2020;**16**(10):1686-1697. DOI: 10.7150/ijbs.45472

[36] Lu H, Stratton CW, Tang Y-W. The outbreak of pneumonia of unknown etiology in Wuhan, China: The mystery and the miracle. Journal of Medical Virology. 2020;**92**. DOI: 10.1002/jmv.25678

[37] Johnson M. Wuhan 2019 novel coronavirus – 2019-nCoV. Materials

and Methods. 2020;**10**(Jan):1-5. DOI: 10.13070/mm.en.10.2867

[38] Wu F, Zhao S, Yu B, Chen Y-M, Wang W, Song Z-G, et al. A new coronavirus associated with human respiratory disease in China. Nature. 2020;**579**. DOI: 10.1038/s41586-020-2008-3

[39] Chan JFW, Yuan S, Kok KH, To KKW, Chu H, Yang J, et al. A familial cluster of pneumonia associated with the 2019 novel coronavirus indicating person-to-person transmission: A study of a family cluster. The Lancet. 2020;**395**(10223):514-523. DOI: 10.1016/S0140-6736(20)30154-9

[40] Zhang N, Wang L, Deng X, Liang R, Su M, He C, et al. Recent advances in the detection of respiratory virus infection in humans. Journal of Medical Virology. 2020:92. DOI: 10.1002/jmv.25674

[41] Wan Y, Shang J, Graham R, Baric RS, Li F, Wan CY. Receptor recognition by the novel coronavirus from Wuhan: An analysis based on decade-long structural studies of SARS coronavirus. Journal of Virology. 2020;**94**:127-147. DOI: 10.1128/JVI.00127-20

[42] Wrap D, Wang N, Corbett KS, Goldsmith JA, Hsieh CL, Abiona O, Graham BS, Mclellan JS. Cryo-EM Structure of the 2019-nCoV Spike in the Profusion Conformation. 2019. Available from: http://science.sciencemag.org/

[43] Bertholom C. Évolution génétique du Sars-CoV-2 et ses conséquences. Option/Bio. 2021;**32**(639):22. DOI: 10.1016/S0992-5945(21)00197-5

[44] Candido DS, Claro IM, Jésus JG, Souza WM, Moreira FRR, Dellicour S, et al. Evolution and epidemic spread of SARS-CoV-2 in Brazil. Science, New York. 2020;**369**(6508):1255-1260. DOI: 10.1126/SCIENCE.ABD2161

[45] van Dorp L, Ackman M, Richard D, Shaw LP, Ford CE, Ormond L, et al. Emergence of genomic diversity and recurrent mutations in SARS-CoV-2. Infection, Genetics, and Evolution. 2020;**83**:104351. DOI: 10.1016/J.MEEGID. 2020.104351

[46] SARS-CoV-2 variants of concern as of 11 May 2021. 2021. Available from: https://www.ecdc.europa.eu/en/covid-19/variants-concern

[47] Binn LN, Lazar EC, Keenan KP, Huxsoll DL, Marchwicki RH, Strano AJ. Recovery and characterization of a coronavirus from military dogs with diarrhea. Proceedings, Annual Meeting of the United States Animal Health Association. 1974;**78**:359-366

[48] King AM, Lefkowitz E, Adams MJ, Carstens EB. Virus Taxonomy: Ninth Report of the International Committee on Taxonomy of Viruses. Vol. 9. Elsevier; 2011

[49] Erles K, Toomey C, Brooks HW, Brownlie J. Detection of a group 2 coronavirus in dogs with canine infectious respiratory disease. Virology. 2003;**310**(2):216-223. DOI: 10.1016/s0042-6822(03)00160-0

[50] Erles K, Shiu KB, Brownlie J. Isolation and sequence analysis of canine respiratory coronavirus. Virus Research. 2007;**124**(1-2):78-87. DOI: 10.1016/j.virusres.2006.10.004

[51] Decaro N, Buonavoglia C. An update on canine coronaviruses: Viral evolution and pathobiology. Veternary Microbiology. 2008;**132**(3-4):221-234. DOI: 10.1016/j.vetmic.2008.06.007

[52] Le Poder S. Feline and canine coronaviruses: Common genetic and pathobiological features. Advances in Virology. 2011;**2011**:609465. DOI: 10.1155/2011/609465

[53] Angot JL, Brugère-Picoux J. Introduction générale sur les coronavirus animaux et humains [General introduction to animal and human coronaviruses]. Bulletin de l'Académie Nationale de Médecine. 2021;**205**(7):719-725. DOI: 10.1016/j.banm.2021.05.011

[54] Saif LJ. Bovine respiratory coronavirus. The Veterinary Clinics of North America. Food Animal Practice. 2010;**26**(2):349-364. DOI: 10.1016/j.cvfa.2010.04.005

[55] Laude H. Two newly identified enteropathogenic coronaviruses in swine are related to bat or bird viruses. Bulletin de l'Académie Vétérinaire de France. 2020. DOI: 10.4267/2042/70845

[56] Pan Y, Tian X, Qin P, Wang B, Zhao P, Yang YL. Discovery of a novel swine enteric alpha coronavirus (SeACoV) in southern China. Veterinary Microbiology. 2017;**211**:15-21

[57] Jindal N, Mor SK, Goyal SM. Enteric viruses in Turkey enteritis. Virus Diseases. 2014;**25**:173-185. DOI: 10.1007/s13337-014-0198-8

[58] Guy JS. Turkey coronavirus enteritis. In: Swayne DE, Boulianne M, McDougald LR, Nair V, Suarez DL, editors. Diseases of Poultry. Vol. I. 2020. pp. 402-408

[59] JJ (Sjaak) de Wit & Jane KA Cook. Spotlight on avian coronaviruses. Avian Pathology. 2020;**49**(313-316). DOI: 10.1080/03079457.2020.1761010

[60] Liais E, Croville G, Mariette J, Delverdier M, Lucas MN, Klopp C, et al. Novel avian coronavirus and fulminating disease in guinea fowl, France. Emerging infectious diseases. 2014;**20**:105-108. DOI: 10.3201/eid2001.130774

[61] Plateforme ESA. 2020. COVID-19 et animaux. Available from: https://www.plateforme-esa.fr/

[62] Fritz M, Rosolen B, Krafft E, Becquart P, Elguero E, Vratskikh O, et al. High prevalence of SARS-CoV-2 antibodies in pets from COVID-19+ households. One health (Amsterdam, Netherlands). 2021;**11**:100192. DOI: 10.1016/j.onehlt.2020.100192

[63] Schlottau K, Rissmann M, Graaf A, Schön J, Sehl J, Wylezich C, et al. SARS-CoV-2 in fruit bats, ferrets, pigs, and chickens: An experimental transmission study. The Lancet Microbe. 2020;**1**(5):e218-e225. DOI: 10.1016/S2666-5247(20)30089-6

[64] Mykytyn AZ, Lamers MM, Okba N, Breugem TI, Schipper D, van den Doel PB, et al. Susceptibility of rabbits to SARS-CoV-2. Emerging Microbes & Infections. 2021;**10**(1):1-7. DOI: 10.1080/22221751.2020.1868951

[65] Munster VJ, Feldmann F, Williamson BN, van Doremalen N, Pérez-Pérez L, Schulz J, et al. Respiratory disease in rhesus macaques inoculated with SARS-CoV-2. Nature. 2020;**585**(7824):268-272

[66] Cohen J. Mice, hamsters, ferrets, monkeys. Which lab animals can help defeat the new coronavirus? Science| AAAS. 2020;**13**

[67] Yves Sciama. Wild and domestic animals, they also catch the Covid. Reporterre. 2021. Available from: https://reporterre.net/Animaux-sauvages-et-domestiques-ils-attrapent-aussi-le-Covid

[68] McAloose D, Laverack M, Wang L, Killian ML, Caserta LC, Yuan F, et al. From people to Panthera: Natural SARS-CoV-2 infection in tigers and lions at the Bronx zoo. MBio. 2020;**11**(5):e02220-e02220. DOI: 10.1128/mBio.02220-20

[69] Freuling CM, Breithaupt A, Müller T, Sehl J, Balkema-Buschmann A, Rissmann M, et al. Susceptibility of raccoon dogs for experimental SARS-CoV-2 infection. Emerging Infectious Diseases. 2020;**26**(12):2982-2985. DOI: 10.3201/eid2612.203733

[70] Munnink BBO, Sikkema RS, Nieuwenhuijse DF, Molenaar RJ, Munger E, Molenkamp R, et al. Jumping back and forth: Anthropozoonotic and zoonotic transmission of SARS-CoV-2 on mink farms. BioRxiv. 2020

[71] European Centre for Disease Prevention and Control. Detection of New SARS-CoV-2 Variants Related to Mink. ECDC: Stockholm; 2020

[72] Hellmér M, Paxéus N, Magnius L, Enache L, Arnholm B, Johansson A, et al. Detection of pathogenic viruses in sewage provided early warnings of hepatitis A virus and norovirus outbreaks. Applied and Environmental Microbiology. 2014;**80**(21):6771-6781. DOI: 10.1128/AEM.01981-14

[73] Sinclair RG, Choi CY, Riley MR, Gerba CP. Pathogen surveillance through monitoring of sewer systems. Advances in Applied Microbiology. 2008;**65**:249-269. DOI: 10.1016/S0065-2164 (08)00609-6

[74] Sims N, Kasprzyk-Hordern B. Future perspectives of wastewater-based epidemiology: Monitoring infectious disease spread and resistance to the community level. Environment International. 2020;**139**:105689. DOI: 10.1016/j.envint.2020.105689

[75] Hamouda M, Mustafa F, Maraqa M, Rizvi T, Aly HA. Wastewater surveillance

for SARS-CoV-2: Lessons learned from recent studies to define future applications. Science of the Total Environment. 2021;**759**(143493). DOI: 10.1016/j.scitotenv.2020.143493

[76] Mao K, Zhang H, Yang Z. Can a paper-based device trace COVID-19 sources with wastewater-based epidemiology? Environmental Science Technology. 2020;**54**(7):3733-3735. DOI: 10.1021/acs.est.0c01174

[77] Qi F, Qian S, Zhang S, Zhang Z. Single-cell RNA sequencing of 13 human tissues identify cell types and receptors of human coronaviruses. Biochemical and Biophysical Research Communications. 2020;**526**(1):135-140. DOI: 10.1016/j.bbrc.2020.03.044

[78] Gonzalez R, Curtis K, Bivins A, Bibby K, Weir MH, Yetka K, et al. COVID-19 surveillance in Southeastern Virginia using wastewater-based epidemiology. Water Research. 2020;**186**:116296. DOI: 10.1016/j.waters.2020.116296

[79] Medema G, Heijnen L, Elsinga G, Italiaander R, Brouwer A. Presence of SARS-Coronavirus-2 RNA in wastewater and correlation with prevalence of COVID-19 reported in the early stage of the outbreak in the Netherlands. Letters in Environmental Science and Technology. 2020;7(7):11-516

[80] Randazzo W, Truchado P, Cuevas-Ferrando E, Simón P, Allende A, Sánchez G. SARS-CoV-2 RNA in wastewater anticipated COVID-19 occurrence in a low prevalence area. Water Research. 2020;**181**. DOI: 10.1016/j.watres.2020.115942

[81] Gupta S, Parker J, Smits S, Underwood J, Dolwani S. Excrétion virale persistante du SRAS-CoV-2 dans les fèces – un examen rapide. Dis

colorectal. 2020;**22**(6):611-620. DOI: 10.1111/codi.15138

[82] Jones DL, Baluja MQ, Graham DW, et al. Shedding of SARS-CoV-2 in feces and urine and its potential role in the person-to-person transmission and the environment-based spread of COVID-19. Science of the Total Environment. 2020;**749**:141364. DOI: 10.1016/j.scitotenv.2020.141364

[83] CDC. Real-Time RT-PCR Diagnostic Panel. Centers for Disease Control and Prevention, CDC-006-00. 2020:1-80. Available from: https://www.fda.gov/media/134922/download

[84] Corman VM, Landt O, Kaiser M, Molenkamp R, Meijer A, Chu DK, et al. Detection of 2019 novel coronavirus (2019-nCoV) by real-time RT-PCR. Eurosurveillance. 2020;**25**(3):23. DOI: 10.2807/1560-7917.ES.2020.25.3.2000045

[85] Kitajima M, Ahmed W, Bibby K, Carducci A, Gerba CP, Hamilton KA, et al. SARS-CoV-2 in wastewater: State of the knowledge and research needs. Science of the Total Environment. 2020;**739**:139076. DOI: 10.1016/j.scitotenv.2020.139076

[86] Pecson BM, Darby E, Haas CN, Amha YM, Bartolo M, Danielson R, et al. SARS-CoV-2, interlaboratory consortium. Reproducibility and sensitivity of 36 methods to quantify the SARS-CoV-2 genetic signal in raw wastewater: Findings from an interlaboratory methods evaluation in the US. Environmental Science: Water Research & Technology. 2021;7:504-520. DOI: 10.1039/d0ew00946f

[87] Green H, Wilder M, Collins M, Fenty A, Gentile K, Kmush BL, et al. Quantification of SARS-CoV-2 and cross-assembly phage (crAssphage) from wastewater to monitor coronavirus

transmission within communities. MedRxiv. 2020;**2**:1-18. DOI: 10.1101/ 2020.05.21.20109181

[88] Ahmed W, Angel N, Edson J, Bibby K, Bivins A, O'Brien JW, et al. First confirmed detection of SARS-CoV-2 in untreated wastewater in Australia: A proof of concept for the wastewater surveillance of COVID-19 in the community. Science of the Total Environment. 2020a;**728**:138764. DOI: 10.1016/j.scitotenv.2020.138764

[89] Dumke R, de la Cruz Barron M, Oertel R, Helm B, Kallies R, Berendonk TU, et al. Evaluation of two methods to concentrate SARS-CoV-2 from untreated wastewater. Pathogens. 2021;**10**(2):195. DOI: 10.3390/ pathogens10020195

[90] Kumar M, Kuroda K, Patel AK, Patel N, Bhattacharya P, Joshi M, et al. Decay of SARS-CoV-2 RNA along the wastewater treatment outfitted with up-flow anaerobic sludge blanket (UASB) system evaluated through two sample concentration techniques. Science of the Total Environment. 2021;**754**:142329. DOI: 10.1016/j.scitotenv.2020.142329

[91] Ahmed W, Bertsch PM, Bivins A, Bibby K, Farkas K, Gathercole A, et al. Comparison of virus concentration methods for the RT-qPCR-based recovery of murine hepatitis virus, a surrogate for SARS-CoV-2 from untreated wastewater. Science of the Total Environment. 2020;**739**(June):139960. DOI: 10.1016/j. scitotenv.2020.139960

[92] Haramoto E, Malla B, Thakali O, Kitajima M. First environmental surveillance for the presence of SARS-CoV-2 RNA in wastewater and river water in Japan. Science of the Total Environment. 2020;**737**:140405. DOI: 10.1016/j.scitotenv.2020.140405

[93] Wurtzer S, Marechal V, Mouchel JM, Maday Y, Teyssou R, Richard E, et al. Evaluation of lockdown impact on SARS-CoV-2 dynamics through viral genome quantification in Paris wastewater. MedRxiv. 2020. DOI: 10.1101/2020.04.12.20062679

[94] Wang XW, Li JS, Guo TK, Zhen B, Kong QX, Yi B, et al. Concentration and detection of SARS coronavirus in sewage from Xiao Tang Shan hospital and the 309th hospital. Journal of Virological Methods. 2005;**128**(1-2):156-161. DOI: 10.1016/j.jviromet.2005.03.022

[95] Arora S, Nag A, Sethi J, Rajvanshi J, Saxena S, Shrivastava SK, et al. Sewage surveillance for the presence of SARS-CoV-2 genome as a useful wastewater-based epidemiology (WBE) tracking tool in India. Water Science and Technology. 2020;**82**(12):2823-2836. DOI: 10.2166/ wst.2020.540

[96] Kumar M, Patel AK, Shah AV, Raval J, Rajpara N, Joshi M, et al. First proof of the capability of wastewater surveillance for COVID-19 in India through detection of the genetic material of SARS-CoV-2. Science of the Total Environment. 2020;**746**:141326. DOI: 10.1016/j.scitotenv.2020.141326

[97] Rimoldi SG, Stefani F, Gigantiello A, Polesello S, Comandatore F, Mileto D, et al. Presence and infectivity of SARS-CoV-2 virus in wastewater and rivers. Science of the Total Environment. 2020;**744**:140911. DOI: 10.1016/j. scitotenv.2020.140911

[98] Amoah ID, Kumari S, Bux F. Coronaviruses in wastewater processes: Source, fate and potential risks Environment International.May 2020;**143**:105962. 10.1016/ j.envint.2020.105962.

[99] Lu R, Wang J, Li M, Wang Y, Dong J, Cai W. SARS-CoV-2 detection using

digital PCR for COVID-19 diagnosis, treatment monitoring and criteria for discharge. MedRxiv. 2020. DOI: 10.1101/2020.03.24.20042689

[100] Dong L, Wang X, Wang S, Du M, Niu C, Yang J, et al. Interlaboratory assessment of droplet digital PCR for quantification of BRAF V600E mutation using a novel DNA reference material. Talanta. 2020;**207**. DOI: 10.1016/j. talanta.2019.120293

[101] Khan K, Tighe SW, Badireddy AR. Facteurs influençant la récupération de l'ARN du SRAS-CoV-2 dans les eaux usées brutes et les boues d'épuration à l'aide de la méthode de concentration à base de polyéthylène glycol. Journal of Biomolecular Techniques. 2021;**32**(3):172-179. DOI: 10.7171/ jbt.21-3203-012

[102] La Rosa G, Bonadonna L, Lucentini L, Kenmoe S, Suffredini E. Coronavirus in water environments: Occurrence, persistence and concentration methods—A scoping review. Water Research. 2020;**179**:115899. DOI: 10.1016/j.watres.2020.115899

[103] Fears AC, Klimstra WB, Duprex P, Hartman A, Weaver SC, Plante KS, et al. Persistence of severe acute respiratory syndrome coronavirus 2 in aerosol suspensions. Emerging Infectious Diseases. 2020;**26**(9):2168-2171. DOI: 10.3201/eid2609.201806

[104] Mukherjee A, Babu SS, Ghosh S. Thinking about water and air to attain sustainable development goals during times of COVID-19 pandemic. Journal of Earth System Science. 2020;**129**(1). DOI: 10.1007/s12040-020-01475-0

[105] Sivakumar B. COVID-19 and water. Stochastic Environmental Research and Risk Assessment. 2020;**2020**:1-4. DOI: 10.1007/s00477-020-01837-6

[106] Sichewo PR, Vander Kelen C, Thys S, Michel AL. Risk practices for bovine tuberculosis transmission to cattle and livestock farming communities living at the wildlife-livestock-human interface in northern KwaZulu Natal, South Africa. PLoS Neglected Tropical Diseases. 2020;**14**(3):e0007618. DOI: 10.1371/journal.pntd.0007618

[107] Sidhu JPS, Sena K, Hodgers L, Palmer A, Toze S. Comparative enteric viruses and coliphage removal during wastewater treatment processes in a subtropical environment. Science of the Total Environment. 2018;**616-617**:669-677. DOI: 10.1016/j.scitotenv.2017.10.265

[108] Kitajima M, Iker BC, Pepper IL, Gerba CP. Relative abundance and treatment reduction of viruses during wastewater treatment processes—Identification of potential viral indicators. Science of the Total Environment. 2014;**488-489**(1):290-296. DOI: 10.1016/j.scitotenv.2014.04.087

[109] Simmons FJ, Xagoraraki I. Release of infectious human enteric viruses by full-scale wastewater utilities. Water Research. 2011;**45**(12):3590-3598. DOI: 10.1016/j.watres.2011.04.001

[110] Nordgren J, Matussek A, Mattsson A, Svensson L, Lindgren PE. Prevalence of norovirus and factors influencing virus concentrations during one year in a full-scale wastewater treatment plant. Water Research. 2009;**43**(4):1117-1125. DOI: 10.1016/j. watres.2008.11.053

[111] Keegan AR, Robinson B, Monis P, Biebrick M, Liston C. Validation of activated sludge plant performance for virus and protozoan reduction. Journal of Water Reuse and Desalination. 2013;**3**(2):140-147. DOI: 10.2166/ wrd.2013.032

[112] Ye Y, Ellenberg RM, Graham KE, Wigginton KR. Survivability, partitioning, and recovery of enveloped viruses in untreated municipal wastewater. Environmental Science & Technology. 2016;**50**(10):5077-5085. DOI: 10.1021/acs.est.6b00876

[113] Verbyla ME, Mihelcic JR. A review of virus removal in wastewater treatment pond systems. Water Research. 2015;**71**:107-124. DOI: 10.1016/j.watres. 2014.12.031

[114] Nqombolo A, Mpupa A, Moutloali RM, Nomngongo PN. Wastewater treatment using membrane technology. In: Wastewater and Water Quality. Rijeka: InTech; 2018. DOI: 10.5772/intechopen.76624

[115] Neuman BW, Buchmeier MJ. Supramolecular architecture of the coronavirus particle. Advances in Virus Research. 2016;**96**:1-27. DOI: 10.1016/ bs.aivir.2016.08.005

[116] Wigginton KR, Pecson BM, Sigstam T, Bosshard F, Kohn T. Virus inactivation mechanisms: Impact of disinfectants on virus function and structural integrity. Environmental Science and Technology. 2012;**46**(21): 12069-12078. DOI: 10.1021/es3029473

[117] Page MA, Shisler JL, Mariñas BJ. Mechanistic aspects of adenovirus serotype 2 inactivation with free chlorine. Applied and Environmental Microbiology. 2010;**76**(9):2946-2954. DOI: 10.1128/AEM.02267-09

[118] Shirbandi K, Barghandan S, Mobinfar O, Rahim F. Inactivation of coronavirus with ultraviolet irradiation: What? How? Why? SSRN Electronic Journal. 2020. DOI: 10.2139/ssrn.3571418

[119] Kim J, Jang J. Inactivation of airborne viruses using vacuum ultraviolet photocatalysis for a flow-through indoor air purifier with short irradiation time. Aerosol Science and Technology. 2018;**52**(5):557-566. DOI: 10.1080/02786826.2018.14313

[120] Casanova L, Rutala WA, Weber DJ, Sobsey MD. Survival of surrogate coronaviruses in water. Water Research. 2009;**43**(7):1893-1898. DOI: 10.1016/j.watres.2009.02.002

[121] Smith EC, Denison MR. Coronaviruses as DNA wannabes: A new model for the regulation of RNA virus replication fidelity. PLoS Pathogens. 2013;**9**(12):e1003760. DOI: 10.1371/ journal.ppat.1003760